DESIGN OF PLUMBING AND DRAINAGE SYSTEMS

Design of Plumbing and Drainage Systems

By

Louis Blendermann

Vice-President, Research and Engineering, Josam Manufacturing Company, and member of The American Society of Sanitary Engineering

Second Edition

Fifth Printing

INDUSTRIAL PRESS INC., 200 Madison Ave., New York, N.Y.

Library of Congress Catalog Card No. 59-7600

PRINTED IN THE UNITED STATES OF AMERICA

SECOND EDITION

CONTENTS

PART 1 — SITE WORK

PART 2 — BUILDING DRAINAGE SYSTEMS

PART 3 — BUILDING WATER SYSTEMS

PART 4 — BUILDING FIRE PROTECTION

PART 5 — PLUMBING FIXTURES

PART 6 — LABORATORY SERVICES AND FACILITIES

Part 1
Site Work

CHAPTER 1

Location and Construction of Dry Wells

Every so often the engineer and designer encounters a problem concerning the disposal of rain water that will fall on the roof of a proposed building. This problem is a difficult one, especially when there are no available storm sewers, brooks, streams, or other points of discharge at the site to which the rain water may be conducted, and it is further complicated when the nature of the area or the ground is such that this water must be disposed within the property limits.

While there are several solutions, in this chapter we shall consider only the method that uses dry wells. The function of a dry well is, therefore, that of an underground receiving tank into which the rain water flows and from which, by reason of its construction, this water is recharged into the earth. Features of design in location and construction will be described that will enable the dry well to provide many years of trouble-free service and efficiency in operation.

Location of Dry Wells

Before the dry wells are located, several factors must be considered, as follows:

1. Rainfall and total roof area to be drained, which will determine the size and number of required dry wells.
2. Proposed grouping of leaders for the purpose of leaving the building with as few storm sewers as possible.
3. Characteristics of the soil and the level of the possible ground water, both of which will affect the location of the dry wells. Often when ground water is a condition, it will be necessary to construct two or more shallow type dry wells which are located above the water level.

4. Future construction of a public storm sewer near the property, into which the rain water may then be diverted, is a matter which should be discussed with the local sewer or highway departments. If the construction of such a sewer is proposed at a future date, the engineer should, if possible, so locate the dry wells, that when the public storm sewer is installed, the dry wells could be filled with earth and the storm water lines extended to the public storm sewer system.

These considerations have a definite bearing on the design and location of the proposed rain water disposal system. Once the location of the dry wells has been determined, it is important to see that they will operate at full capacity. Using one dry well as an example, a recommended location and installation is shown in Fig. 1.

As shown in Fig. 1, the dry well is preferably installed in sloping ground. If the grade is level or slightly depressed, then the grade over the dry well should be raised and sloped away as indicated. This practice will prevent surface water from accumulating and seeping into the dry well thereby reducing its effective capacity. Many dry well installations of adequate size and construction have failed in their purpose because surface water accumulation had not been taken into account. To avoid this occurrence, it is necessary

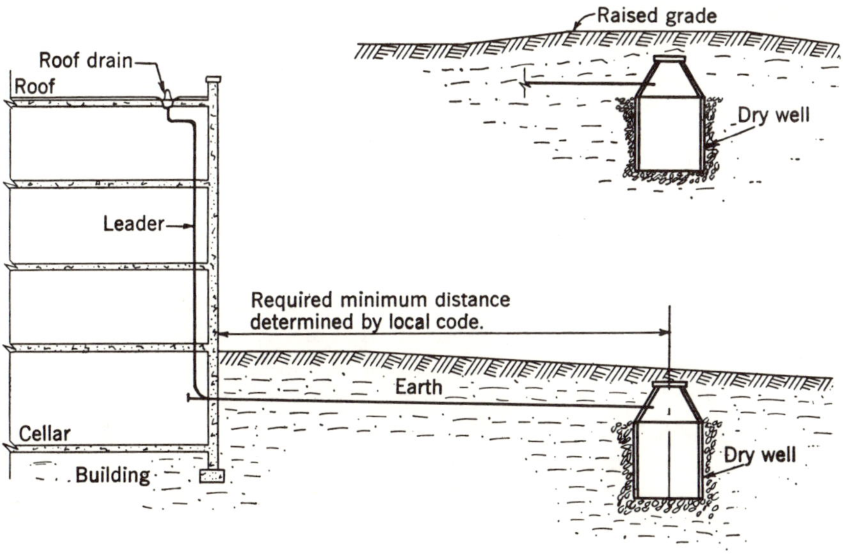

Fig. 1. Typical location of a dry well. Where ground level is depressed, it must be raised as indicated.

to inform the prospective contractor of the desired procedure to be followed with regard to the treatment of the grade over the dry well as shown in Fig. 1. This can be accomplished by means of a brief note of explanation added to the detail of the dry well on the drawings or by a short paragraph covering the installation of the dry well, included in the specifications.

Determining Size of Dry Well

The size of a proposed dry well is governed by the following:

1. Rate of rainfall prevalent in the area.
2. Total roof area to be drained.

The rate of rainfall prevalent in the area, based on the intensity of rainfall in inches per hour, and the time of concentration in minutes, will determine the required size of dry well for a given roof area. The heaviest rainfall will occur during a fractional part of a storm, and the dry well is usually sized for this concentrated flow. Tables of the records of rainfall for any given area may be obtained at local United States Weather Bureau stations. Since the rainfall rates given in the tables are based on the greatest intensity over a number of years, it is safe to assume that the dry well will be of more than ample size.

Once the rainfall rate has been decided upon, it must be applied to the total roof area to be drained. For very large roofs, several dry wells may be required; for small roofs, one dry well may suffice. Using a small roof as an example, by applying the rainfall rate to the total area, the resultant flow of water can be quickly determined. The dry well should be of such capacity to receive this entire flow. If the flow is such, that the size of the dry well would be enormous, then it is recommended that an additional dry well be considered for the purpose of maintaining moderate sizes.

The water in the dry well will drain into the earth at a rate dependent upon the drainage qualities of the soil. In sand or gravel the rate of drainage is rapid. In other soil which is not as porous, the rate of drainage is somewhat reduced. Often, when tight soil conditions exist, the dry well is increased slightly in size to compensate for the additional number of hours required for drainage. It is good practice, when the drainage rate of the soil must be known, to make a percolation test. If the soil is so tight that drainage is almost impossible, then dry wells or other similar means of disposal may not be used to their best advantage. In this case the rainwater must be disposed of in another manner. Often, test borings are made in

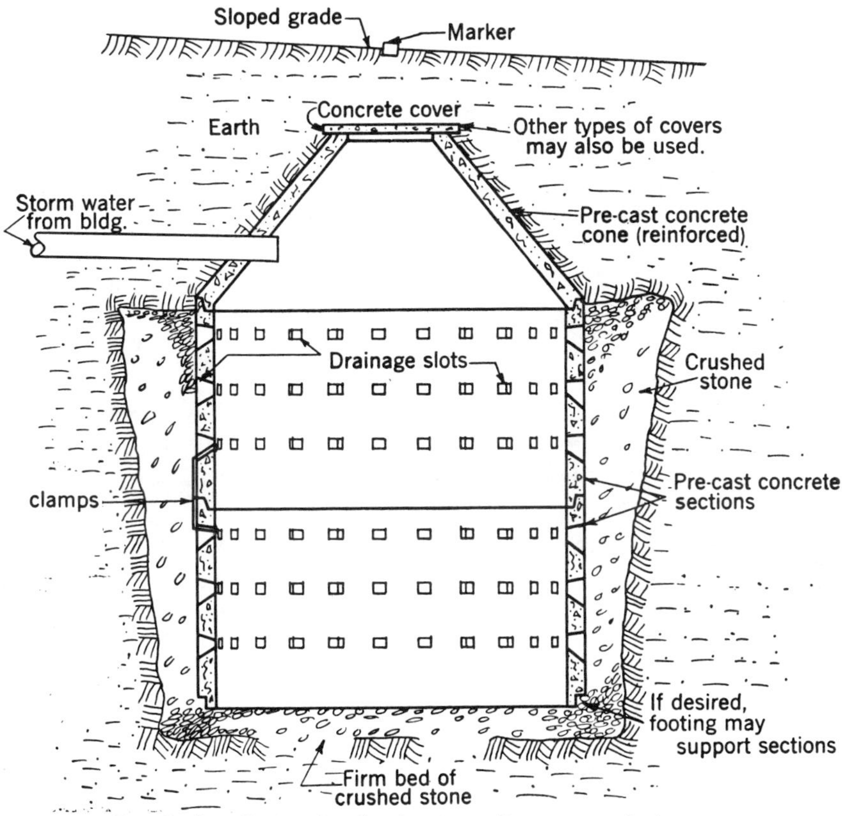

Fig. 2. Installation details of a dry well constructed of pre-cast concrete sections.

an effort to locate soil that is more porous, into which the water may then be disposed, by means of deep drainage pipes. However, in this article we are concerned only with normal soil conditions.

Construction Details

Dry wells may be constructed of cement block, brick, stone and other material so laid with open joints that the water can freely drain into the earth. Also available are various types of pre-cast concrete dry wells with holes or slots for drainage. Regardless of the materials used, it is of prime importance that the dry well be carefully constructed to withstand normal, structural and surface loads and to provide maximum efficiency. Since a dry well will be in operation for a good number of years, and because of its diameter and depth, every precaution should be taken in design and construction to prevent structural failure. In addition, as mentioned earlier, the dry well should be located above the mean high level of the ground water table. A typical dry well installation is illustrated.

The dry well used as an example in Fig. 2, is of the reinforced precast concrete type. The space around the dry well has been filled with crushed stone for the following reasons:

1. Crushed stone allows for free and easy drainage.
2. Soot and other matter that settles on the roof of a building in dry weather is washed into the dry well during a storm where it will in time, form an impervious layer on the surface of the close grained earth in contact with the exterior of the drainage slots, thus greatly reducing the effective rate of drainage. The crushed stone will prevent this action, by allowing the small particles of foreign matter to disperse. Normally years will elapse before a sufficient accumulation of silt could retard the drainage.

These points when considered during the design period, will assure a dry well system of adequate size and complete fulfillment of efficiency in function. The same considerations may also be applied to existing dry wells in order to increase their drainage capacity. When the drainage capacity of a dry well that has been in operation for a number of years becomes troublesome, it is reasonable to assume that the porous quality of the soil immediately surrounding the dry well has become clogged with impervious foreign matter. To correct this condition, it is necessary to dig out this earth and replace it with fresh sand or gravel, and crushed stone in a manner shown in Fig. 2, and the dry well, will again function properly.

CHAPTER 2

Function of Drainage Basins

A necessary design consideration for every new construction project is the system of storm water piping and appurtenances that will receive and conduct the storm water to suitable points of disposal. This system must be designed to handle the roof water as well as the water that falls on the paved and unpaved areas, and ground surfaces. For a project consisting of many thousand square feet of roof and paved surfaces, the amount of water to be handled reaches tremendous proportions.

Rainfall Intensity

The amount of rain that will fall on a given area in a prescribed time can be calculated by means of rainfall data tables compiled by the United States Department of Agriculture. These tables represent the amount of rain that has fallen in a given area over the past 2, 5, 10, 15, 25, 50, and 100-year periods. Such data have been further simplified to reflect the intensity of rainfall in inches during 10 or 15 minutes or other period of minutes. This same type of information is often available in the local plumbing code or can be obtained from local plumbing officials in the area where the project is to be erected.

Rainfall Disposal

After the expected quantity of water to be handled has been determined, it is then necessary to consider the proper means of disposal. If storm water sewer facilities are available in the street, the problem is somewhat simplified since it is only necessary to check the capacity of the existing street sewer to see if it can receive the additional storm water load. The same applies to an existing, combined sanitary and storm sewer. If neither of these is available, it becomes necessary to discharge the water into a nearby brook, lake or any natural body of water.

If the site area is so located that none of these natural discharge means are at hand, the storm water will have to be discharged elsewhere. When this condition is encountered, dry wells are usually employed. Dry wells are perforated concrete structures of a predetermined capacity into which the rain water is conducted. Sufficient dry wells of proper capacity are included to handle the total rain water that will accumulate on the site. The dry wells serve their purpose by receiving and holding the rainfall which then drains into the earth as fast as the percolation rate of the soil will allow.

When the square feet of roof and paved area on a site is so extensive that dry wells alone will not suffice, then other means of disposal must be considered. Roof and paved areas constitute the biggest rain water load to be handled. The ground surface will absorb a certain per cent of the water and therefore it does not present the same problem as non-porous surfaces. However, even the ground surfaces must be drained by means of catch basins. Thus if the site is level or slightly depressed, flooding of the area will result during a heavy storm. To avoid this occurrence, attention is focused on drainage basins to receive and dispose of the storm water.

Drainage Basins

A drainage basin, or recharge basin as it is sometimes called, is nothing more than a hole or depression dug into the earth on the site at a lower elevation into which the storm water from the buildings, paved areas and general site will flow. The function of a drainage basin is to receive and retain this water, but to permit it to drain into the earth as fast as the percolation rate of the soil will allow. The word recharge originated with the ability of this pit to replenish ground water streams. Ground water in many areas is a primary source of water supply. Paved areas and increased water consumption deplete such supplies. When the water is permitted to drain back into the earth, the ground water certainly benefits. Recharge basins, however, have a specific application in certain parts of the country. The function of the basin in this discussion is that of a drainage basin only and a typical application is shown in Fig. 1.

In this figure, it will be noted that the site used for illustrative purposes is that of a good sized manufacturing plant, consisting of a main building, numerous manufacturing buildings as well as a boiler house, parking areas, and grounds around the buildings. The

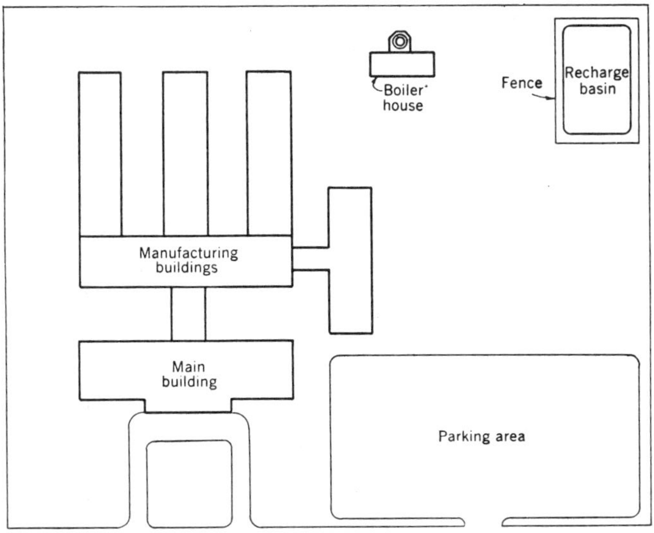

Fig. 1. Typical application of drainage basin.

drainage basin is located at the corner of the property, which happens to be best suited for the grade of the property.

All the rain falling on the buildings, paved areas and general surface area is intercepted by proper drains and conducted to the drainage basin. Manholes for cleanout purposes should be placed at suitable intervals along the storm water sewer, particularly at the intersection of two or more drainage lines as well as at proper spacing along the storm water line to ensure adequate and convenient clearing of the drainage lines.

A fence is shown around the drainage basin to prevent persons from falling in. This provision is especially important as protection for children.

Construction

A drainage basin is usually constructed or formed by means of a bulldozer. The surplus earth can be used for fill in grading the site. A typical section of this basin is shown in Fig. 2 and shows how the sides of the basin are sloped at an angle that best suits the depth of the basin and the stability of the soil. This prevents the sides of the basin from sliding into the basin. Sometimes, crushed stone is placed at the bottom of the basin and partly up the side to facilitate drain-

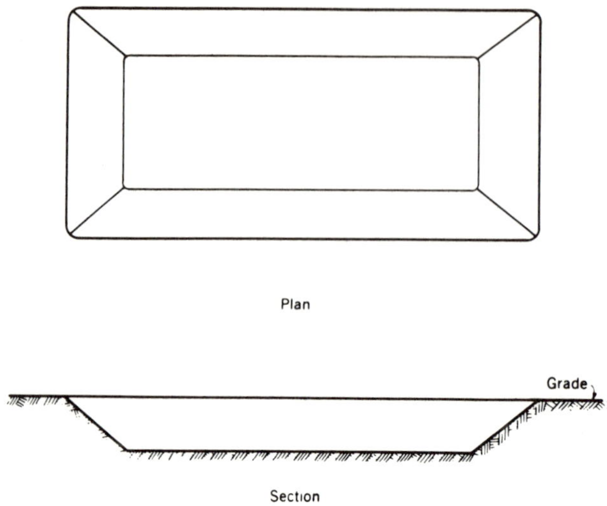

Fig. 2. Typical section of drainage basin.

age. However, the addition of crushed stone depends entirely upon the particular design.

The size of any drainage basin is related to the amount of water to be retained. The percolation rate of the soil is also an important factor. Should percolation tests taken of the soil indicate a sluggish drainage rate, it is then advisable to increase the size of the drainage basin to accommodate an extra quantity of water in the event of an extended rainfall.

Outfall Structures

Since the underground storm water lines will discharge into the drainage basin, it is necessary to erect some sort of a structure that will prevent the flowing water from washing away the earth at the point of its contact in the basin. There are many different types of structures that can be used for this purpose. A typical outfall structure is detailed in Fig. 3.

The outfall structure, Fig. 3, is constructed entirely of reinforced concrete and allows the water to splash on the concrete pad and spread out evenly along the bottom of the basin. This prevents the damage to the bottom and sides of the basin that could otherwise be caused by the water. The overall size of the structure is determined by the diameter of the drainage line.

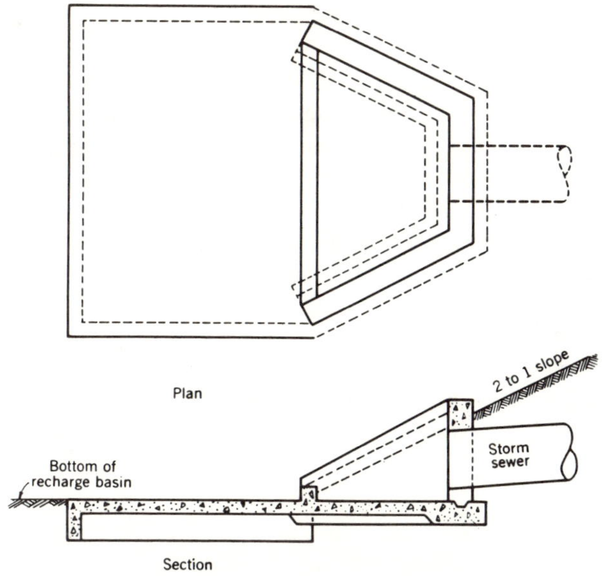

Fig. 3. Typical outfall structure.

Maintenance of a Drainage Basin

Like very other appurtenance, a drainage basin must be maintained periodically to obtain the greatest efficiency in its use. Since water will more or less always be contained in the basin, attention must be given to appropriate mosquito control. Proper care of course will prevent the breeding of mosquitoes.

Silt and other fine foreign matter will clog the surface layer of the depression, which in time prolongs the time required for the water to seep into the earth. To rectify this condition, it is often necessary to scrape the surface of the basin to expose the porous earth again. This is generally done every two or three years, depending upon the intensity of such clogging impurities. This task is generally performed during the dry seasons of the year.

Benefits

In addition to serving the drainage needs of the property, drainage basins very often can be used to further advantage in beautifying the property itself. By developing the area immediately surrounding the basin, it is possible to transform a necessary drainage facility into a garden spot. Trees, flowers, shrubs, walkways and benches certainly enhance the area and provide a spot of relaxation for employees.

Moreover, the water contained in the basin offers an excellent secondary reserve for fire protection purposes. In addition, by means of pumps and an adequate strainer on the intake line, the water can be used to sprinkle the lawn and flowers. There are perhaps many other uses for which this water can be employed. These benefits certainly compensate for any loss of property value or valuable space that the owner normally hesitates to relinquish.

Other Applications

Drainage basins need not only be used for a private building enterprise but can be used for communities, small home developments, schools, hospitals or other groups of buildings.

CHAPTER 3

Protection of Site Utility Lines

New buildings that are now being erected, are completed in relatively short periods of time. Careful planning, improved methods of construction and modern equipment have made this possible.

In fact, the construction work on a project generally proceeds at such a rapid pace that it is not unusual to see many of the component parts of a building progressing at the same time. Even though the work proceeds at this rate, one must remember that the project is intended to endure for a good number of years.

Therefore, the engineer designing the plumbing systems must take every precaution to ensure that his services and equipment are properly installed. This means, that not only should his work be adequately designed and specified, but in addition the components should be carefully protected during the installation.

Utilities for the Project

Figure 1 illustrates a typical arrangement of utilities for a project containing a group of six apartment house buildings. The utilities, consisting of sanitary, storm and water lines were installed between the buildings and they connect with existing street utilities. Only these particular three lines have been shown in order to simplify the discussion.

The proper protection of the site or project utilities is a serious problem, since often the utilities are installed in the early stages of construction in order to provide water and drainage facilities required by the construction work. Although the piping is tested and inspected after it is installed, and despite the fact that the contractor provides a workmanship guarantee for a period of a year or more after installation, leaks can still develop. After the backfill has been applied to the pipe trench, it is often impossible to detect leakage caused by pipe settlement, due to heavy loads passing over the trench.

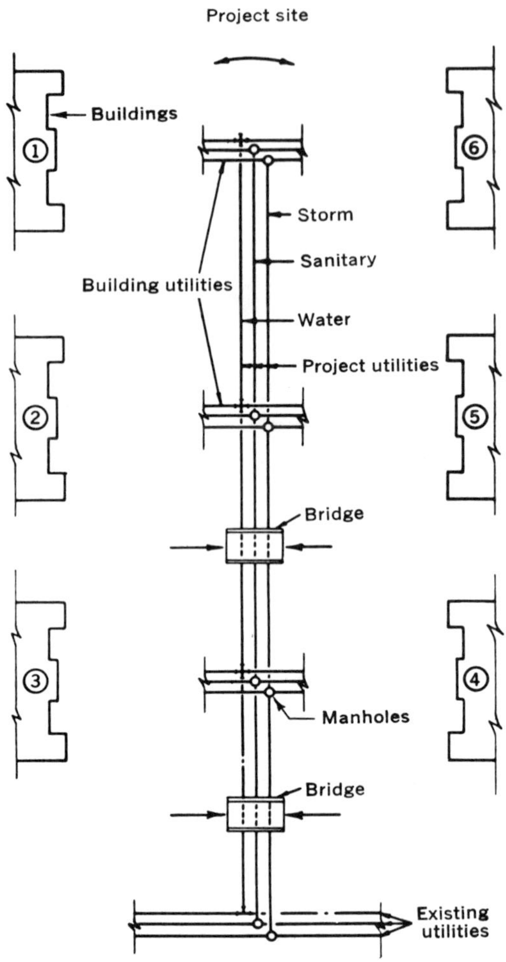

Fig. 1. Typical layout of site utilities and location of bridges. Valves and other equipment are not shown.

It may take several years to discover that a serious leak is existing, but such a leak can cause buildings to settle and crack and finished grade to sag or wash away. In addition, the discharge of raw sewage into the ground can create a menace to public health. While good design and the proper selection of materials will assure a sound system, unless the piping is protected after installation, the careful consideration given, will have been wasted.

Measure of Protection

Assuming that fill is not encountered at the site and that the piping is installed in trenches which have been excavated in natural earth. If the trench bottom is accurately graded for the slope of the piping, the utilities will function properly. Furthermore, if bell holes are cut in the trench bottom for the hubs of the piping and if the backfill is carefully handled, the utilities may be considered properly supported and well installed.

Figure 2, indicates a typical arrangement of the underground utilities and the manner in which they are supported. As shown in this illustration, the utilities are installed on the bottom of a trench which has been carefully excavated with a minimum of disturbance to the natural earth, thereby assuring a good firm support for the piping. The storm and sanitary lines were installed with the proper slope to satisfy the elevations of the buildings and the finished grading. Care has been exercised in excavating the trench to provide a shelf or ledge of natural earth for the support of the water line. This procedure lessens the possibility of pipe settlement with its resulting damage.

Dangers to Underground Utilities

If the wall thickness of the piping shown in Fig. 2 has been adequately specified for the working pressure, size and function of the lines, and if the backfill has been carefully handled, the lines should withstand reasonable amounts of weight or pressure exerted on the grade, directly over the piping.

However, at a site similar to the one shown in Fig. 1, there is a continual movement of heavy trucks, delivering the necessary construction material for the buildings. In addition, cranes, bulldozers and other types of heavy equipment are also being moved over the trenches and present a serious danger to the utilities. The result is that the piping will be subjected to unusual loads and strains for which it has not been designed. This can crush the pipe or cause the lines to settle to the extent that leaks will develop.

Bridges to Span Trenches

To avoid this, it is necessary to take certain steps to protect the utilities. One such precaution is to provide bridges that span the width of the trenches. The bridges are located at points over the trench, best suited for movement of heavy equipment as determined by the needs of the various contractors. The bridges are

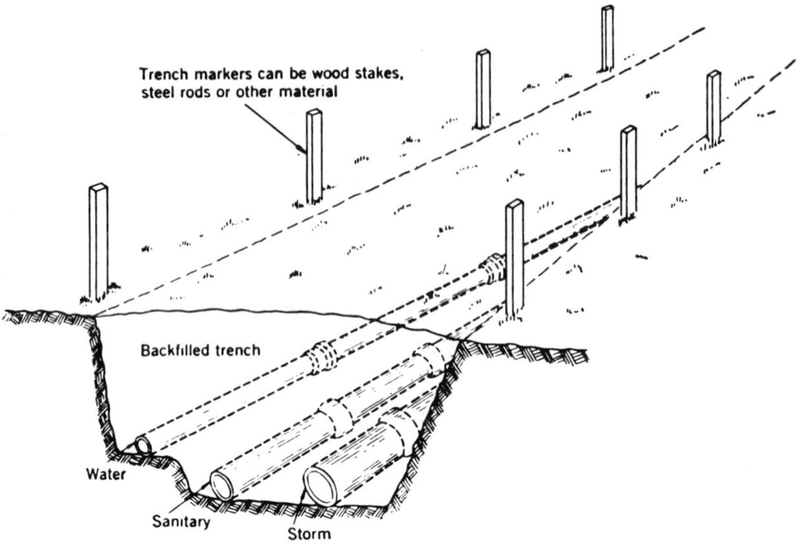

Fig. 2. Installation of utility lines in a trench. As a protective measure, trench markers are used to show trench limits.

generally built and maintained by the general contractor. A typical location of bridges is shown in Fig. 1. Two bridges are required and have been so located to provide the best access to either side of the trench. A third bridge was not considered since access to either side of the trench is available in the area of the curved arrow. However, Fig. 1 is only an example, for other projects may require different and more numerous locations. The required width of the bridge is dependent upon the size of the largest piece of equipment to be moved. This is usually determined by the general contractor.

The bridge itself may consist of timber planks or steel plates, supported on a framework of heavy timber and piling at either side of the trench. In addition, underground supports are usually extended from the piles to the center of the span.

The building utilities should be provided with similar bridging protection if the utilities are installed early during the construction stages. Often these building utilities are not installed until the construction is well advanced.

The purpose of the bridge is to distribute a heavy load passing over the trench away to the framework rather than directly over the piping. The material used for the bridge and the method of construction is the responsibility of the general contractor. It is the responsibility of the engineer to make certain by provisions on drawings or in the specifications, that the plumbing contractor will know this protection is called for so that he can bring it to the attention of the general contractor.

Trench Markers

In addition to providing bridges over the pipe trenches at specific locations, it is also necessary to make certain that the bridges will be used and that equipment will not be moved over other portions of the trench.

This can be accomplished by setting stakes or markers along the sides of the trench, denoting the location of the backfilled trench. The stakes will allow workmen to cross the trench, but will indicate the location of the trench which trucks and other necessary heavy equipment are to avoid.

Figure 2 illustrates such a system of trench markers. The stakes are easily set and can be removed when it is certain that the heavy equipment will no longer be moved over the trench. The trench markers will also discourage the stockpiling of heavy construction materials over the pipe trench.

These precautions are just a few of the necessary steps to be taken for the protection of the site utilities. If the piping systems are expected to last for a good number of years, then they should at least be installed with care at the start.

It is fortunate on most projects that both the plumbing contractor and the general contractor give this problem adequate consideration, and take the necessary precaution to protect the piping. However, this is not always the case. Therefore, any consideration the engineer can give this matter, either by instructions on the drawings or in the specifications, will assure the necessary protection of the installed piping.

CHAPTER 4

Design and Construction of Manholes

Manholes are important. They are provided so that the sewer line, whether it be sanitary, storm or combined sewer, can be easily inspected and cleaned. Manholes are placed at suitable intervals along the sewer and permit two important functions: (1) they serve as junctions to join together one or more sewer lines; (2) they offer the means of access to inspect and clear the lines of any accumulation or stoppages that may occur.

When adequate attention is given to the many details connected with the location and construction of manholes, they can be counted upon for long life and years of trouble-free operation.

Location of Manholes

The location of the manholes for a particular site is dependent upon several factors including site topography, quality of soil, points of sewer connection and invert elevation of street sewers. Figure 1 is an arrangement for the sewer lines and manholes on a typical site.

Referring to Fig. 1, it will be noted that the manholes have been placed at locations best suited to receive the building sewers from the building. For the purpose of this design we shall consider that each of the building sewers joins at a manhole junction and continues from that point directly to the street sewer. For small sewers up to 24-in. diameter, the manholes should be placed approximately 300 ft. apart, since this is the distance most practical for scraping the sewer lines or clearing stoppages. On sewers over 24 in., the manholes can be placed approximately 500 ft. apart.

In general, manholes should be placed at every change in grade, at each change in direction of the sewer, and at every junction of two or more sewers. Where possible, the sewers should run in a straight line from manhole to manhole, and the changes in line or grade should occur at or in the manholes.

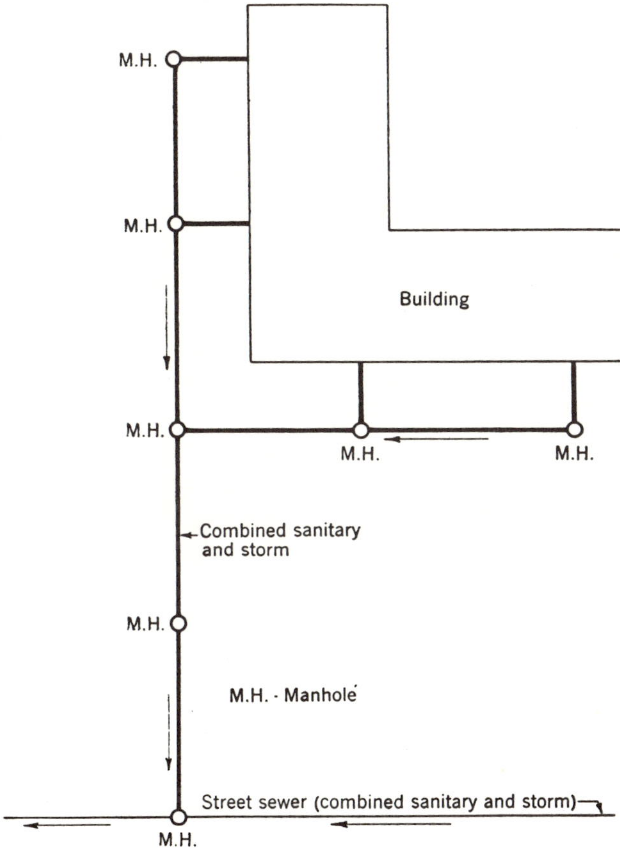

Fig. 1. Plan of typical layout showing location of manholes.

Shape of Manholes

Manholes can be erected in a number of shapes to suit the particular needs of the sewer line of which it is to become a component part. Some of the different shapes more commonly used are illustrated in Fig. 2.

As shown in Fig. 2, the manholes range from a simple shape as shown by detail *A*, to the more complicated *E* type. Following is a brief description of each type.

A — simply constructed manhole of cylindrical shape, more commonly used for small diameter domestic sewers.

B — bottle shaped manhole for sewers up to 24 in. diameter.

C — manhole with straight cylindrical walls and tapered cone for sewers up to 24 in. diameter. This manhole is fully detailed in Fig. 3.

D — manhole erected as a component part of a large diameter sewer.

E — manhole erected over sewers of very large diameters.

Materials Used in Constructing Manholes

The base is the first step in constructing a manhole. Since the base is required to support the walls and upper structure of the manhole it should be quite substantial. The width and thickness of the base is determined by the height and weight of the manhole to be supported as well as the particular quality of the ground in which it is to be placed. Quite often, reinforcing rods are used in the construction of the base to ensure a durable support.

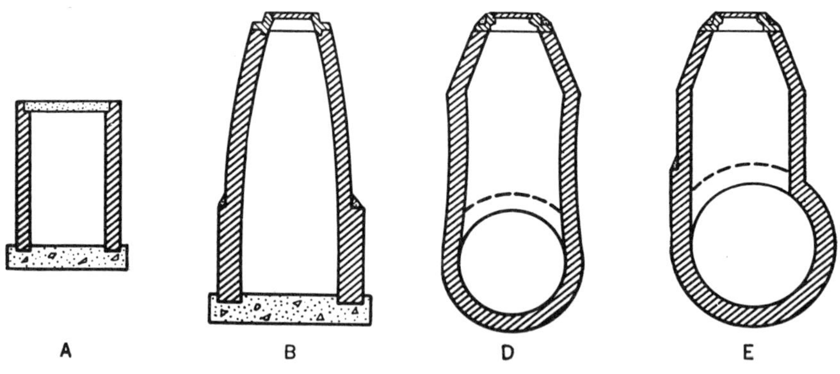

Fig. 2. Cross sections to show various shapes of manholes.

The walls of a manhole are generally constructed of brick, concrete block or poured concrete. Another popular type of construction is that of reinforced concrete sections that are assembled at the site. The sections come in prescribed lengths and may be fitted into one another by means of fitted joints. Information regarding the type and number of sections that can be used as well as their structural strength may be obtained from the manufacturers of the pre-cast sectional manholes.

There is no place for an inadequately constructed manhole in modern construction. A faulty manhole is a constant threat because of the possibility of a cave-in with its attendant hazard to life.

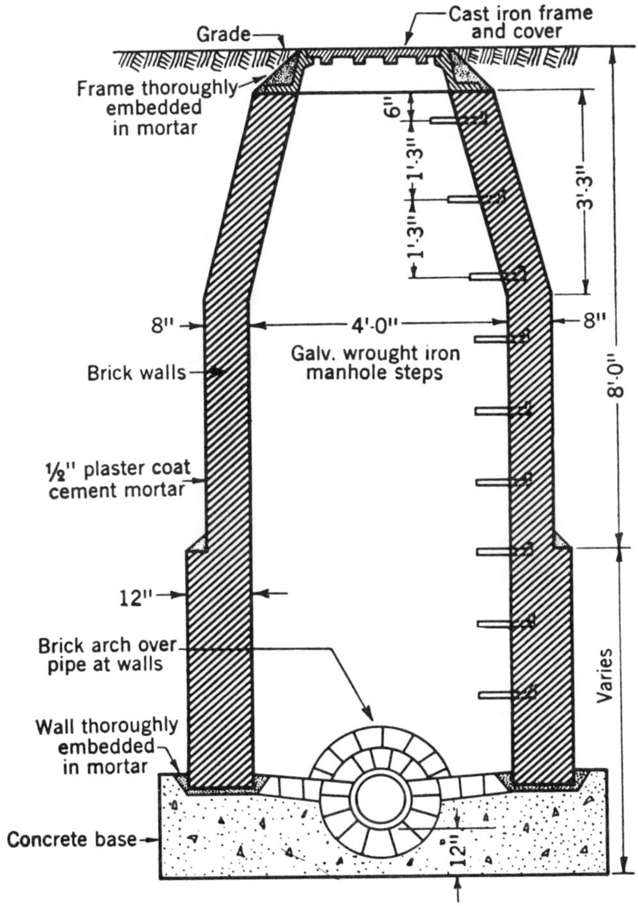

Fig. 3. Detail of typical manhole.

Assurance for Proper Construction

The engineer of a proposed project can be sure that proper manholes of adequate strength will be provided on his job if he includes the necessary details on his drawings and complete instructions in his specifications. The details will advise the plumbing or the general contractor of the type of manhole expected. Instructions in the specifications will require the contractors to assume the responsibility of furnishing a well-constructed manhole.

In addition, the engineer should require from the responsible contractor, a guarantee that the manhole will be erected in stable earth, otherwise that piling be provided under the base to support the manhole in a rigid manner. Any and every precaution taken in the design and construction of the manholes will pay dividends during the many years of expected use.

Typical Manhole

When manholes are required for small size sewer lines up to 24 in. diameter, a recommended construction would be similar to that shown in Fig. 3. It is used, for example, because such sizes are encountered more frequently in project site work.

Figure 3 shows that a good solid base is poured to support the heavy upper wall construction. For the walls, it is recommended that for the first 8 ft. below grade, the walls be at least 8 in. thick. Should the depth exceed this measurement, the walls should be at least 1 ft. thick as shown. Brick is the wall construction material that has been used in the design illustrated.

If ground water is a condition prevalent at the site, or if a heavy lateral pressure is expected, additional study should be given to the design of the manhole. Extra strength is always required to compensate for the exterior pressure exerted by the ground conditions.

For such conditions, the outside walls of the manhole is plastered with cement mortar to provide an encasing shell that gives additional strength to the manhole. In addition, the mortar offers full protection against ground water or the corrosive action of the soil.

Galvanized wrought iron rungs are embedded in the walls of the manhole to permit ready access into the manhole. The inside diameter of the manhole should always be at least 4 ft., since this is the amount of space required for a man to perform the necessary inspection, scraping or clearing of the sewer lines.

The frame and cover is usually of cast iron of the type and strength to withstand the loads exerted on the grade directly over the manhole. There are many shapes and sizes in frames and covers available to serve the specific needs of the manhole.

Drop Manhole

When sewers are connected to a manhole at about the same elevation, the piping connections are easily made. Frequently, because of topography or length of run, the main sewer is installed at a deep elevation whereas the branch sewer may be installed at an elevation only a few feet under the surface of the grade.

A drop manhole serves as a junction and receives sewer lines

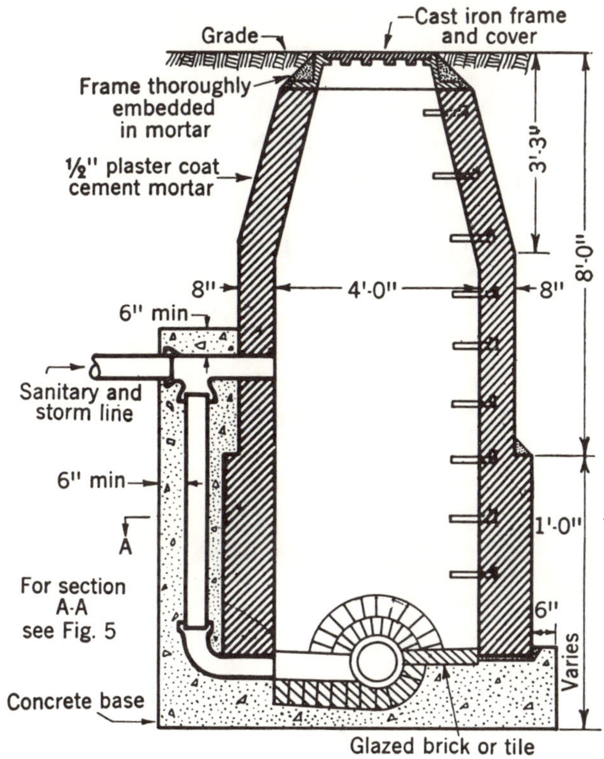

Fig. 4. Design details of a drop manhole.

that are installed in the ground at two different elevations. The lower sewer line enters the manhole at the low level. The higher sewer line, for the reason of short length of run or topography, is installed at a higher level in the ground. The vertical pipe receives the normal waste flow and conducts it to the bottom of the manhole at an elevation equal to that of the other connecting sewer line. However, when the flow of water or waste is greatly increased in the upper sewer line it will by virtue of its velocity discharge into the upper part of the manhole with only part of the flow dropping into the vertical pipe.

This arrangement prevents the water from continually splashing into the manhole which is rather deep and avoids the erosion effect on the concrete that follows the constant waste splashing action. If the waste passes through the manhole as a smooth and gentle flow, this action will be reduced. The bottom of the manhole is generally lined with glazed tile or glazed brick to further withstand the wearing action of the water passing through. Figure 4 pictures a typical drop manhole. The construction of the drop manhole is very similar to that of a regular manhole except for the vertical piping. The vertical piping is encased in concrete to prevent movement or leakage in the line. When a drop manhole is contemplated for a particular project, the engineer should include all the same considerations in the design and specifications that would generally be included for the regular type manhole. The bottom of the manhole would appear something similar to that shown in Fig. 5. This

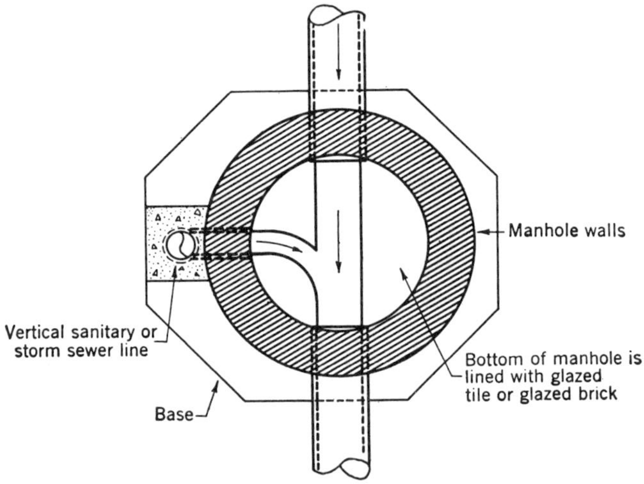

Fig. 5. Section A-A of Fig. 4 to show bottom of drop manhole.

figure shows that the waste passes through an open trough in the bottom of the manhole. Both the trough and the bottom of the manhole is lined with glazed tile or glazed brick. Although, the base is of a hexagon shape, it can also be square or any other shape as desired for the particular installation.

Other Applications

Manholes are often used for purposes other than that required for sewer lines. They may be used to provide access to underground water control valves and on occasion, for gas or other special service installations that are designed for the particular requirements of the buildings on the site.

There are a number of varied applications for the underground installation of steam or other heating lines. Regardless of the varied application for which the manhole may be employed, it is of prime importance that it be of adequate strength and construction to fulfill its purpose.

CHAPTER 5

How to Install Exterior Cleanouts

Cleanouts are a necessary component of sanitary, storm water or other types of drainage systems. Cleanouts are the means by which obstructions or deposits of sand and grit may be cleared in an underground drainage line thereby preventing the backflow or flooding that might otherwise result. With the provision of cleanouts spaced at prescribed intervals, it would not be necessary to excavate in order to gain access to an underground line in the event of a stoppage.

Cleanouts serve an important purpose in exterior drainage lines as well as for the drainage lines contained within a building. This is especially true for those drainage lines that are installed at a shallow depth within the limits of the property site. Manholes are generally used to aid a means of cleanout in exterior drainage lines. However, for those lines installed only a few feet below the surface, regular cleanouts are often used. Some of the cleanouts that may be utilized for the different conditions encountered in outside installations will be described.

Ground Settlement

Unless exterior cleanouts are properly installed, they will settle and lose their usefulness. More important, unless the normal settlement of the ground surface is taken into consideration, rupture of the underground piping can result.

After an underground piping system has been installed and the trench is backfilled, it will take a period of several months before the fill over the piping becomes as compact as natural earth. Compacting the backfill by physical means is an expedient but it still requires a period of time for the earth to become stable again. Even after the earth has finally compacted to a natural degree, there are a number of factors that can cause further settlement. Vibration of trucks passing in the street, increased foot traffic or the settlement due to the holes burrowed in the earth by moles or other underground animals.

If a cleanout were extended to the surface of the grade in a rigid manner, any ground settlement would either expose the cleanout or would place a damaging strain on the underground piping to the extent that the piping might rupture or sag. This development could create serious trouble and involve the owner of the building in the additional expense of repairing the damage. A leaking sanitary line is a potential threat to health because it contaminates the ground area into which it discharges.

It is, therefore, necessary that the cleanouts be installed in such a manner that they will give or settle with the movement of the earth and yet not place any strain on the underground piping. Following, are some of the methods that may be employed with good success to safeguard the piping.

Grass Surfaces

For sections of drainage lines installed at shallow bury under areas surfaced with grass and not subject to heavy foot traffic, a cleanout arrangement similar to that shown in Fig. 1, may be used.

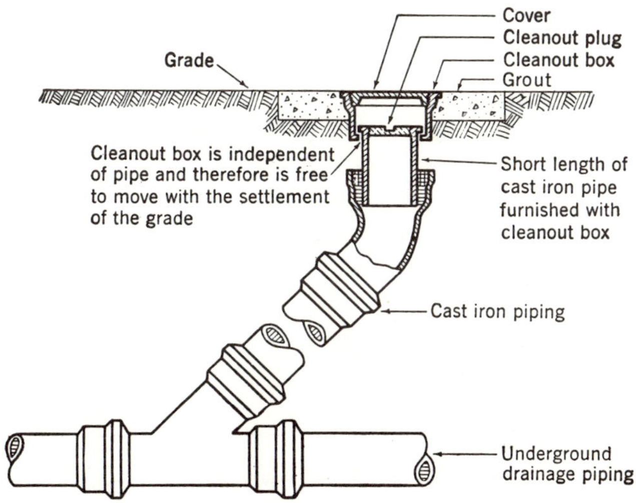

Fig. 1. Typical arrangement of cleanout and piping for areas not subject to heavy foot traffic.

As illustrated in Fig. 1, a cast iron cleanout box is grouted in place at surface level of the grade. This box is manufactured in various types and includes either a cast iron cover that can be lifted out or one that is hinged and can be locked with a key. The unit also includes a short length of cast iron pipe, provided with a threaded connection at one end into which an adequate cleanout plug is placed. This piece of pipe has cutoff sections spaced about one inch apart to offer flexibility in meeting any length required by the installation.

Should there be any settlement of the grade, the cleanout box will settle too. Since the cleanout box is independent of the short length of piping, no strain is placed on the piping itself. Instead, the connection is like a loose sleeve and the box will slide a little further down the length of the cast iron pipe. Sufficient space has been provided between the top of the cleanout plug and the cover of the box to allow for more than the maximum amount of settlement expected.

This method will provide for years of trouble-free service and removes any danger of rupture or sagging to the underground piping.

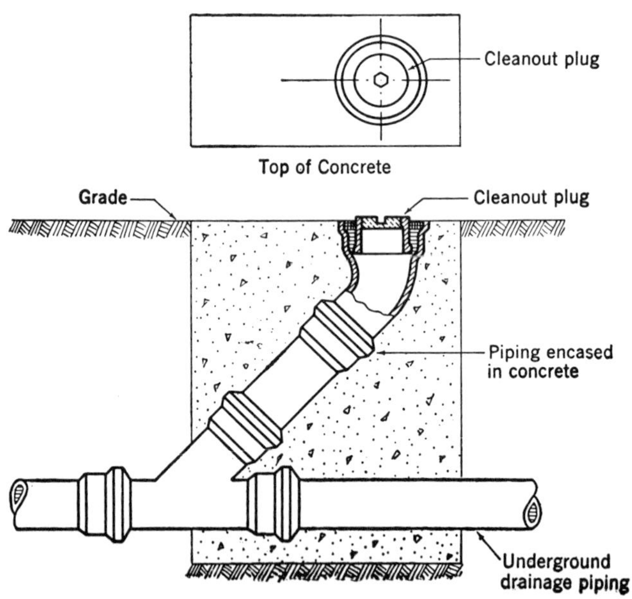

Fig. 2. Arrangement of cleanout and piping for areas subject to constant foot traffic and occasional heavier loads.

Plain Surfaces

For sections of drainage lines installed at shallow bury under unfinished surface areas, subject to constant foot traffic and occasional heavier loads, a cleanout arrangement similar to that shown in Fig. 2 can be used. In this design the cleanout piping extends up to the surface of the grade with a cleanout plug installed in the top of the piping. The entire section of cleanout piping is encased in concrete as shown, to prevent any damaging strain, either by surface traffic or grade settlement from bearing on the underground piping. This is a positive manner of protecting the piping.

At a later date if the grade recedes away from the top of the concrete, it is an easy matter to fill in with additional earth to make the surface of the grade level again. This is only one of similar methods that may be used.

Paved Top

When sections of the drainage lines installed at shallow bury are contained under black top or other paved surfaces, a method similar to that shown in Fig. 3, may be employed.

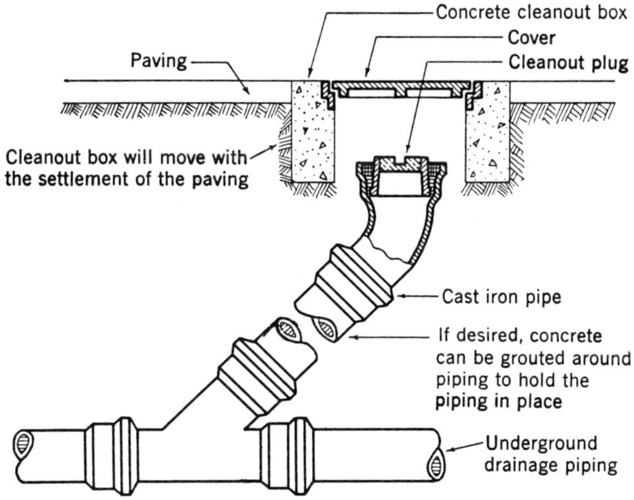

Fig. 3. Cleanout and piping installed under areas that are paved.

As detailed in Fig. 3, a cleanout box, constructed of concrete, is placed flush with the surface of the paving. This box can be round or square as desired and is fitted with an extra heavy cover to withstand the surface loads imposed. The cleanout plug is installed in the hub of the fitting, as illustrated.

In the event that the pavement settles, the unit will also settle without placing any strain on the underground piping. Here again, sufficient space is provided between the top of the cleanout plug and the underside of the cover to allow for more than the maximum amount of settlement expected.

Playgrounds

For sections of drainage lines installed at shallow bury under playgrounds or similar areas where concrete is the surfacing material, a method equal to that shown in Fig. 4, may be considered.

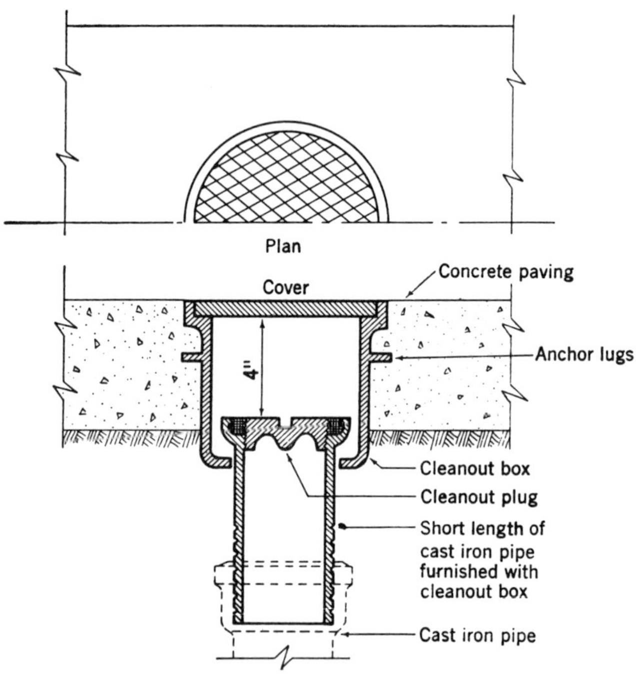

Fig. 4. Typical arrangement of cleanout and piping for areas surfaced with concrete.

Although the surface of the playground is concrete, it is still subject to a certain amount of settlement over the period of years. Therefore, careful attention must be given the installation of the cleanouts to prevent any possibility of damaging strain being placed on the underground piping.

A durable cast iron cleanout box is used to provide access to the cleanout plug. The top of the box should be installed flush with the surface of the concrete. If the box projects above the top of the concrete it could cause children to trip. The exposed surface of the cover is scored to prevent slipping. If tamper-proof protection is desired, a hinged cover equipped with a key and lock can be furnished. A number of different types of boxes and covers are manufactured for this purpose.

When this arrangement is carefully installed it can be counted upon for many years of reliable service.

Many types of units are manufactured that will serve the same general purpose as the units illustrated. With this availability of products to choose from, one is sure to find a unit to fit the most specific requirements.

CHAPTER 6

Parking Area Drainage

When a building is located on level or raised ground there is usually no problem in draining the rain water that falls on the site in the vicinity of the building. In fact, the grade immediately surrounding the building is generally sloped away from the building to allow the water to follow its normal path along the terrain and to seep into the earth in a natural way. Thus, for buildings so situated, rainfall drainage is no great concern.

However, a building often will be erected on a site that will include a paved area for parking cars or for other purposes. If the surface of this area is paved with asphalt, concrete or other surfacing materials, only a small per cent of the rain water will be absorbed. Therefore, it is necessary to slope this area to enable the water to run off into the street or to other portions of the property where it will drain into the earth.

Methods for Drainage

When it is not feasible for the water to drain into the earth because of the location of the site and its particular grade slope, then attention must be given to other means to drain the area of its expected rainfall. Other means may consist of connections to storm or sanitary gravity systems within the building, pumping systems, or dry wells.

Every now and then, a condition will be encountered where, because of certain site limitations or other reasons, the parking area is located in a depression. Because of its negative slope, water cannot drain away from the building itself.

Refer to the key plan in Fig. 1. The parking area, which is located behind the building, is in a grade depression with even the driveway pitching toward the parking area. Unless proper steps are taken to ensure adequate drainage, the entire area or at least a portion thereof will remain flooded after a storm. This is not a desirable condition

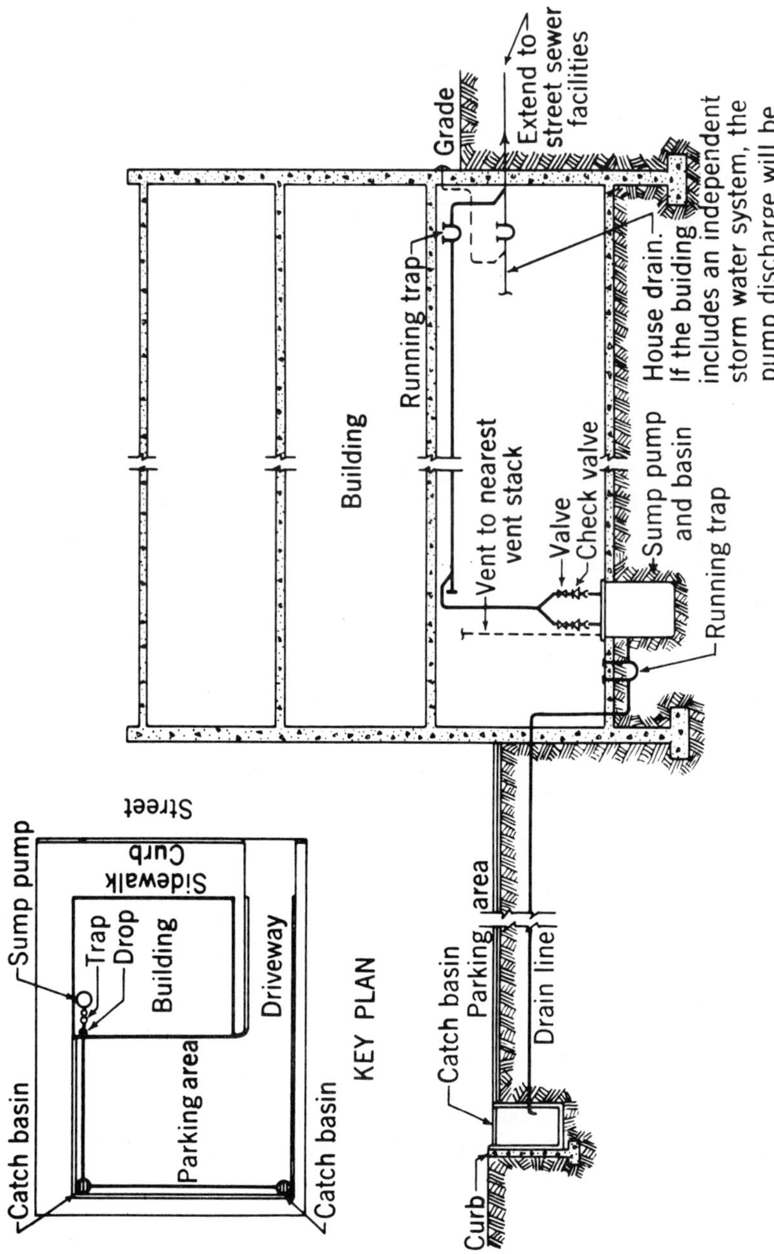

Fig. 1. Method of draining parking area by means of a sump pump installed within the building. Key plan shows the layout of building in respect to the grounds.

and can only result in considerable annoyance to the owner of the building and his tenants or customers. It will result in criticism of the design.

Dependent upon certain factors, pertinent to the site, there are several methods that may be utilized to rectify this condition.

Sump Pumps

If the parking area is so located with proximity to the building that the space by means of underground piping can be drained into the gravity storm or combined sanitary or storm system within the building, then the disposal of the rain water is greatly simplified. If, however, the parking area is at such an elevation that it is not possible to utilize the building gravity systems, then some thought must be given to the application of a sump pump. Figure 1 shows an installation where a sump pump is employed.

As detailed in this figure, the storm water runs into the catch basin from which it is conducted, via the underground piping, into the building and to the sump basin. The sump pump should be of the duplex type and of a capacity to handle the flow during a peak storm. The capacity of the basin is normally determined on the basis of the pumping cycle desired for the pumps. The pumps lift the water to an elevation that will allow the water to drain by gravity into the interior storm water system or combined facilities and thus discharge into the exterior street facilities. If neither storm or sanitary facilities are available in the street, then the water must be pumped to the curb or other points of disposal.

Dry Wells

If the soil characteristics at the site are porous enough to permit a rapid absorption of the rainfall, dry wells may be employed. Figure 2 is a typical example showing how and where the dry wells are placed.

As shown in Fig. 2, catch basins are set at the low point or points of the parking area. Rainfall runs into the catch basins from which, by means of underground piping, the water is conducted to the dry well. One or more dry wells, as the need may be, are constructed just outside the area of the parking space shown; the dry well should be of sufficient size and capacity to receive the quantity of rain that will fall on the area in a 24-hour period. Rainfall intensity tables will supply data defining the amount of rain that will fall on a specific area in a given time. This data, as well as the leeching rate of the soil, determine the size and capacity of the dry well.

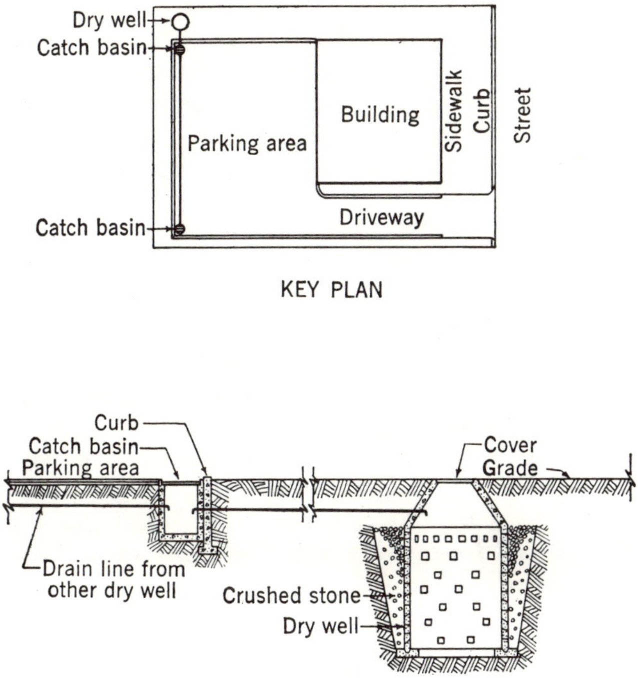

Fig. 2. Method of draining parking area using a dry well.

The dry well shown in Fig. 2, is of the pre-cast type because of the extra strength it offers. Crushed stone is placed around the dry well to increase its drainage capacity. Without the crushed stone, the earth would soon become clogged with the silt and debris that settles on the parking area and washes into the dry well with the rain. The crushed stone prevents this by permitting the silt and debris to filter down to the stone thereby greatly increasing the serviceability of the dry well.

If the earth has a good percolation rate, and the dry wells are placed at an elevation that is high enough above ground water level, the dry well can be counted upon for many years of dependable service.

Submersible Pumps

If the property is so situated that storm or combined sewer facilities do not exist in the street and the ground is not suitable for installing dry wells, then other methods for disposing of the rain water must be explored. The task is further complicated if, because of the construction of the building and arrangement of floor space, it is impossible to conduct the rain water to suitable pumping equipment in the building. When these conditions arise, it may be necessary to give some thought to the type of pumping equipment that can be located at the exterior of the building near the parking area. Figure 3, represents a typical installation of this kind.

Referring to Fig. 3, the rain water is intercepted by the catch basins which, in this case, conduct the water to a pit containing the pumping equipment as shown. A submersible pump is used in this example to show the range of different types of pumping equipment available for this purpose.

As the water runs into the pit, the water level rises in the pit until at a predetermined height it actuates the pump. The water is then discharged through the pump discharge and up the grade slope as illustrated, to a point where this discharge runs by gravity out to the curb.

The reason why it discharges to the curb is that this is the only place on the site left for the disposal of the rain water. When water is discharged at the curb it will follow the slope of the street gutter

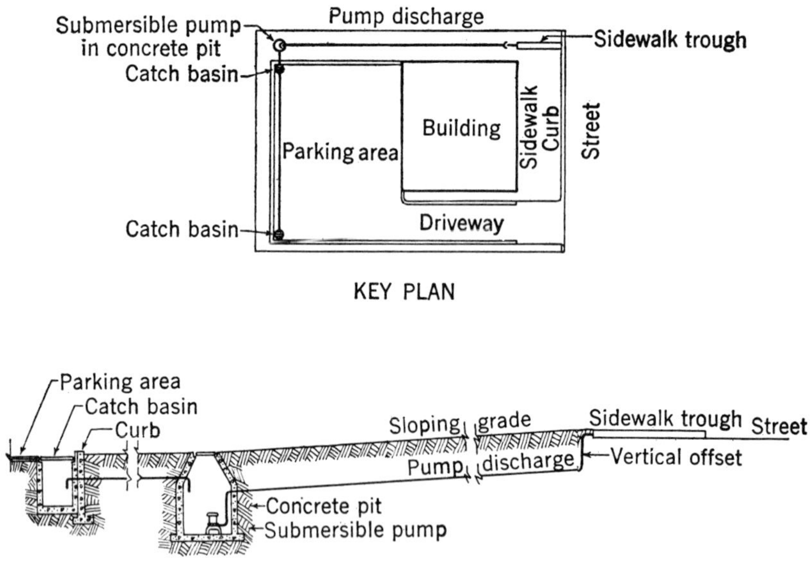

Fig. 3. Draining the parking area with the aid of a submersible pump. Water is discharged at the curb.

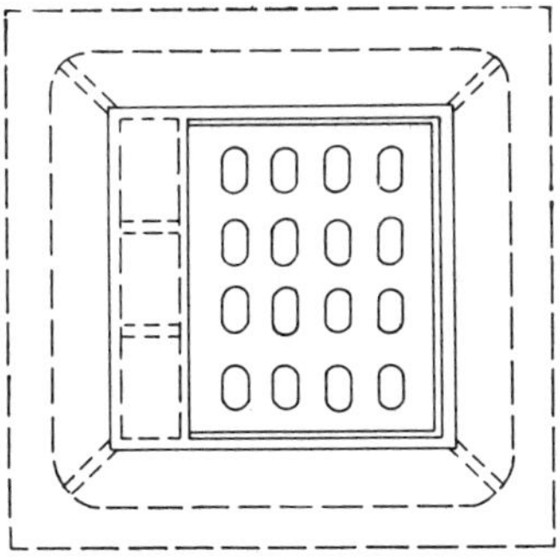

PLAN

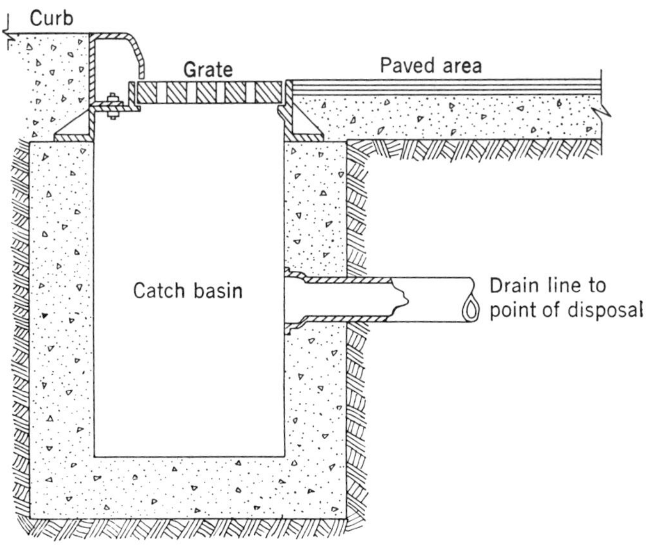

SECTION

Fig. 4. Typical construction of a catch basin.

and eventually reach a natural drainage area, as does the rain that falls on the street. Generally, when such conditions exist at the site, it is usual practice to also conduct the roof leaders of the building to the curb and allow the water to discharge at this point.

The pump pit is of the storage capacity that will permit the frequency of pumping cycles desired. The controls for the submersible pump can be located, either over the pit, in the pit, on the nearest exterior wall of the building or at an interior wall of the building. Many types of controls are available for this purpose. In any event, a disconnect switch should be placed in or above the pit for the safety of the maintenance man before entering the pit.

If a regular sump pump is used, the motors would either have to be exposed or contained in a small pump house erected over the pit. However, either method can be used with a good deal of success. The motors for a regular sump pump could of course be located underground but the building of the watertight compartment needed to house the motors might be too expensive in this particular case.

Catch Basin

Figure 4, presents a detail of the catch basin construction shown in each of the illustrations. Although there are many types available for this purpose, this particular one is shown as example. It is a good design. The space provided below the inlet or outlet of the basin is to intercept and retain the debris that washes into it.

Large Paved Areas

When large paved areas are encountered, the same provisions as outlined can be employed with the exception that all of the equipment used, should be of greater capacity. For example, when only dry wells are used, the prescribed number corresponding to the size of the overall area to be drained should be provided. The same consideration must be applied to equipment in the other methods shown.

This same method of treatment can be applied to all types of buildings under similar circumstances where a paved area is to be included. This covers schools, industrial plants, shopping centers, restaurants, public buildings and all other general types of buildings.

CHAPTER 7

Draining the Planting Area

Planting areas have become more and more popular in all types of buildings. Many new hotels, motels, department stores, apartment houses, shopping centers, banks and numerous other types of buildings will include a planting area as a decorative scheme to enhance either the interior or exterior of a building.

The planting area may consist of a small space, containing but a few flowers and shrubs or it may be an elaborate area devoted to many types of flowers, shrubs or even small size trees.

When such planting areas are utilized outside of a building, special consideration need not be required since the water falling on the planting area will seep through and into the earth. When the planting area is included inside the building or on the roof of a building, then proper drainage becomes a major consideration. The same thought must be applied to planting areas located at the exterior of a building over an underground portion of the building.

Interior Planting Areas

For example, let us assume a planting area provided by an architect for the interior of a bank, as shown in Fig. 1. This area has been gracefully curved to offer a distinctive appearance. Indicated in the figure is the location of the planting area drain which is a vital part of the planting area. The drain permits a natural seepage of water through the earth. Otherwise the soil would become saturated, which is not the best condition for many types of plantings.

Unless a special arrangement is considered for watering the area, a source of water supply must be located near the area so that a hose can be used for the watering.

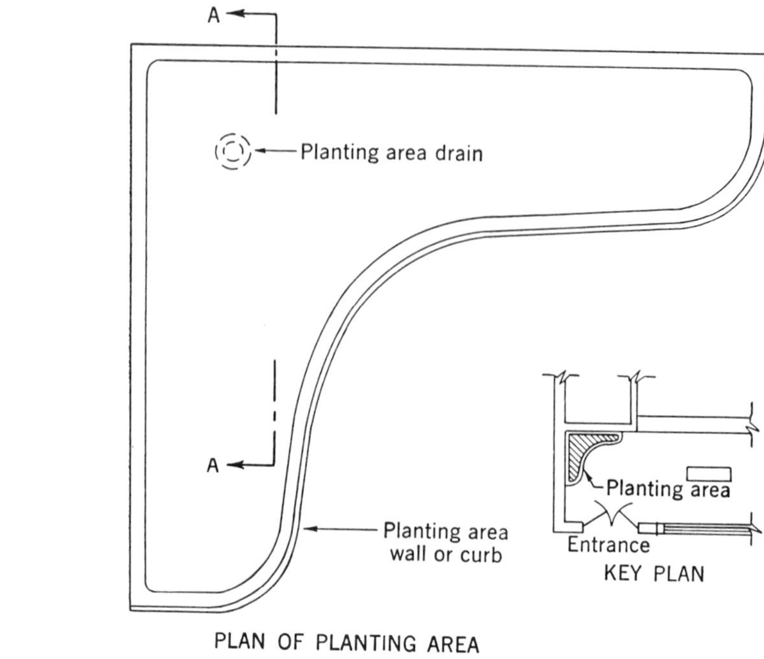

Fig. 1. Layout of a planting area that might be included in a building. Key plan shows location of planting area.

Section A-A through the planting area of Fig. 1, is shown in Fig. 2 and illustrates pertinent details as well as the means for providing drainage.

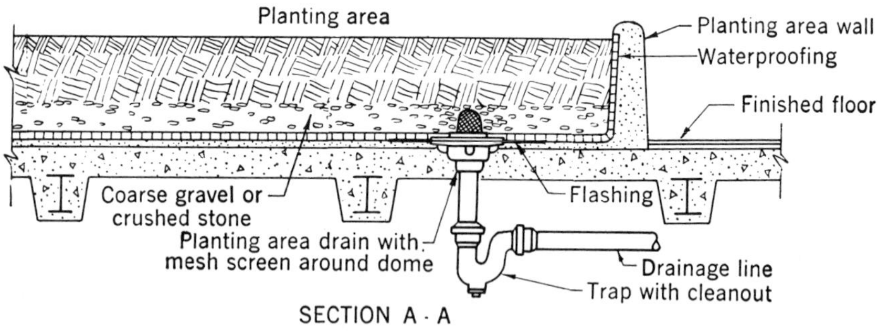

Fig. 2. Section through planting area of Fig. 1, showing arrangement of planting area drain.

The architect may choose to select any number of different types of construction materials for the walls around the planting area. For example, we have indicated concrete walls which can be lined with waterproof material or can be painted with a waterproofing compound. These walls are generally 18 to 24 inches high, depending upon the proposed depth of the soil, the architect or landscape architect wishes to maintain.

The bottom of the planting area is sloped toward the drain. The drain itself, consists of a unit, especially designed for this purpose, which is flashed to the construction in a watertight manner. Lead or copper flashing is used for this purpose. A shallow dome which is slotted or perforated is placed on the drain. This dome includes a bronze mesh screen that fits over it.

Coarse gravel or crushed stone is placed at the bottom of the planting area to a height that covers the dome. The function of this material is to allow the water to seep through the earth in a natural way and then flow freely to the drain. This material also filters out the fine silt that is washed down with the water and prevents it from clogging the fine mesh screen.

The drainage line should not be connected directly to the sanitary system but instead should be discharged as indirect waste into an open sink or vessel that is properly vented.

Roof Planting Areas

The planting area that may be provided for a penthouse garden is generally similar in construction to that of the interior planting area. Here, however, the elements must be taken into consideration.

Figure 3 shows an installation of this type.

The planting area can be of any desired shape or size according to the architect's design. A rather simple arrangement is pictured for the purpose of discussion.

In addition to being watered by a hose, a roof planting area is also subject to natural rainfall. A heavy rainfall will quickly saturate the soil and if not controlled, will flow over the walls of the curb, carrying precious top soil with it. In addition to losing the top soil, this soil also spoils the neat appearance of the roof area. Thus, unless proper means of drainage is provided, the top soil will have to be replaced after every rain storm and the roof area will have to be cleaned after each storm.

To avoid this condition, one or more standpipes with drainage domes must be employed, depending upon the overall size of the

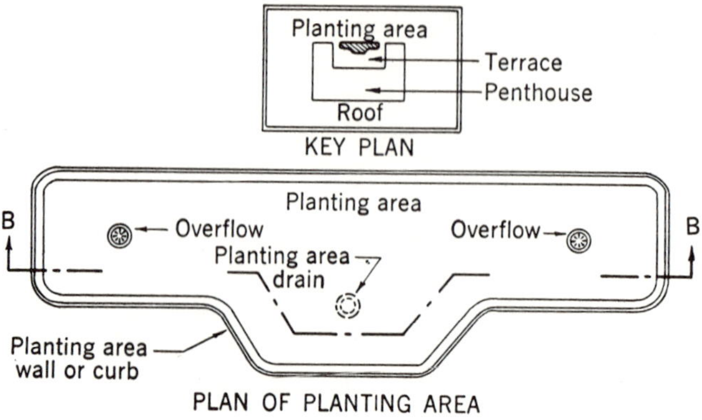

Fig. 3. Layout of a planting area that might be included on the roof of a building. Key plan shows planting area in relation to penthouse.

planting area. These overflow drains are usually so placed that they will be concealed by the flowers or shrubs. A typical example of the drainage devices is shown in Fig. 4.

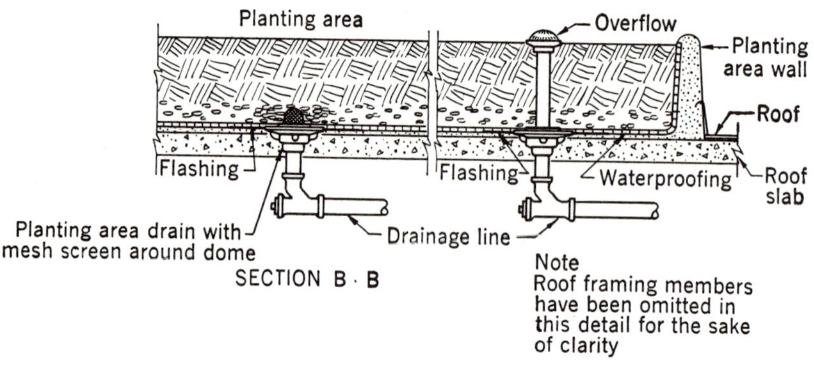

Fig. 4. Section through planting area of Fig. 3, showing planting area drain and overflow drain.

The purpose of the overflow drains is to carry off the rainwater when it exceeds the height of the level of the top soil contained in the planting area. The slotted dome is oversized to allow air pressure to balance the water discharging into this piping.

A planting area drain, with the same function as that explained for the interior planting area, is installed at the bottom of the area.

This drain also functions to allow water to drain through the earth in a natural way.

Every precaution must be exercised when erecting a planting area on a roof to avoid any possibility of rain water seeping under the area and causing leaks. Since this is the architect's problem, it will not be given any further attention at this time.

The planting area drains and the overflow drains can be connected to the interior leader piping. Cleanouts are placed in the drainage line to permit clearing the line of any stoppages that may be caused by accumulation of earth silt. When adequate attention is given to the effect of the elements on a roof planting area, one can be assured of a system that will function properly.

Large Exterior Planting Areas

Often a planting area of enlarged scope will be placed at the exterior of a building. As mentioned earlier, if the water in this instance can seep through the soil and into the earth, no unusual problems will be experienced. There are occasions, however, when such planting areas are located over an underground garage or over an underground portion of a building. When this situation is prevalent, special care must be given the design of the drainage system. The engineer has many methods at his disposal. One treatment is shown in Fig. 5.

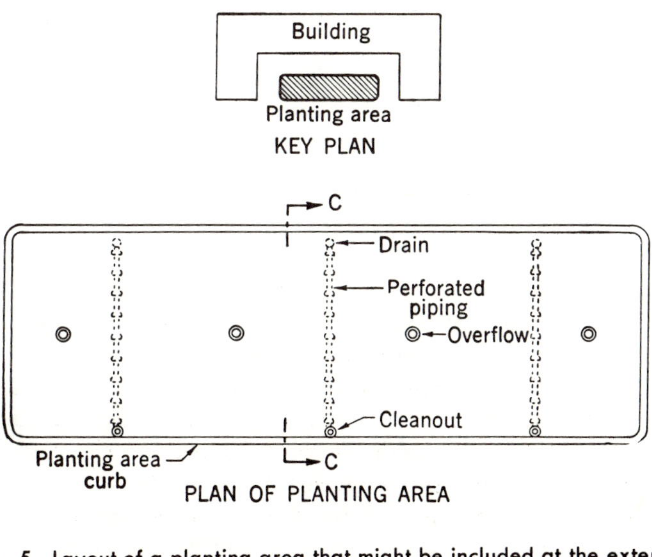

Fig. 5. Layout of a planting area that might be included at the exterior of a building. Key plan shows relation of planting area to building.

The area given, for example, is approximately 20 ft. x 60 ft. We shall assume that it is located over an underground parking garage for the building shown in the key plan. This type of a planting area is quite popular for a number of different types of buildings.

Once again, we have provided overflow drains to prevent the rain water from overflowing the walls of the curb and carrying away the top soil. For draining the planting area we have utilized a system of perforated clay pipe. The clay pipe is installed as shown and connects with a drain that is in turn connected to the drainage piping in the space below. The drainage piping below may be connected indirectly to the sanitary system or directly connected to the system of storm water piping that is run in the space below. The favored means of connection must be determined by the engineer. A section through this planting area is shown in Fig. 6.

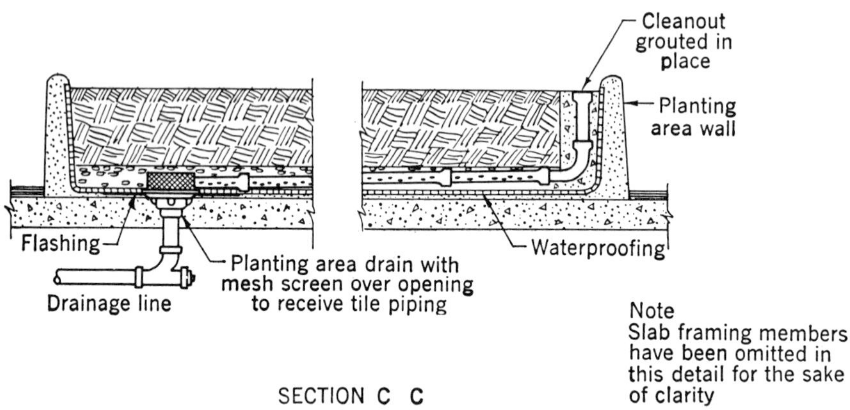

Fig. 6. Section through planting area of Fig. 5, showing drain and arrangement of clay drainage piping.

The overflow drains are installed in the same manner as that shown in Fig. 4. The clay pipe is installed to follow the slope or valley that has been provided in the bottom of the planting area. This slope is in the direction toward the drain.

Coarse gravel or crushed stone is again used to offer a path of travel for the water that seeps through the earth. A cleanout is provided at the end of the clay pipe to permit clearing the clay line of any soil or silt that may accumulate. To accomplish this a hose is used and the water is applied in a gentle manner to flush the accumulation out of the line. It is good practice to install the piping in the space below, in such a manner, that it can be equipped with a stand interceptor with easy access for cleaning. The sand interceptor can be a cast iron basin or any other means that the engineer may devise.

A note of caution: Deep rooted shrubs or trees should never be placed near or directly over the planting area drains or drainage piping. Such roots can block up the mesh screen dome of the planting area drain or can plug the perforated clay piping. Hair-like roots of some trees, can enter the pipe perforations, grow inside the pipe, and completely block drainage flow.

In addition to the methods illustrated for draining planting areas, there are other means that can be employed. When special effects are utilized for watering the planting area, it may also be necessary to give some special attention to the system that can be used to drain the water and control its level.

CHAPTER 8

Installation of Exterior Piping

When installation of a sanitary, storm or combined sanitary and storm system is contemplated, it is very important, in addition to proper pipe sizing, to make sure that the piping will be carefully installed in the earth. The type soil existing at a site is of vital concern. Unstable soil conditions require that every precaution be taken to insure that the piping will not sag in the ground after installation.

When piping sags, two dangerous conditions develop: (1) rupture of the piping at the point of sagging, with its resulting washout and (2) reduced area for flow in the piping, which defeats the purpose of an adequately sized line. All such dangers can be avoided when the piping is supported carefully in the first place.

Typical Installation

A typical installation is shown in Fig. 1, where the piping is run to serve a group of apartment houses. The exterior system may be a sanitary, storm or combination sanitary and storm system. Let us assume that a combination sanitary-storm system is utilized. The line is extended from each building to the manhole, as shown. The lines are then combined and run to the point of termination at the outside facilities.

Piping Materials

There are a number of different piping materials that can be used for exterior piping systems. Most frequently used are: asbestos-cement pipe, bituminized-fiber pipe, cast iron soil pipe, concrete sewer pipe, and vitrified-clay pipe.

The choice of material is dependent upon a number of factors and the judgment of the engineer with regard to the particular

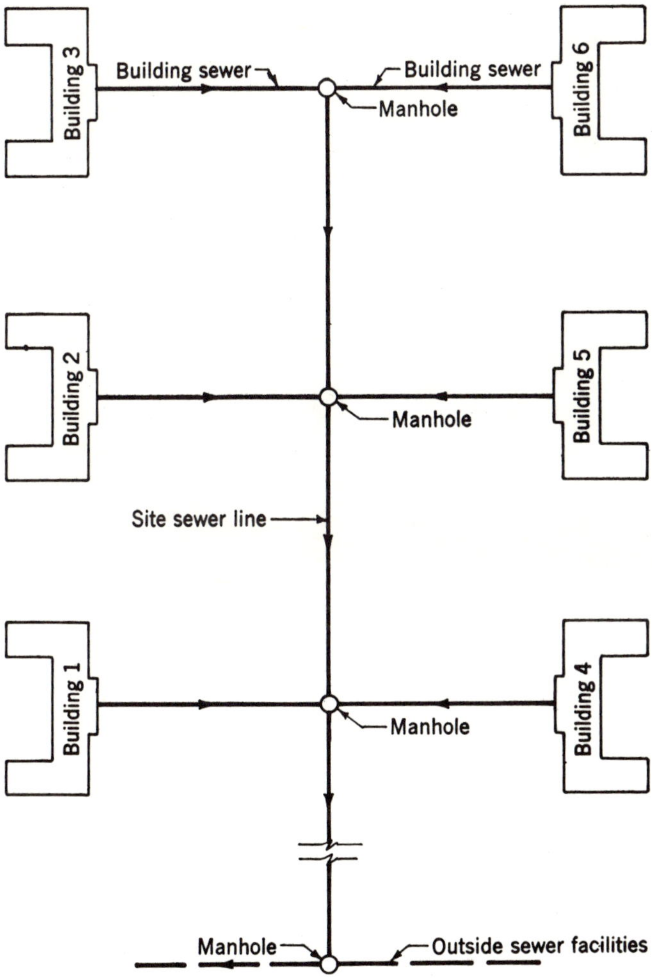

Fig. 1. Typical arrangement of sewer piping that serves a group of apartment houses.

site and the capacity of the required system. In many areas, local plumbing department, sewer department, or highway department will prescribe the type piping that should be used. It is always good practice to get the pipe manufacturers' recommendations concerning use, method of joining, and other helpful data. Also available are handbooks which contain a wealth of information on piping materials. Booklets may be obtained by

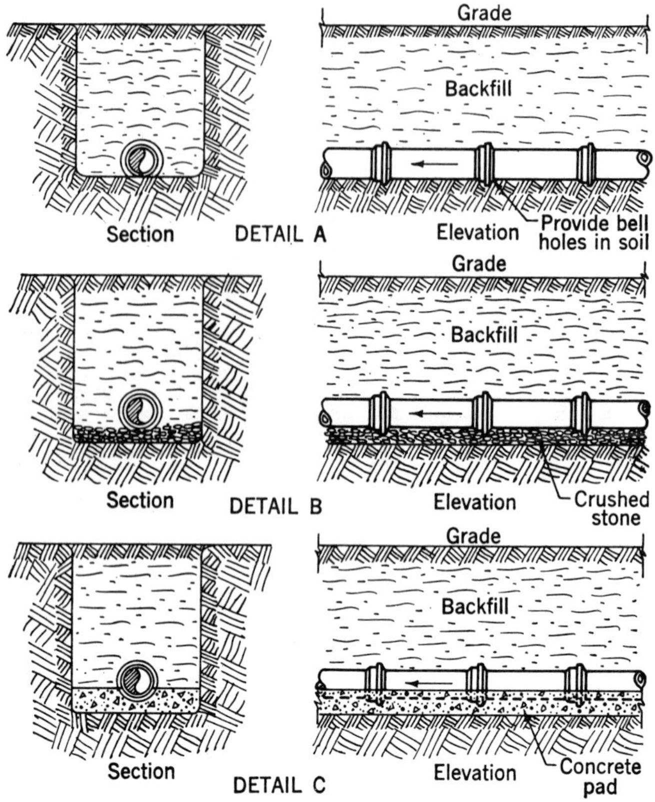

Fig. 2. Three ways to install piping at the bottom of a trench.

writing to a particular manufacturer or by writing to the association of manufacturers for each of the piping materials listed above.

Types of Installations

Under normal conditions, piping can be installed in one of several ways. Fig. 2 shows three of the most common:

Detail *A* shows the piping in a trench that has been carefully dug and graded in stable soil. Bell holes are provided for the hubs so the piping is firmly supported for its entire run. When the backfill in this instance is carefully applied and tamped, a good installation will result.

Detail *B* shows how the piping can be installed in trenches that have been dug deeper than required or are not uniform in depth or grading. In this case, crushed stone is added to the trench bottom and is carefully graded for the proper support of the piping. The crushed stone offers a firm bed for pipe support and is also of value when muddy conditions are encountered at the trench bottom.

Detail *C* illustrates how concrete can be added to the trench bottom for a firm piping support. When required, reinforcing rods are set in the concrete. Also when necessary, the piping is completely encased in concrete.

Fill

Very often, by virtue of the contours of the site, fill is needed in the depressed areas to grade the site to the desired level or slope. Grading is accomplished so quickly nowadays that fill will not have sufficient time to compact. Unless fill is compacted, it cannot support the weight of piping. Figure 3 shows the piping installed in a section of the site where a depressed point in the grading was filled in.

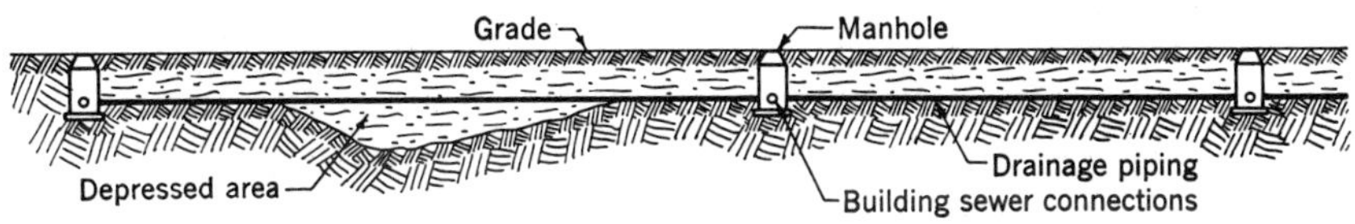

Fig. 3. Depressed areas or voids can affect the installation of underground drainage piping.

Fill requires a number of months to fully compact in a natural way. Since fill placed in this depression does not have sufficient time to compact, it cannot be relied upon to support the piping installed as shown. Unless adequate means are considered for the support of the piping it can sag and rupture. It is needless to mention the health hazard that will result when a line conveying the sanitary waste of the buildings ruptures. Furthermore, a heavy rainfall could cause a washout at the point of rupture.

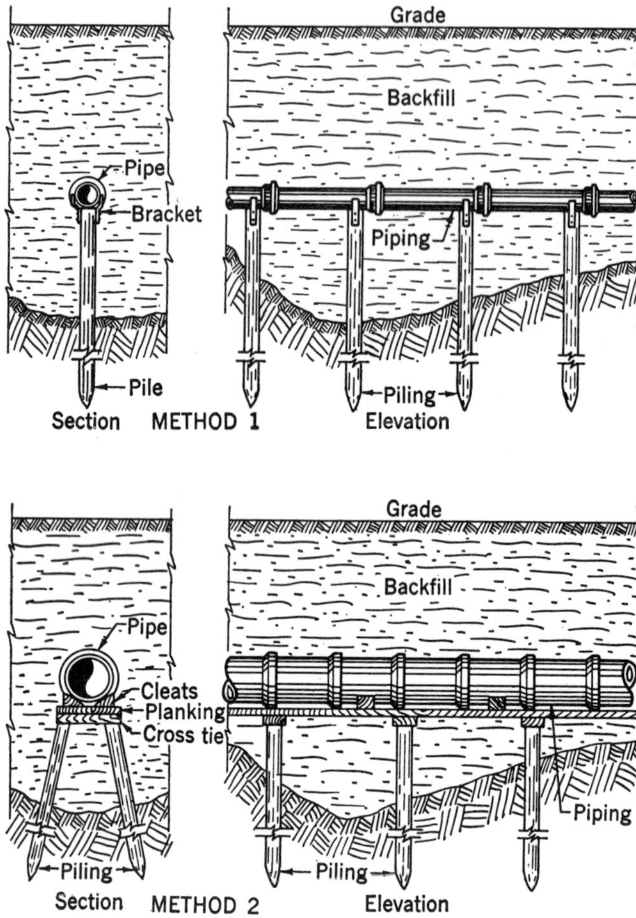

Fig. 4. Two methods of supporting piping in fill.

Piping Supports

When piping is to be supported in fill, there are a number of different methods that may be utilized to support the piping, and two methods that may be used with a good deal of success are shown in Fig. 4.

Method 1 of Fig. 4 shows how piling is driven into the earth for stable footing to support the piping. Brackets are placed on top of the piling to serve as a cradle for the piping. The piling is usually of selected timber and is treated to withstand the type of soil in which it is placed. This method has certain limitations and when larger diameter piping is to be supported, more substantial methods must be considered.

Method 2 shows how large diameter piping can be supported. The piling is driven down into hard pan for adequate footing. A cross member is placed across the top of the piling on top of which timber planking is placed to actually support the piping for the length of the depression into which fill has been placed. Wood cleats are nailed to the planking to afford proper alignment for the piping. Timber used in this arrangement is of selected stock and must be treated to withstand soil action. For both methods, the piling must be spaced in accordance with the weight

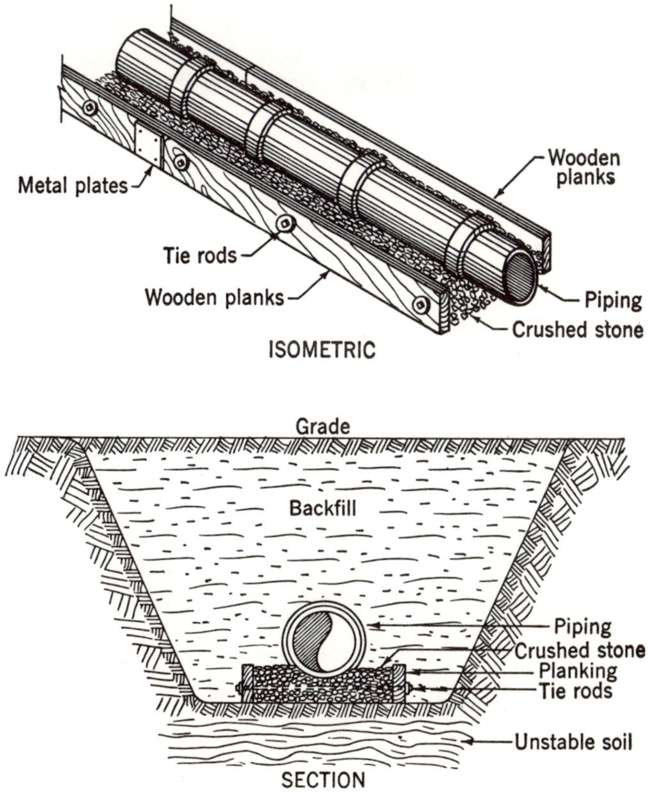

Fig. 5. One arrangement for supporting piping in unstable soil.

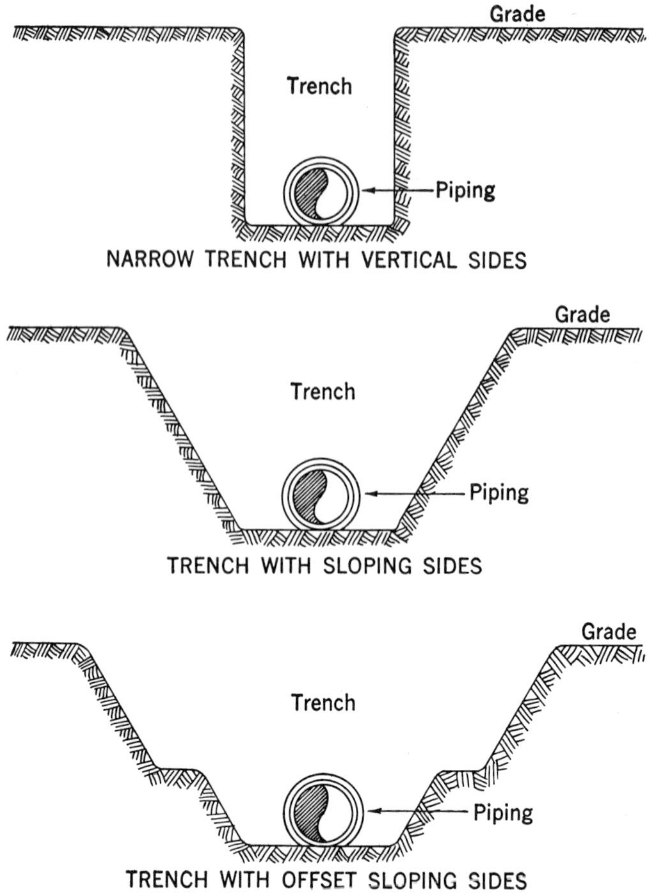

Fig. 6. Types of trenches; use is dependent upon soil conditions.

of the piping. When either method is carefully applied, the danger of line settlement will be completely removed. Both methods of support may be used for any of the piping materials.

Marshy Ground

Sometimes it becomes necessary to run piping in marshy or unstable ground. One way this is handled is shown in Fig. 5. The trench is dug and a cradle of crushed stone provided at the trench bottom to support the piping. This means of support is accomplished in the following manner: A framework of planking is installed at the trench bottom, spaced and held together with tie rods. The planking, of course, is sloped to conform to the designed slope required for the piping. The space between the planking is filled with crushed stone for the actual support of the piping. Other methods are employed with equal success; sometimes timbers are placed like railroad ties to support the piping. It is often necessary to use well points and pumping equipment to keep the trench drained of water while the supporting methods are set in place.

Shape of Trenching

The particular shape of a trench to be dug depends upon the type of soil. Three different trench shapes are shown in cross-section in Fig. 6.

The first, a narrow trench with vertical sides, can only be dug in hard clay. The second method shows how the trench sides are sloped when the trench is to be dug in a soil mixture other than hard, compacted clay. A 2:1 ratio is generally used for the slope of the sides. However, the excavating contractor will utilize a slope that he decides best fits the job conditions. This procedure is necessary to prevent the sides of the trench from caving in and endangering the workman.

The last method shows a trench with sloping sides that were offset to prevent the sides from caving in. This method is needed where a high percentage of sand is present in the soil.

If space is a factor and a narrow trench is to be dug in sandy soil, then it becomes necessary to erect wood siding to prevent the sides of the trench from caving in.

Backfill

Careful placement of backfill is very important to proper installation of underground drainage lines. Backfill placed around the piping itself should be carefully selected and screened to eliminate large objects, as rocks and twigs. All backfill immediately surrounding the piping should be handled with care and carefully tamped to insure that the piping is properly supported. Water, when used to aid soil compacting, should be applied in measured amounts to avoid saturation of backfill around the piping.

The remainder of the trench should then be filled in, with care exercised that large rocks or other heavy objects are screened out of the backfill. When the backfill is replaced in a careful manner, the piping certainly will benefit from the added support the soil affords.

Engineer's Responsibility

Even though soil bearings can be taken along the prescribed route of a drainage line, one never is sure of just what the excavator will find when he digs the trench. Therefore, the engineer is advised to include in his specifications instructions that the contractor furnish and erect the adequate supports needed when the piping is installed in unstable soil. The specifications should also include instructions that piping be carefully sloped as per the calculated pitch to assure that the system will perform in the manner for which it was designed. Many specifications include a section that requests that the contractor present for approval the method of supporting the piping that he proposes to use in the event that unstable soil conditions are encountered. There are a number of ways that this matter can be handled.

Each project must be judged on its own merits. The advice of local plumbing, sewer and highway departments can be of considerable value because of their knowledge of the manner in which similar installations were handled in the area.

CHAPTER 9

Septic Tank Systems

Sewage disposal is an easy matter when street sewer facilities are available, but becomes more difficult when such facilities are not available. When sewer facilities are not available, it becomes necessary to dispose of waste on the property, in a manner approved by local plumbing and health authorities. It is not the intent of this chapter to compare the specific merits of one method of disposal over another, but to review some precautions that must be taken when a private disposal means is contemplated. Let us assume that the site selected is that of a small sized school of the kind being erected in all areas of the country.

School Building

Figure 1 shows the hypothetical school used for this chapter. The building, driveway, paved areas, lawns and play areas are shown. For sewage disposal, a septic tank, distribution box and means of leaching have been selected.

The sewage disposal system is at the front of the building to simplify connection to the street sewer in the event that such facilities are later provided. If the local community or highway department should construct a street sewer in front of the property, it would be an easy matter to disconnect the house sewer from the septic system and extend it out to the street sewer facilities.

If the septic tank is to be located in the area shown, it is first necessary to check with the local authorities and gain their precise requirements concerning the limits within which the system must be placed. Most local plumbing and health authorities have definite regulations concerning the design and installation of the proposed system as well as the required distances between components of the system and the property limits.

In this particular location in front of the building, the system is relatively safe and will not be subjected to surface loads

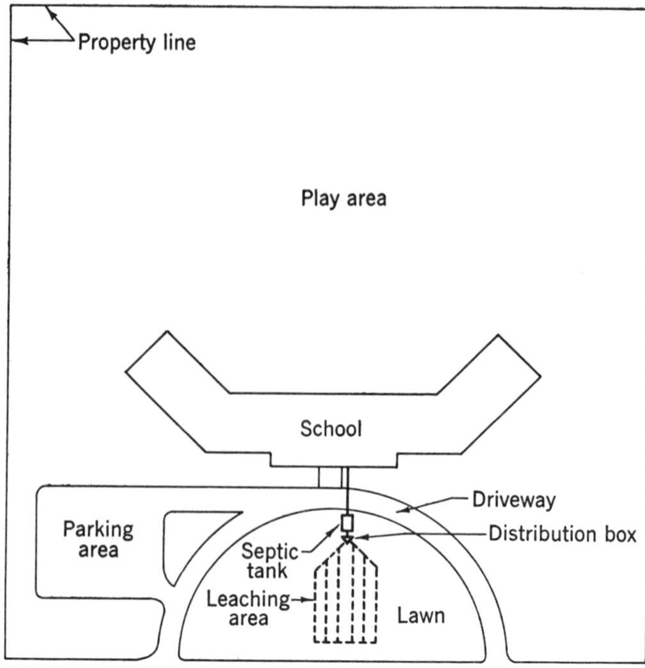

Fig. 1. One possible location of leaching area on the grounds of a small school.

greater than that imposed by foot traffic or grass cutting equipment. As a result, the drainage piping is not likely to settle and will offer many years of dependable service.

Other Locations

Sometimes there is a certainty that street sewer facilities will never be installed. Then, because of the nature of the property, it may be better to place the disposal system at the rear of the school, as in Fig. 2. Here, also, since the system is in a play area, it will not be subject to loads other than that imposed by foot traffic and lawn cutting equipment.

If there is danger of heavy equipment passing over the piping, it may be advisable to erect a fence around the general area of the disposal system. Otherwise, it would be practical to locate the system to one side of the property, or to keep it out of the

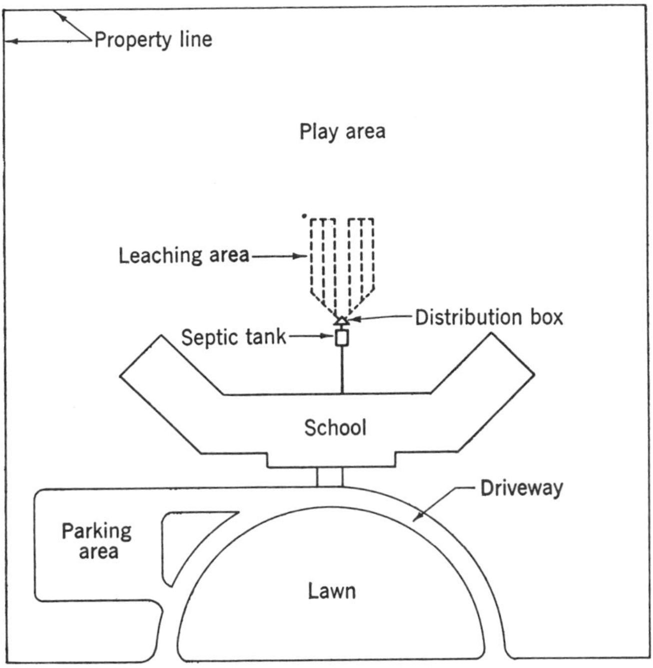

Fig. 2. Septic tank and leaching area are here located at the rear of a school building.

way of areas subjected to heavy loads. It is always good practice to take every precaution to safeguard a sewage disposal system and to anticipate occurrences that could cause piping in the leaching area to sag or break, thus destroying the function of the system.

A further consideration is to so locate the disposal system that, should the school expand its facilities, it can be accomplished with a minimum of alterations to the present system. In planning the disposal system, it is always advisable to anticipate the ultimate growth of the school and to size the septic tank accordingly. Piping in the leaching area is easily extended to accommodate an increase sewage load in the building.

Another important point to remember is that septic tanks must be cleaned periodically and therefore some provision must be made for the entry of cleaning equipment and for ease of

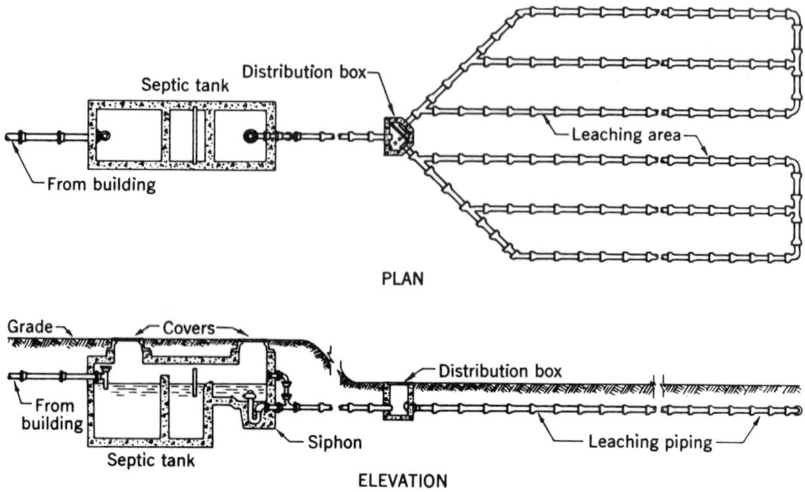

Fig. 3. A common arrangement of septic tank, distribution box and leaching area.

handling. Suction apparatus is generally used for cleaning and the means and method of approach to the septic tank must be given some consideration.

Typical Installation

Figure 3 details one of several types of sewage disposal systems. The septic tank consists of a reinforced concrete structure with access covers extending up to the grade. The outlet connection utilizes a siphon device to dose the drainage piping in the leaching area. The distribution box is a concrete structure with a cover at grade. A loose baffle in the distribution box permits the waste to be directed to one section of the leaching area or the other, allowing one section to be used while the other section drains dry. The baffle position is changed by a maintenance man at regular intervals. Drainage piping used in the leaching area is available in a number of different materials. The particular arrangement of the drainage piping shown in Fig. 3 was arbitrarily chosen for illustrative purposes only. Other arrangements can be used to satisfy the specific needs of the system or the area in which the piping is to be installed.

The supervising authorities must be assured that the soil in which the leaching is to be disposed is capable of handling the

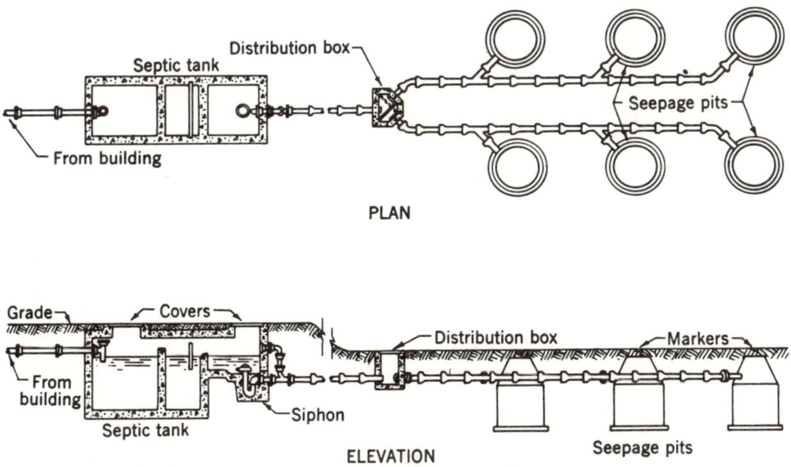

Fig. 4. Arrangement of a septic tank system utilizing a group of seepage pits.

quantity of water that is discharged into it. The engineer must be able to show that the soil is of suitable character and that percolation tests prove that it is of good drainage capacity.

In Fig. 3, the grade slopes from the septic tank, but the design of an actual system must be based on the characteristics of the grade in which the proposed installation is to be made.

Seepage Pits

Another method often used in sewage disposal work is that which includes seepage pits or dry wells, as they are often called. An example of this type of system is illustrated in Fig. 4. The septic tank and distribution box are the same as that shown in Fig. 3. The difference is the means for discharging waste water into the ground. Seepage pits are used instead of the drainage piping.

The number of seepage pits required is determined by the quantities of waste water being disposed. The diameter and depth of the seepage pit is based upon the quantity of water to be handled and disposed of, the number of such pits that must be installed in a given area, and the characteristics of the soil in which the pits are to be installed.

Once again, the baffle in the distribution box permits one group

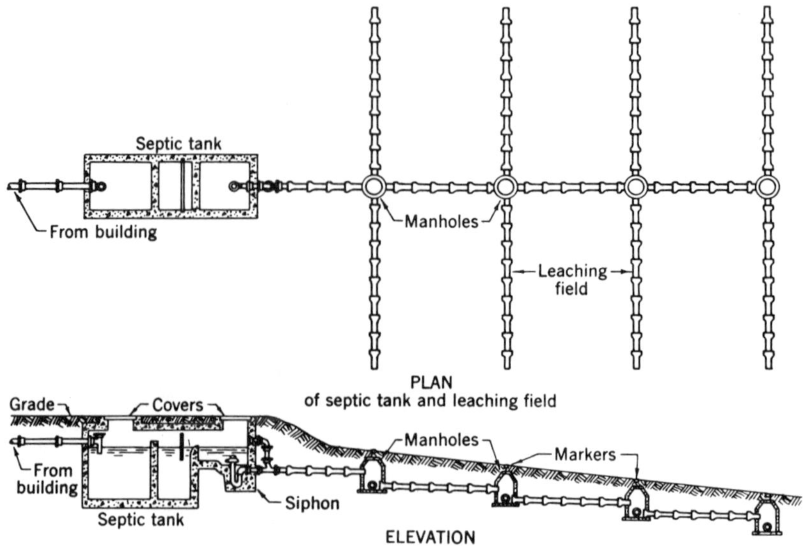

Fig. 5. Typical arrangement of a septic tank system using a leaching field for installations in sloping ground.

of the seepage pits to be used while the other group is draining dry. The design and installation of this method of sewage disposal is also subject to review and approval by the local plumbing and health authorities. The cover on top of the seepage pits offers access for maintenance purposes.

Sloped Grades

Very often, sloping grades are encountered in the leaching area. With such a condition, disposal might be accomplished by the method suggested in Fig. 5. Here the septic tank and distribution box are the same as shown in Fig. 3 and 4. However, the method of disposal varies in that the drainage piping must be so arranged to fully utilize the slope or contour of the grade in which it is to be installed.

This is accomplished by utilizing manholes at prescribed intervals, with each manhole at an elevation that best suits or conforms to the grade slope in which it is erected. Thus the drainage piping extends out from each manhole to serve as the means of disposal. The first manhole is dosed and as the capacity of the drainage piping of the first manhole is exceeded, the waste

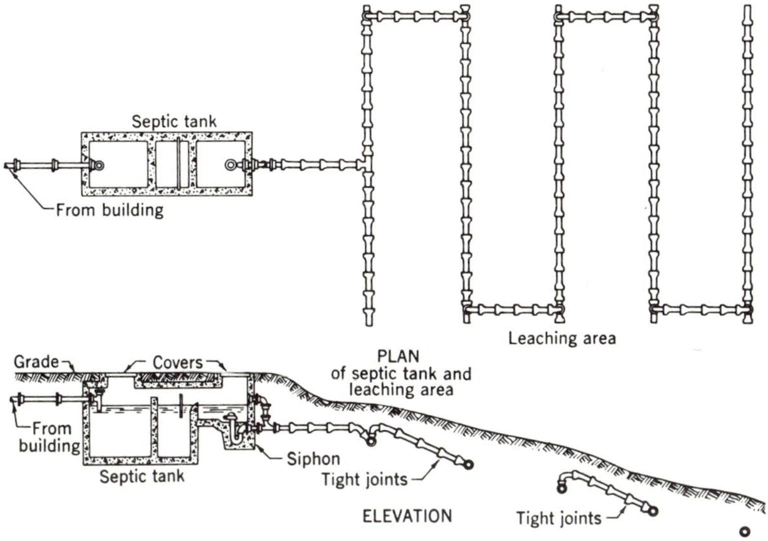

Fig. 6. Typical arrangement of a septic tank system utilizing serial distribution.

water will continue on to the next manhole to be handled by its system of drainage piping. This type of system is well suited to sloping grades where it would be impossible to install the regular kind of disposal piping.

Serial Distribution

In discussing septic tank systems, it is important to mention that comprehensive data are available on this subject in the Manual of Septic Tank Practice, published by the U. S. Public Health Service in 1957. New studies and investigations in this field are continually being made. A recent example is design data on septic tank systems which favor a new method of effluent distribution known as "serial distribution."

Serial distribution uses separate trenches, connected in series. All effluent is discharged into the first trench and the overflow so arranged that when the first trench is sufficiently dosed the effluent will pass on to the next trench. Thus, the entire leaching area will be utilized by the system of overflows. A typical arrangement for an installation with a sloping grade would be similar to that shown in Fig. 6.

CHAPTER 10

Parts of a Septic Tank System

In Chapter 9, the general arrangement of sewage disposal systems, based on the particular location and grade encountered, was discussed. Here, we discuss in more detail the component parts of the systems there described.

Septic Tanks

The prime function of a septic tank is to condition raw sewage so that it readily percolates into the ground. To accomplish this, it provides for the removal of solids, for biological treatment, and for sludge and scum storage.

Removal of solids: Suspended solids in the raw sewage settle to the bottom of the septic tank, thereby allowing the clarified effluent to be discharged for disposal.

Biological treatment: Solids and liquid in raw sewage decompose by bacterial and other natural processes.

Sludge and scum storage: Sludge is the accumulation of solids at the bottom of the septic tank while scum is a mat of floating solids. Both are eventually digested by bacterial action. Nevertheless, the septic tank must be of sufficient capacity and equipped with baffles so that it will retain material which is not fully digested, in the intervals between periodic cleanings.

There is a wide variety of septic tanks, from precast concrete tanks to large, elaborate structures designed to handle specific sewage loads. Furthermore, concrete structures are built in many different shapes and forms to accommodate site conditions. Fabricated steel tanks of varying sizes and shapes are available for septic tank application.

Several types of concrete structures are illustrated in Fig. 1. Detail *A* shows a small septic tank, available pre-cast. Briefly, it consists of an inlet and outlet tee connection and an access cover. A marker at grade level indicates the exact location of the cover.

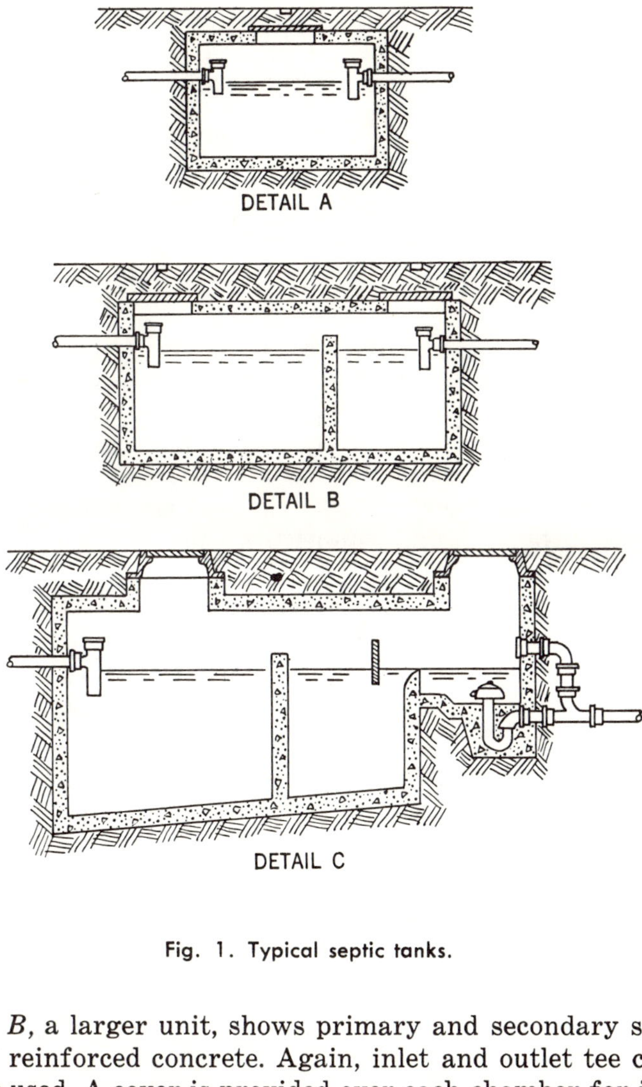

Fig. 1. Typical septic tanks.

Detail *B*, a larger unit, shows primary and secondary settling tanks of reinforced concrete. Again, inlet and outlet tee connections are used. A cover is provided over each chamber for inspection and cleaning purposes. Markers are set in the grade to indicate the exact location of covers.

Detail *C* illustrates a still larger septic tank, including primary and secondary settling chambers. There are an inlet connection, baffling arrangement, a dosing chamber with a siphon device, and access covers at grade. The settling chambers are sloped to allow the sludge to accumulate at the low point in the tank. This facilitates the cleaning operation. The sloping bottom is also desirable when a pumping arrangement is utilized with suction lines connected under the septic tank at the low point.

Concrete septic tank structures should be carefully designed and built to carry the slab across the top of the tank and, in addition, sustain a prescribed surface loading to which the structure may be subjected. With proper care and maintenance, a septic tank will provide many years of dependable service.

Typical Septic Tank

Figure 2 shows the typical septic tank that has been used for illustrative purposes in this and the preceding chapter. It is a reinforced concrete structure, although the reinforcing has not been detailed. The structure consists of a primary and secondary settling tank. An outlet tee is used to baffle the flow of raw sewage into the tank. A baffle in the secondary compartment controls turbulence and retains the scum in the tank.

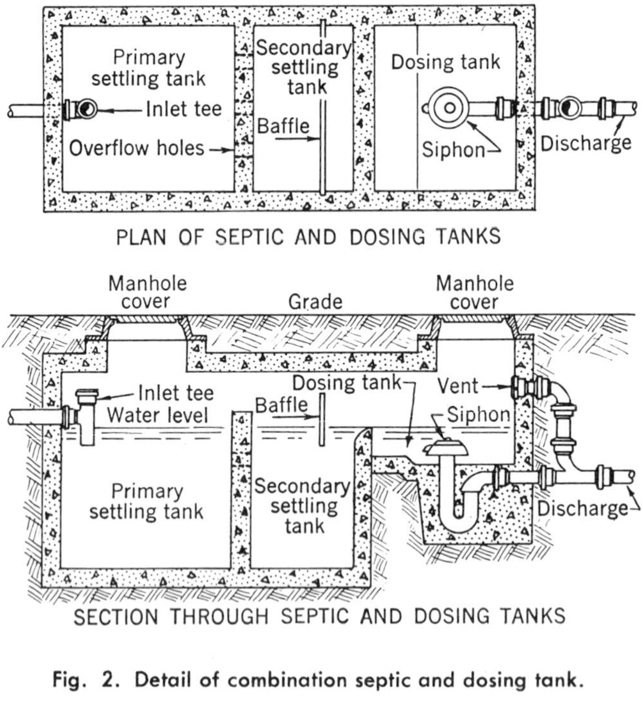

Fig. 2. Detail of combination septic and dosing tank.

As effluent passes into the dosing tank, a prescribed depth of water activates the siphon device and the quantity of waste water is discharged to the means of disposal. A vent connection at the outlet provides the atmospheric pressure needed to assure free flow of the waste water to the leaching means. Covers that extend up to the grade offer access to each compartment.

Many designs and shapes of septic tanks are used with equal success. Moreover, septic tanks constructed of other materials can also be used. It is important to point out however, that the design and construction of the septic tank system must conform to the requirements of the local plumbing or health authority, whichever has jurisdiction.

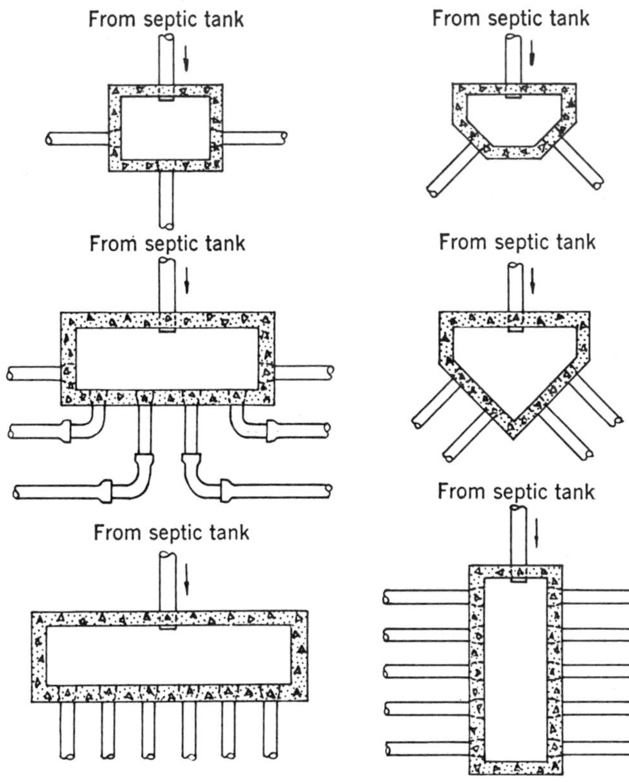

Fig. 3. Plan views of several of the many differently shaped distribution boxes.

Distribution Boxes

A distribution box is a concrete structure between septic tank and leaching area. Its prime function is to ensure an even distribution of effluent to the leaching area for better handling of effluent. There are a number of different types of distribution boxes that will satisfy a specific arrangement of drainage piping or seepage pits. Figure 3 illustrates some of the different shapes and sizes. It is the proposed piping arrangements that dictate the different shapes.

Typical Distribution Box

Figure 4 shows the type of distribution box that has been used for illustrative purposes in this and the preceding chapter. The small scale plan shows the general application of the distribution box with a septic tank and leaching area.

Referring to the plan of the distribution box, note that a baffle is used to divert the effluent to either one section of the piping in the leaching area or the other. The baffle, either wood or steel, may have a handle on top for easy removal. The frequency in diverting the flow of effluent is determined by the maintenance personnel and is usually determined after the system is in operation for some time and its needs have become evident.

The section taken through the distribution box shows the piping connections. The line from the dosing tank is set at a higher elevation than the discharge lines to the leaching area.

The distribution box is equipped with a cover at grade for easy access. The structure itself is of concrete and when surface loads warrant, is constructed of reinforced concrete.

When a serial distribution system is used, a distribution box is not required. When it is used, however, the distribution box is subject to the regulations of the local plumbing and health authorities.

Drainage Piping

To dispose of the effluent into the ground, a system of piping or other means must be provided to distribute the waste water uniformly into the leaching area. Several types of piping materials may be used. Figure 5 shows three types, including a popular type known as farm tile, short lengths of clay pipe laid end-to-end in the drainage trenches; semi-perforated clay

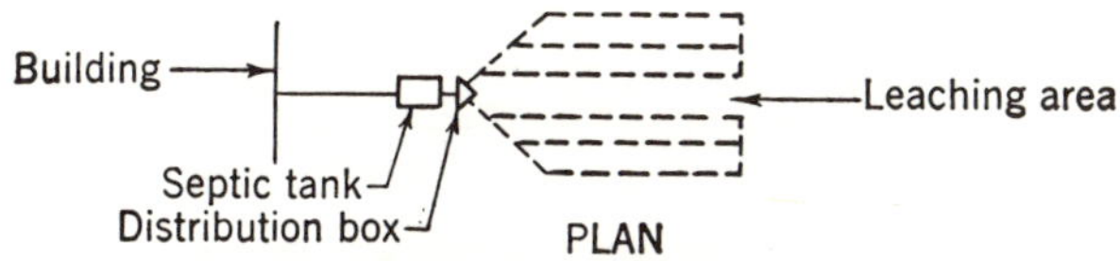

PLAN
Typical layout of septic tank,
distribution box, and leaching area

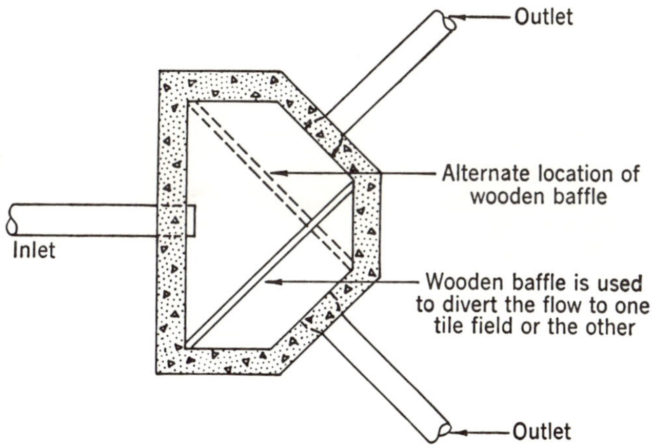

PLAN
OF DISTRIBUTION BOX

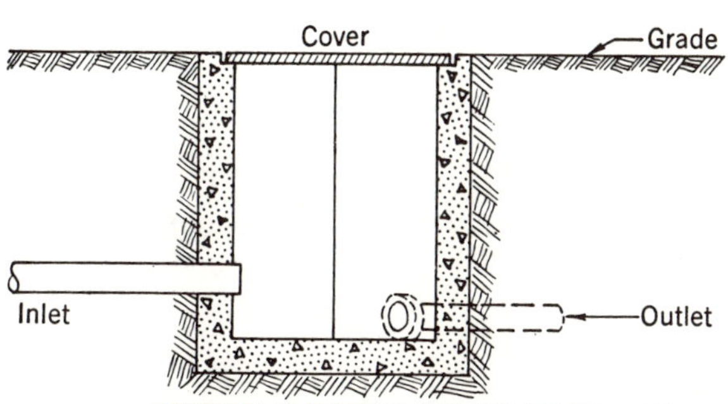

SECTION THROUGH DISTRIBUTION BOX

Fig. 4. Details of a distribution box and plan of its layout in disposal system.

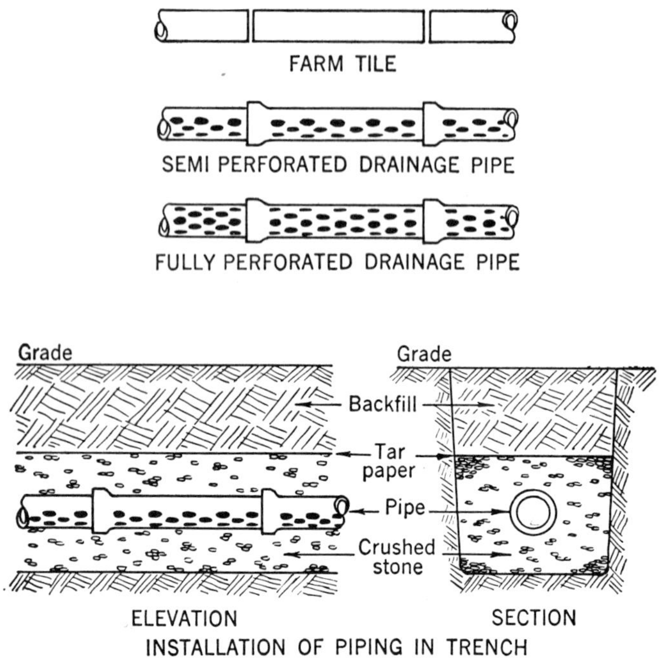

Fig. 5. Details of drainage piping and the typical manner of installation.

piping, available in 2- to 3-ft. lengths and placed together with the joints left open; and fully perforated clay pipe of the same length. Joints of this piping are also left open to facilitate rapid drainage.

Although clay has been shown here, there are other materials available. Also, clay piping is available in several different strengths.

Drainage piping is set in shallow trenches as illustrated in Fig. 5. The trenches must be a prescribed distance apart, dependent upon either the drainage capacity of the soil, the requirements of the local plumbing or health authorities, or the engineers' favored design.

The trenches are carefully excavated, a firm bed of crushed stone applied, and the piping installed. Crushed stone is added under, over and around the piping to assure unrestricted drainage. A layer of tar paper is often applied to the top of the crushed stone to prevent backfill from clogging and diminishing the drainage capacity of the crushed stone. Then backfill is gently added to slightly higher than grade level to complete the installation.

The total length of drainage piping depends upon the absorption capacity of the soil. The better the drainage qualities, the fewer feet of drainage piping required. The local plumbing or health authorities will always request test data on the percolation rate of the soil in which the leaching area is to be located.

Seepage Pits

Another method for disposing of effluent is by seepage pits, or drywells, constructed of brick, concrete blocks, poured concrete, among other materials. A typical seepage pit is detailed in Fig. 6, in this instance, a pre-cast one. This pre-cast unit is avail-

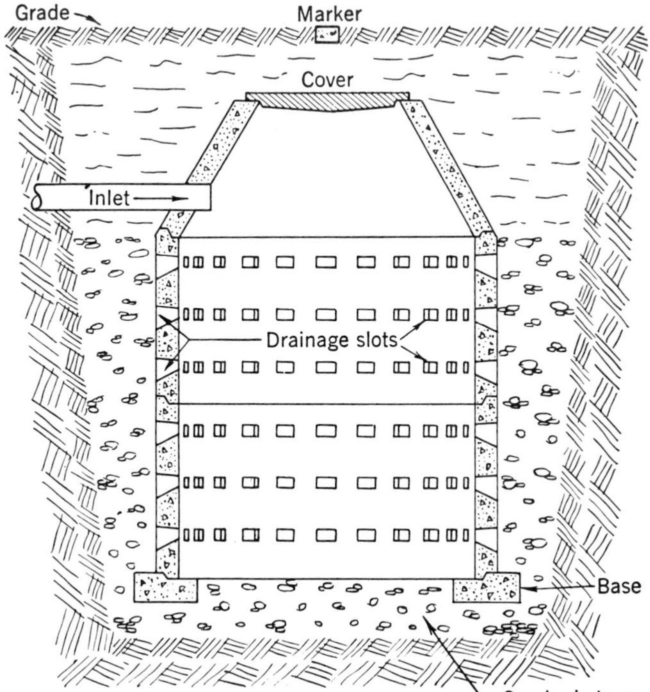

Fig. 6. Detail of a seepage pit fabricated from pre-cast reinforced concrete sections.

able in certain desired heights and diameters. Its sections fit together as shown, secured by clamps. This structure is a very durable type and is reinforced with steel mesh to provide extra strength. Seepage holes are formed when the structure is poured.

To install the seepage pit, a hole is dug, crushed stone applied to the bottom of the hole and then a concrete base poured to support the structure. Next, the sections are lowered and clamped. A pre-cast cone is placed on top of the section. Removal of a cover on top provides access to the interior, and a marker at grade indicates the location of this cover.

Crushed stone is added around the exterior of the seepage pit to increase the drainage capacity of the pit. Backfill is then carefully applied up to grade and the seepage pit is ready for operation.

Seepage pits are usually considered less desirable than the leaching area or absorption field. Seepage pits should never be used where there is a likelihood of the liquid contaminating underground waters and it should not be used when adequate subsurface disposal fields can be provided. Pit excavation should terminate at least 4 ft above the ground water table. The local authorities should always be consulted for their regulations governing the design and construction of seepage pits as well as for the limits of application.

CHAPTER 11

Exterior Sand Traps

Sand or sediment, entering exterior storm sewer or drainage lines over a period of time, can cause serious stoppages or reduce effective flow capacity.

Sand and sediment are carried into exterior piping by rain. Since rain water is often only a gentle flow, sediment will settle at the bottom of the piping and slowly accumulate until it reaches troublesome proportions. During dry periods, sediment hardens like cement. Once this hardening occurs, it is difficult to wash away the sediment, even with a heavy flow of rain water. Rodding may then be required to dislodge the material.

Parking areas, school yards, general site areas, walkways and roadways are all exterior surfaces which must be drained. Such surfaces contribute very heavily to the sediment that enters the underground drainage lines. Therefore it is necessary to provide the means or appurtenances that will intercept and retain the sediment and thus prevent it from entering the drainage lines. Some of the exterior sand traps that may be used for this purpose are described in this chapter.

Types of Structures

There are many structure types that can be used to intercept sand and sediment. Several of such structures, each with a specific purpose, are shown in Fig. 1. These structures include area drains, catch basins, clay pipe interceptors, cast iron basin interceptors, concrete pits and manholes.

Catch Basin

The catch basin offers drainage facilities for streets, roadways, walkways and parking areas. Details are shown in Fig. 2. It is usually constructed of concrete or brick. When required, reinforced concrete may be used. It is equipped with a frame and

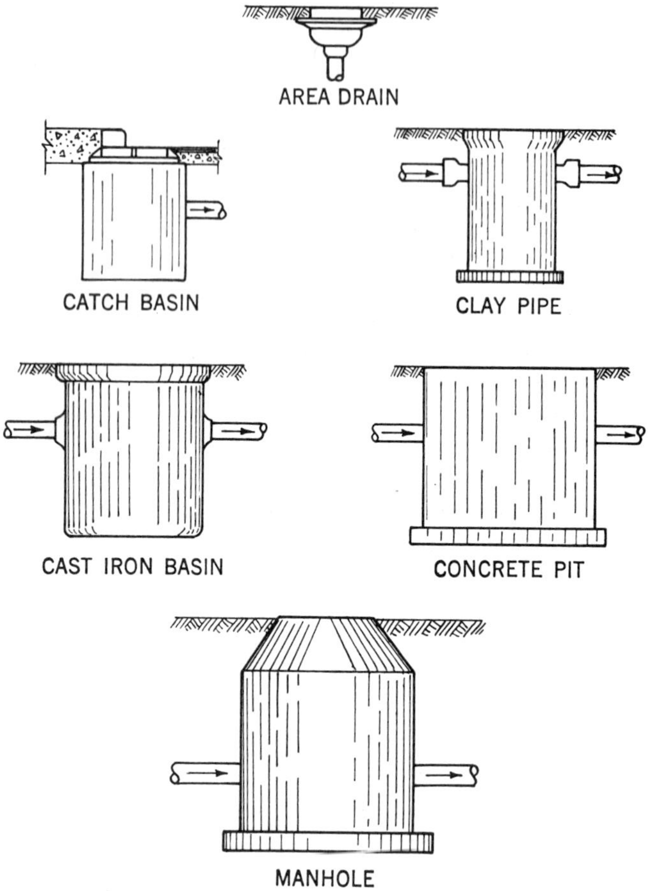

Fig. 1. Five types of sand traps.

cover to withstand traffic loads. A grate with slots or square holes prevents entry of large objects. When the catch basin is to be a component part of the curb, the frame is constructed as suggested by the detail.

The catch basin often must conform to the standards of the local highway or plumbing departments. Depth of the unit is important, since this is the pocket that will retain the accumulation of sediment that is intercepted.

Sediment washed into the pit settles at the bottom while clear

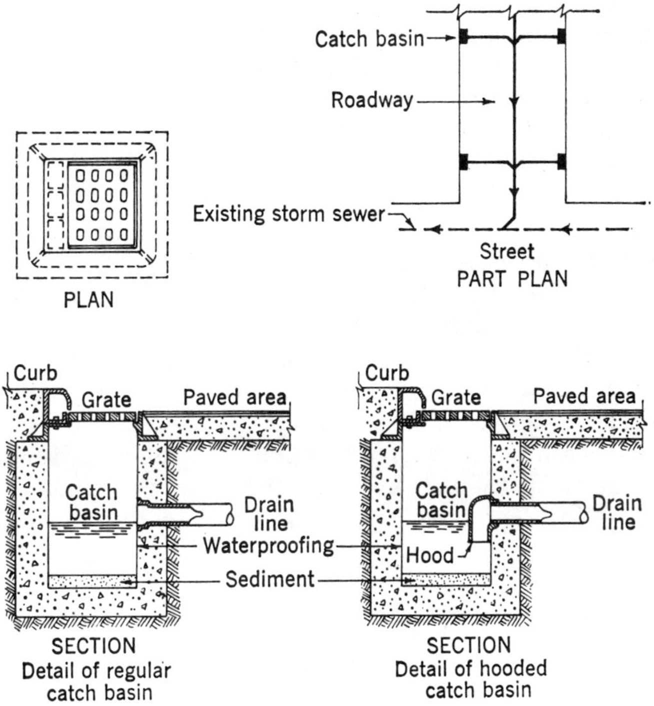

Fig. 2. Two types of catch basins that serve as sand traps.

water passes on into the drainage piping. Periodically, the pit must be cleaned out.

When a hood is required, it may be a fitting or a manufactured device. See Fig. 2. The hood (1) prevents floating debris in the basin from washing into the drainage line and (2) serves as a trap seal in preventing foul odors in the drainage line from relieving at the catch basin.

Where winters are severe, attention must be given to the design of the catch basin. It is often necessary to omit the pocket and to intercept the sediment at an in-line sand trap where the water, due to its depth in ground, will not freeze.

The part plan of Fig. 2, shows a typical installation of catch basins to serve a roadway. Note that the basins connect with a common drainage line which, in turn, join with the street storm water facilities.

Area Drains

In parking areas it is necessary to provide a system of drains and drainage piping if the rain water does not run off, by reasons of slope, toward a natural point of discharge. In many areas, the local codes require that drainage be provided for the parking area regardless of its natural runoff. The requirements must be checked with the local authorities.

A typical area drain, Fig. 3, is constructed entirely of cast iron and is equipped with a heavy duty grate. The grate must withstand heavy traffic loads imposed by automobiles and trucks. The grate in this instance features small square drainage holes to repel the entry of large objects. Also, the small holes prevent persons, especially children, from wedging their foot in the holes. Area drains may also be constructed of concrete or brick with a cast iron frame and cover fitted on top.

The area drain is so constructed that the sand, sediment and debris will settle at the bottom while the clear water passes on to the drainage line. The hood serves as a trap seal and is equipped

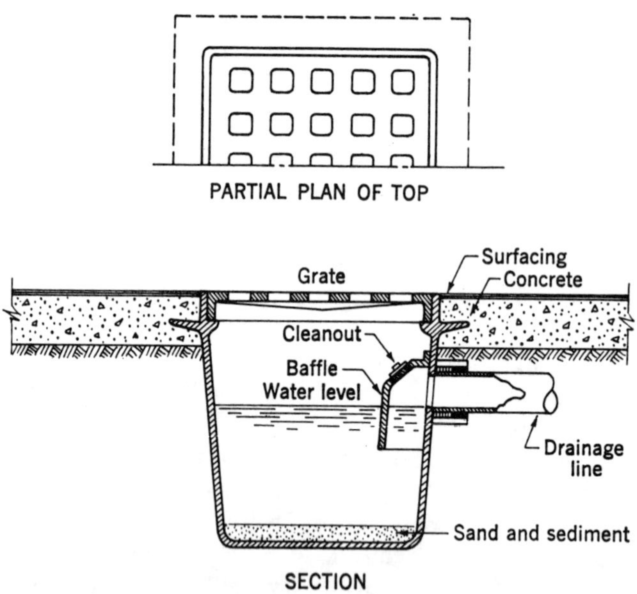

Fig. 3. How a cast iron sump drain can be used to intercept sand and sediment.

with a cleanout to offer access to the drainage line beyond the drain. Many manufactured drains are available.

Regular Area Drains

When the area drain is not required to serve as a sand trap, a regular area drain, which does not include the pocket below the drain outlet for the retention of sand, sediment and debris, is employed. See Fig. 4.

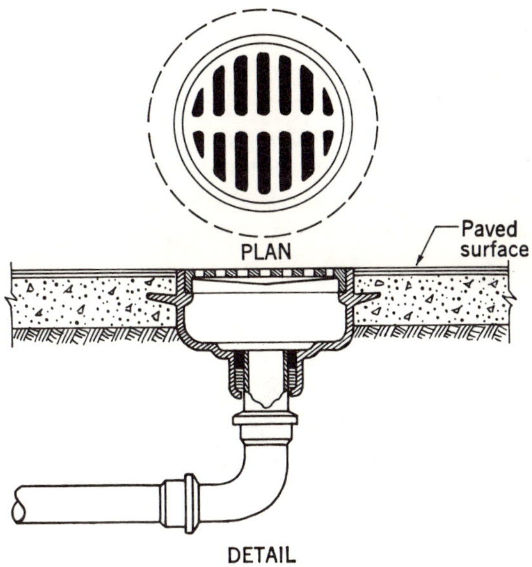

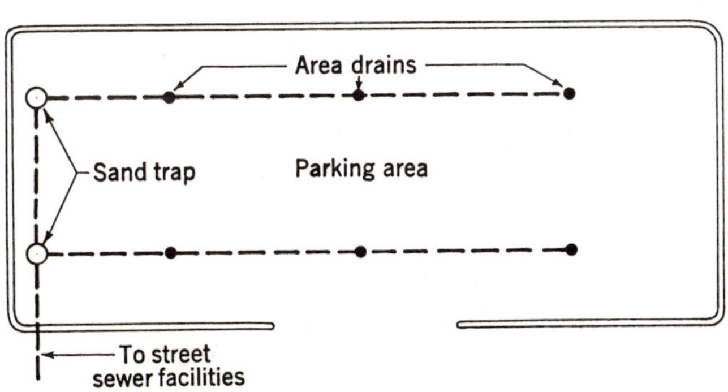

Fig. 4. How sand traps can be installed to serve a system of area drains.

The drain in Fig. 4 is a bottom outlet type. Different sizes and shapes are available. This drain also must be of heavy construction. The grate will be subject to heavy traffic loads.

The small scale plan shows how the area drains might be arranged for a parking area. The surface of the parking area is sloped toward the drains. The drains are connected to underground piping which, in turn, connects with sand traps that are placed at the edge of the parking field. Many types of sand traps are available and a few will be described.

Clay Pipe Sand Traps

The clay pipe sand trap is quite popular and fulfills a broad application need. Shown in Fig. 5, it consists of a length of clay pipe either set on a slab of concrete and grouted in place or set into the concrete directly after it is poured. Clay pipe is available in several sizes to fit any precise need. The cover can be constructed of cast iron, concrete or wood.

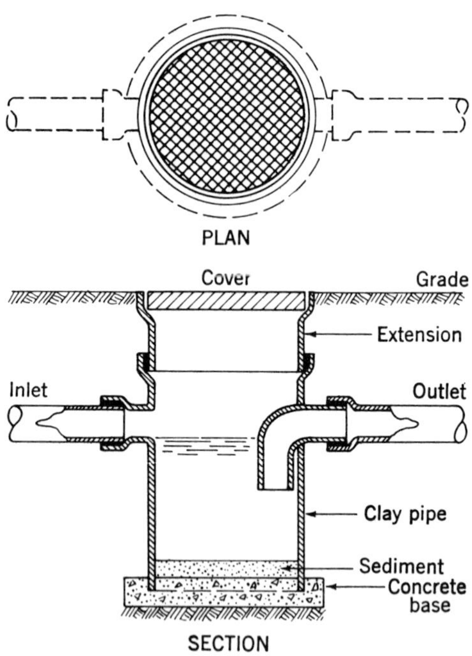

Fig. 5. Use of clay pipe in sand trap design.

Extensions are easily made. A length of clay pipe is cut to a prescribed length and grouted into the hub of the sand trap with concrete. This arrangement will satisfy any arrangement or elevation of underground drainage piping. When a hood is required, the clay pipe sand trap may be ordered that way, and the benefit of a trap seal is gained.

Cast Iron and Concrete Sand Traps

Sand traps of cast iron or concrete, similar to those shown in Fig. 1, may also be used. Cast iron units are available in different sizes to handle any drainage load concentration. Concrete sand traps must be sized properly and equipped with a cast iron frame and cover. Hoods to offer trap seal protection may be incorporated into either the cast iron basin or the concrete pit.

Manholes

When extensive exterior drainage systems are contemplated and the drainage piping will be run rather deep in the ground, the manhole type interceptor is very practical. Figure 6 is a manhole usually constructed of concrete, block or brick. When

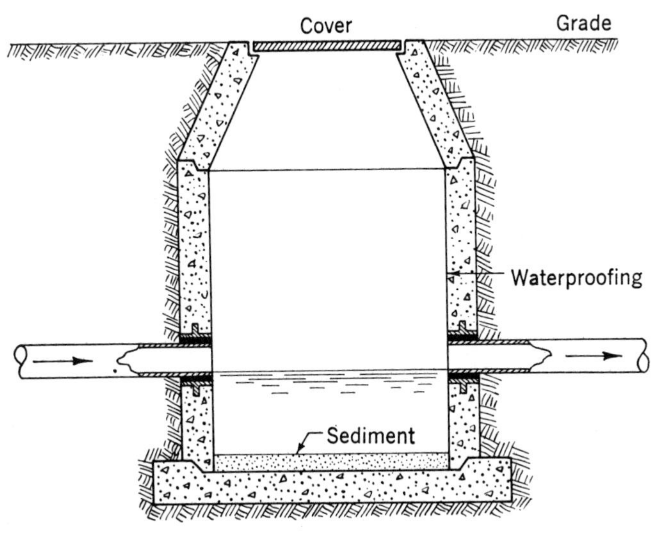

Fig. 6. Installation of a manhole to serve as a sand trap.

required, reinforced concrete may be used. Its size must be determined by the engineer. Wall thickness, too, must be given careful consideration with regard to the surface loads to which the structure will be subjected. The cover is cast iron or steel. The pocket needed for the retention of the sediment should be at least 18 to 24 inches in depth. If the engineer so desires, he can make it deeper. Although a hood has not been shown, it can be included if trap seal protection is desired.

Site Drains

Graded areas and depressed graded areas generally need some means that will drain water accumulations. Area drains may be used for this purpose provided that a sand trap is placed in the underground drainage line. A much more practical approach is to construct a manhole and equip it with a slotted grate to enable this structure to function as an area drain.

This structure is rather similar in construction to the unit shown in Fig. 6, with the exception that a cast iron frame and grate has been placed at the top of the structure. The grate is equipped with narrow slots to prevent the entry of large foreign objects. Here again, a pocket should be provided at the bottom of the site drain to retain debris washed into it with the rainwater. If desired, a hood can be used.

Water accumulations are drained from a site by locating site drains as one would area drains. If the area is rather extensive, several rows of site drains may be required.

Special Sand Traps

Occasionally a special type of sand trap is required to meet unusual needs. An example is an industrial plant where certain waste water is discharged into a stream. Similar applications may also demand the need for the special structure shown in Fig. 7, which is constructed of reinforced concrete. It consists of two compartments: In the first, waste water is received and most of the sediment will settle to the bottom where it can be easily cleaned; in the second compartment, any fine matter still suspended in the water will settle. The purpose of the baffle is to reduce the turbulence of the waste water as it enters the compartment and thus allow the matter to settle more easily.

The unit shown is open at the top, which would indicate that it is installed in warm areas. When installed in cold areas, the

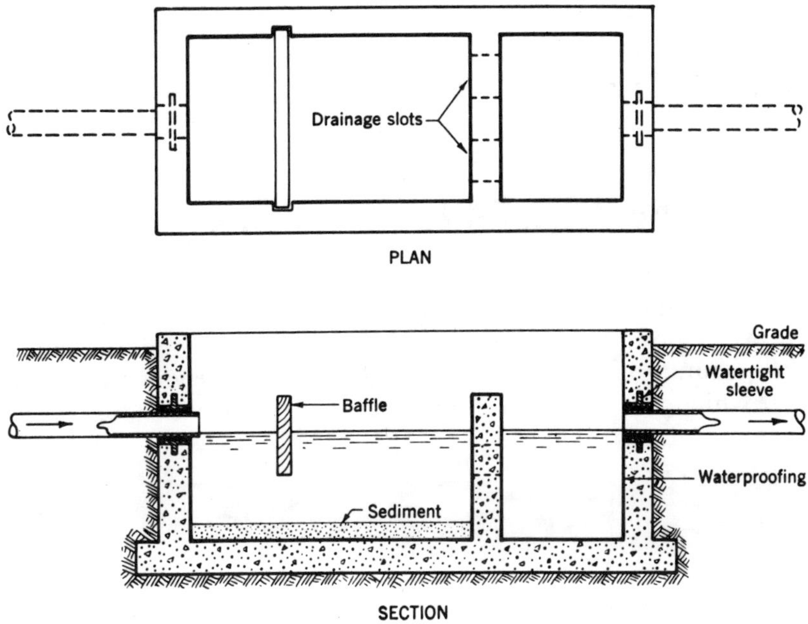

Fig. 7. Details of a large sand trap for interior or for exterior applications.

structure is generally placed in an enclosed, heated building. Quite often the greenhouse type of building is used. Although not shown in detail, a hood can be installed on the outlet connection in the second compartment, if required.

This type of sand trap is of value when heavy concentrations of matter are handled and it is not desired to have this matter enter the exterior street sewer facilities. Moreover, the structure has application when solids resulting from industrial or commercial processes are involved. The solids are intercepted and retained by the structure for periodic cleaning.

Regardless of the specific application, there is a structure that can serve the intended purpose. Before starting the proposed design of such a system, it is always good practice to consult with the local plumbing or highway departments for their requirements or suggestions concerning installations. They will also advise on the depth of trap seal to be maintained when hoods are involved.

CHAPTER 12

Lawn Sprinkler Systems

New buildings erected on spacious grounds in many suburbs are carefully designed to harmonize with the surrounding area and the grounds are attractively landscaped to produce a building plot which is an asset to the community. The light industrial type

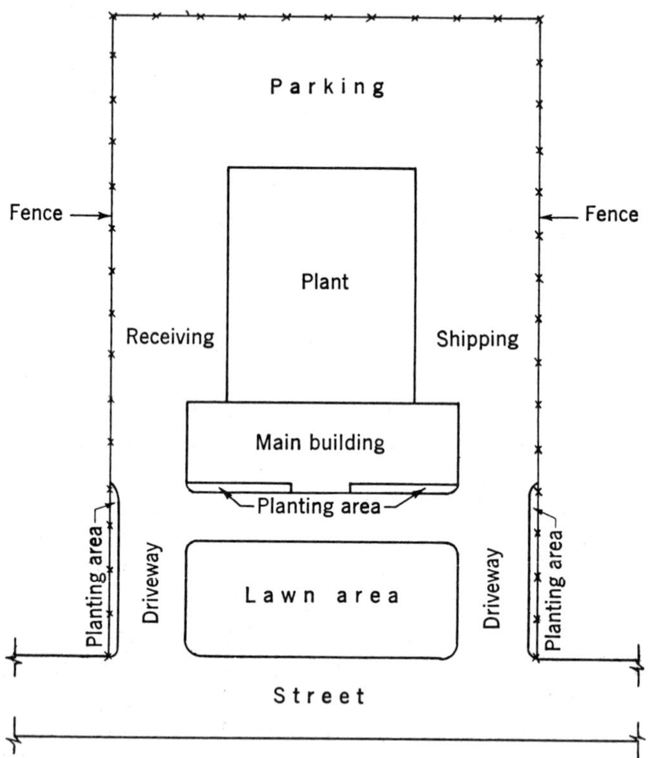

Fig. 1. Relation of lawn area to industrial plant site.

building is typical, where landscaping and site hide most of the plant operations and help to retain the peaceful atmosphere of the area. The installation of a lawn sprinkler system is therefore considered important.

We shall consider a typical industrial plant site for illustrative purposes in this chapter as shown in Fig. 1. It is situated on a large plot. Adequate space has been provided for shipping and receiving as well as for employee parking. The portion of the site facing the street is well landscaped and includes a spacious lawn area. When properly maintained, this area will enhance the property and present a neat, attractive appearance to the passing traffic. A single lawn area of medium size has been used for illustration.

Spacing of Lawn Sprinklers

To achieve an even distribution of water, the lawn sprinklers must be so placed that they will completely cover the lawn area, as shown in Fig. 2.

The spray pattern is so arranged that each sprinkler spray will overlap the other. This consideration will provide overall coverage despite slight water pressure variations that may occur from time to time. Water accumulation at the edge of the lawn will take care of the small voids near the curb.

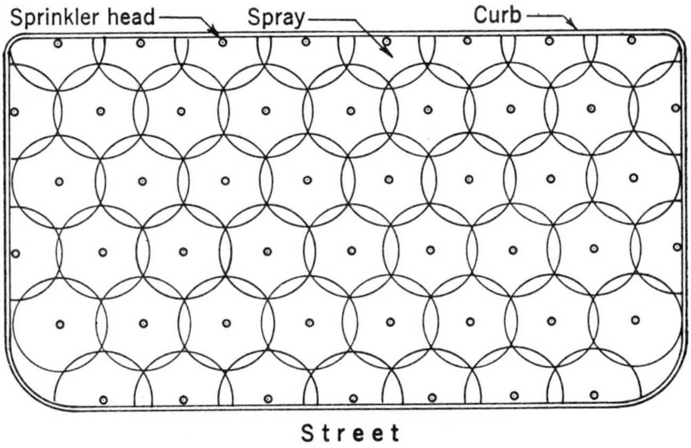

Fig. 2. Typical spray pattern of sprinklers for a lawn area.

The maximum coverage offered by each sprinkler head is dependent upon the design of the head and the water pressure. The manufacturer's reference material will give minimum and maximum effective spray patterns for each of the heads at varying water pressures.

A regular spray head covers an area of 360° and the heads at the curb, an area of 180°. Other spray patterns for specific needs can be worked out with the manufacturer. It is important that the spray does not exceed the limits of the lawn area, otherwise the overall spray pattern will lose much of its decorative beauty and may create a nuisance.

While a lawn sprinkler system assures a healthy lawn, it also provides one of the most distinctive means to enhance a building site. It is economical in supplying the prescribed amount of water, while manual hosing of lawns can be most wasteful.

Arrangement of Piping

Sprinkler piping must be arranged to satisfy the desired spray pattern. A typical piping arrangement is shown in Fig. 3.

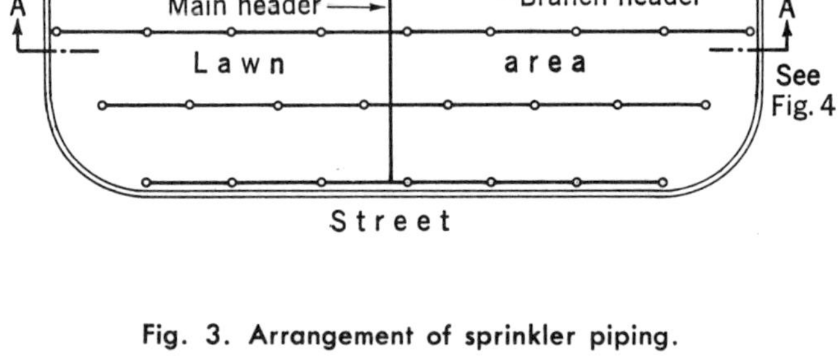

Fig. 3. Arrangement of sprinkler piping.

This is a rather simple arrangement but it illustrates a method utilizing all the sprinklers at one time. The piping consists of a water supply from the building, a main and branch headers. The main header is an equalizing line serving the branch header. The water supply is connected to the main header at the mid-point between the several rows of branch headers. In this manner, a uniform pressure is provided all sprinkler heads and the spray pattern will be of uniform intensity. A lawn sprinkler system loses some of its decorative effect when the overall spray pattern varies in volume.

Several factors determine the best piping arrangement. The first criterion is to provide a complete system with the least piping. This can be accomplished by plotting several arrangements on paper and scaling the total lineal feet of piping for each method.

Required trenching is a factor, for excavation can be costly. It is suggested that this factor be considered when the systems are plotted on paper. A piping arrangement should be found that requires the least amount of excavation.

Assuming that building supply water pressure is adequate, care must be exercised in calling for the proper slope of piping so that water remaining in the piping can be drained off into the building from sections of the system where the water is subject to freezing in winter. All piping must be properly sized and the entire system balanced if the system is to fulfill its intended function.

Installation of Sprinkler Heads

Details concerning installation of sprinkler heads are shown in Fig. 4.

Section A-A represents a typical arrangement of branch headers on either side of the main header. Branch headers are normally installed 12 to 24 inches below the surface of the lawn.

Several piping materials may be considered. The characteristics of the water and soil are factors in this choice. Copper tubing is, perhaps, the most generally used material for this purpose although plastic piping is being used in increasing amounts.

Many types of sprinkler heads are available, ranging from inexpensive to deluxe models. The pop-up type spray head is most popular. Water pressure lifts the head to spray position, as shown in Detail *B* of Fig. 4. Crushed stone covers the trench bottom, as illustrated in Detail *B*, especially when positive slope of piping is to be maintained.

Detail *A* shows a trimmer device for removing the grass around the sprinkler head. If the grass growth is not controlled,

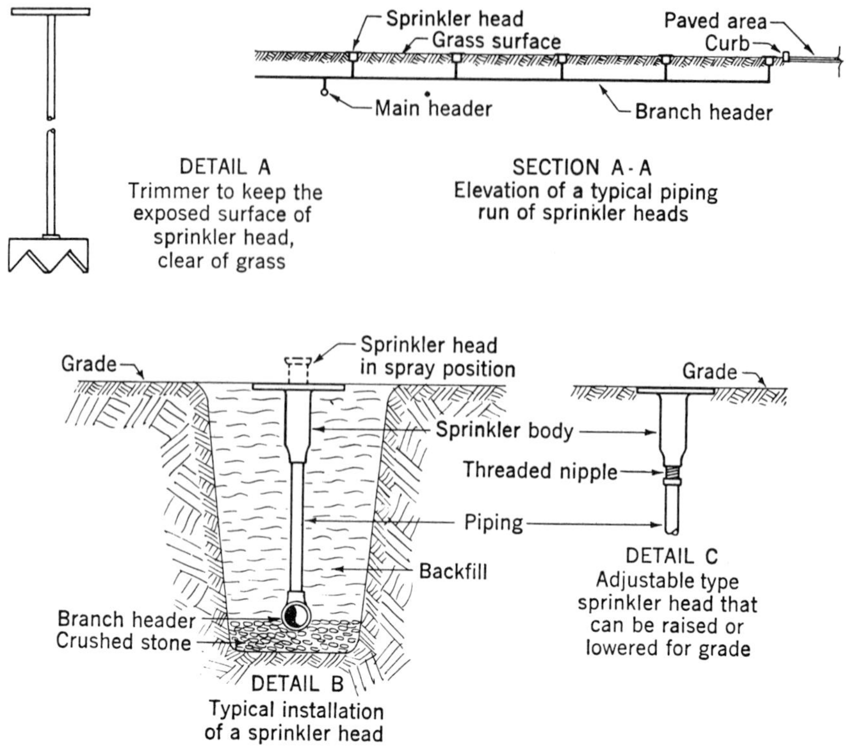

Fig. 4. Details pertaining to the installation of lawn sprinkler heads.

it can move the sprinkler head out of alignment, interfering with the action of the pop-up head.

Since the grading will settle due to (1) foot traffic, (2) the weight of grass cutting machines and (3) because the branch header is subject to some settlement or movement, it is advisable to consider a means of installation that will provide some extent of flexibility. Detail *C* shows an adjustable head that can be raised or lowered to satisfy the changing conditions of the grade, providing that it may always be flush with the grade elevation.

Sectionalization of Sprinkler Systems

Very often, due to the size of the lawn or the lack of water pressure, it is necessary to water the lawn in sections. Of the several ways in which it can be accomplished, Fig. 5 shows a

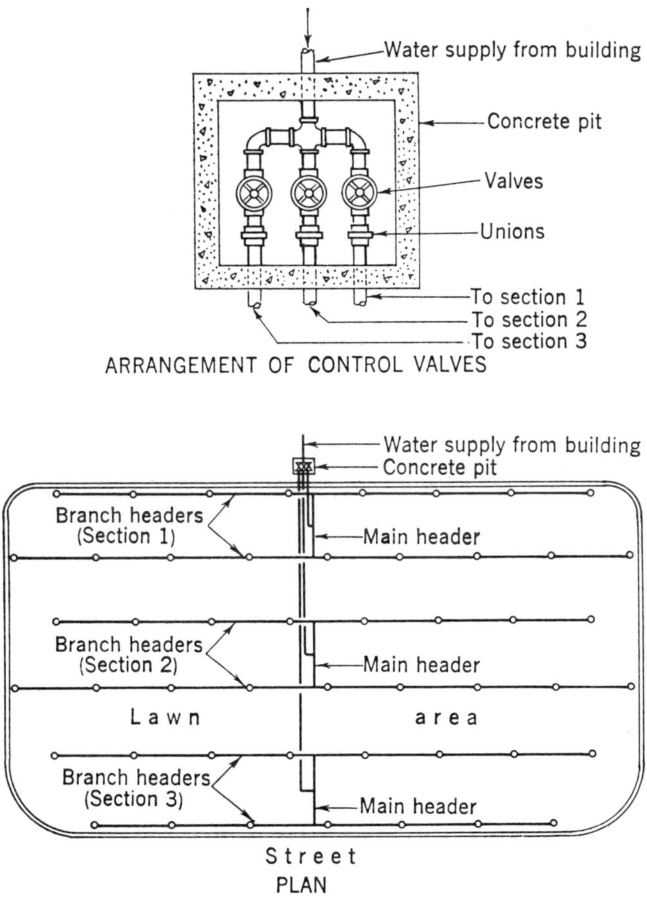

Fig. 5. How arrangement of sprinkler piping can be subdivided when there is insufficient pressure for the entire system.

method that may be utilized with the particular lawn area shown.

In this case, the lawn area has been divided into three sections. A control valve for each section has been placed in a concrete pit outside of the lawn area. The operator will water one section of the lawn for a prescribed period of time and then shut it off to water another section. The valves are outside of the lawn area so the operator can work the valves without getting wet. The detail shows how the valves can be placed in the pit for ease of handling.

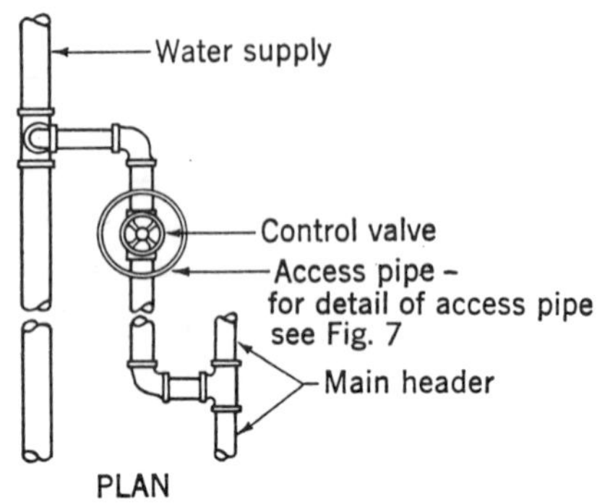

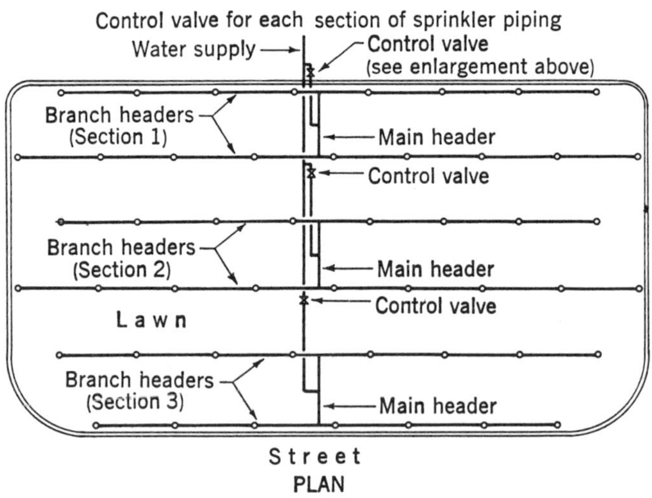

Fig. 6. Another way to sub-divide when there is insufficient pressure for an entire system.

Another method is shown in Fig. 6. The control valve is near the section it serves, as may be required in some installations. Access to the underground valve is by means of a length of clay pipe placed in the ground, flush with the grade. The operator removes the cover to turn on the valve. In each case, the valve is installed out of range of the water spray for the section it controls.

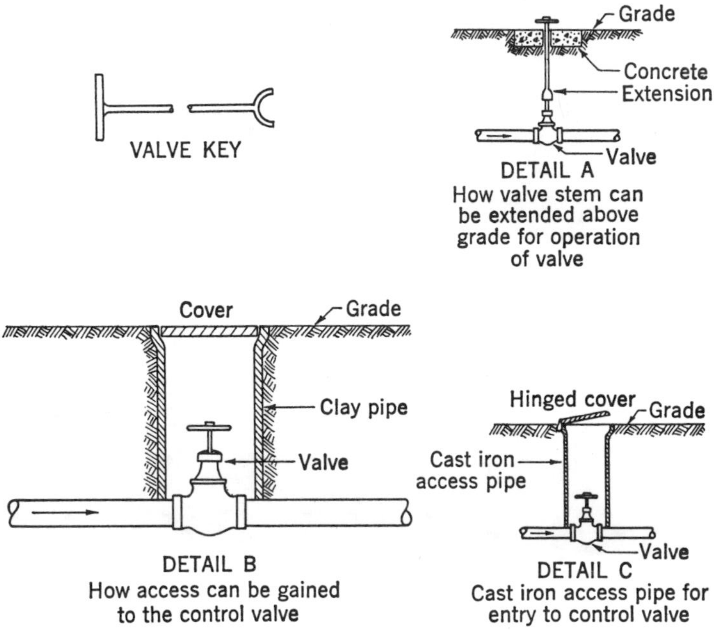

Fig. 7. Methods which may be used to operate underground control valves.

Access to Control Valves

Since control valves are placed on the underground piping, proper access must be provided. Several methods are detailed in Fig. 7:

Detail *A* shows a way to provide a means of control at grade level. The extension, affixed to the valve stem, is extended through a small slab of concrete and above grade. This method provides ease of access for the operator. Detail *B* shows how a piece of clay pipe can be used. It is necessary only to remove the cover to gain access to the control valve. Detail *C* consists of a manufactured cast iron access pipe with a hinged cover. The cast iron pipe can be cut to any desired length.

Drainage of Lawn Sprinkler Systems

In cold climates, it is most urgent that the water be drained from the underground piping system at the end of the season.

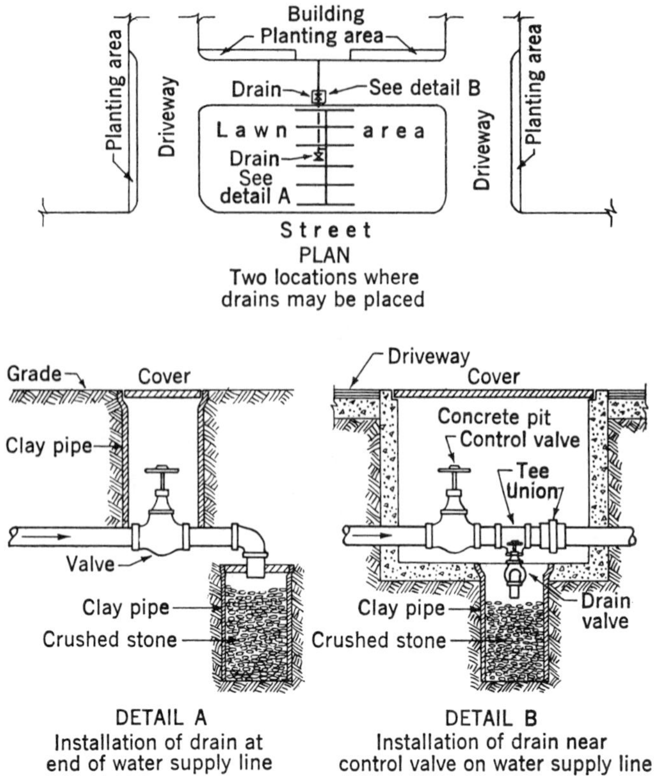

Fig. 8. Methods that may be used to drain lawn sprinkler piping.

Otherwise, the underground piping, buried in shallow fill, can freeze and rupture or sections of it moved out of alignment.

Whenever possible, it is desirable to slope the underground piping so that the water in the piping will drain back into the building where it can be drawn off with relative ease. However, when large lawn areas are involved, consideration may have to be given to other locations.

These methods are illustrated in Fig. 8. The plan shows a typical site and indicates just where the points of drainage might be applied.

Detail *A* shows how a drain may be installed at the end of the water supply line. The piping in this case must be sloped in the direction of the drain. A valve, installed at the end of the piping, when opened will drain into the clay pipe filled with crushed stone. Another length of clay pipe has been used at grade to offer access to the valve. In Detail *B* the system is drained near the valve that controls the supply of water to the sprinkler system. The means consists of a drain valve which, when opened, permits the water to drain into a piece of clay pipe, again filled with crushed stone. The entire assembly is housed in the concrete pit, as shown.

Part 2
Building Drainage Systems

CHAPTER 13

Footing Drains Around Buildings

Occasionally, the site conditions for a proposed project are such that the building, by nature of the ground contour, will be located in a low or depressed area where surface water accumulation during a rainstorm presents a serious problem for the foundation wall and the earth support for the footings. Since this is a matter of grave concern to the structural engineer, he usually requests that a system of subsoil drainage piping be installed to prevent damage from accumulated water. The proper design of such a system then becomes the responsibility of the engineer designing the plumbing work.

Subsoil drainage systems have varied applications. In this chapter we shall consider only the system that may be used successfully for a condition similar to that previously mentioned — a system generally referred to as footing drains. These footing drains are designed to intercept the surface water before it can reach the foundation wall or earth support under the footings, and to conduct this water to a suitable point of discharge. A typical system of footing drains with cleanouts and sand interceptor is shown in Fig. 1.

Footing drains, illustrated in Fig. 1, are installed around the perimeter of the building. The water is conducted to the sand interceptor and then to the point of discharge which in this case is assumed to be a storm water drainage line from the building. The system includes the drainage piping, cleanouts and a sand interceptor, each of which has an important role in the satisfactory operation of the system.

Although there are several different piping arrangements that may be used with this system, the arrangement shown in Fig. 1 is basic and serves as the example in this case. We will discuss the function and installation of the component parts of the footing drainage system shown.

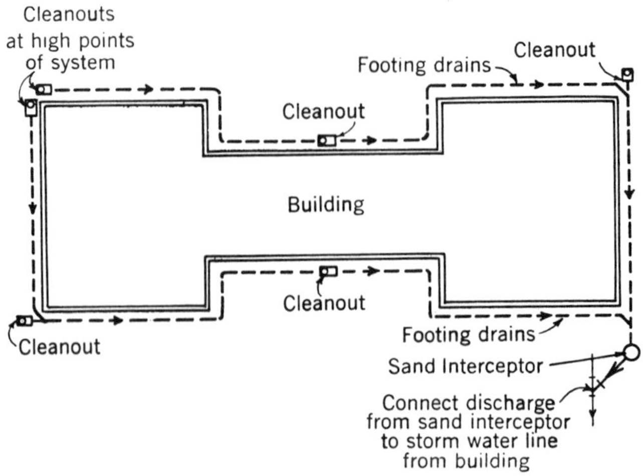

Fig. 1. Layout of a footing drainage system.

Installation of Footing Drains

Clay pipe or porous concrete pipe is used for footing drains of which the clay pipe is more generally used. Clay pipe is available in two to three-foot lengths and 4-inch diameter is the size usually selected for a system of this type. The piping is also obtainable in regular lengths or in lengths that are fully perforated or partially perforated.

As indicated in Fig. 2, the drainage line is installed near but above the bottom of the footing. If regular piping is used, a covering of tar paper or other manufactured material available for this purpose, is placed over the open joints to prevent sand or earth from entering the line. If perforated piping is used, it is of the partially perforated type with the perforations set at the bottom of the line. Since water will follow the easiest path, it will flow through the drainage piping to the point of discharge. A firm bed of crushed stone is placed under the piping for support. Crushed stone is also placed around and above the piping to facilitate easy drainage. In addition, the piping is installed with a slope of approximately one inch to every 25 feet, toward the sand interceptor. For extra long runs of piping, the slope may be even less than this figure.

It is important to mention that consideration be given to the area around the building to the extent that area drains be provided to draw away as much as possible of the surface water. Area drains will

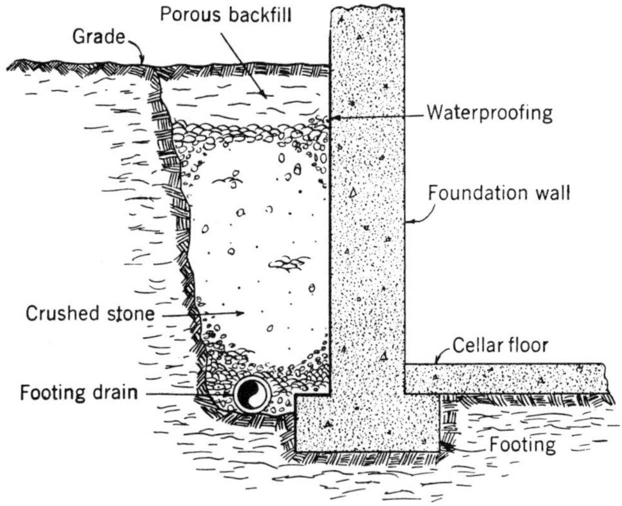

Fig. 2. Detail of a footing drain installation.

lessen the load otherwise placed on the footing drains. It is good practice to slope the grade away from the building as this prevents heavy concentrations of water over the drainage trench and allows for a more natural seepage of water to the footing drains.

The system shown in Fig. 2 is intended only for the drainage of surface water seepage and is not intended for the drainage of a ground water condition. The system required for a ground water condition is a different problem and will be discussed in one of the later chapters.

In brief, the drainage trench and piping functions as though it were an open trench around the building which would intercept the surface water and surface water seepage before it could reach the building, and conduct it to a suitable point of discharge.

Cleanouts

A system of footing drains, like any other system of drainage piping, should be equipped with cleanouts to allow for the clearing of any stoppages that would tend to interfere with the flow of water through the system. Cleanouts for this system have an important part in the proper operation of the system and they should be placed at the ends or high points of the system and in other locations which provide for easy access to the entire system. A typical detail of a cleanout is shown in Fig. 3.

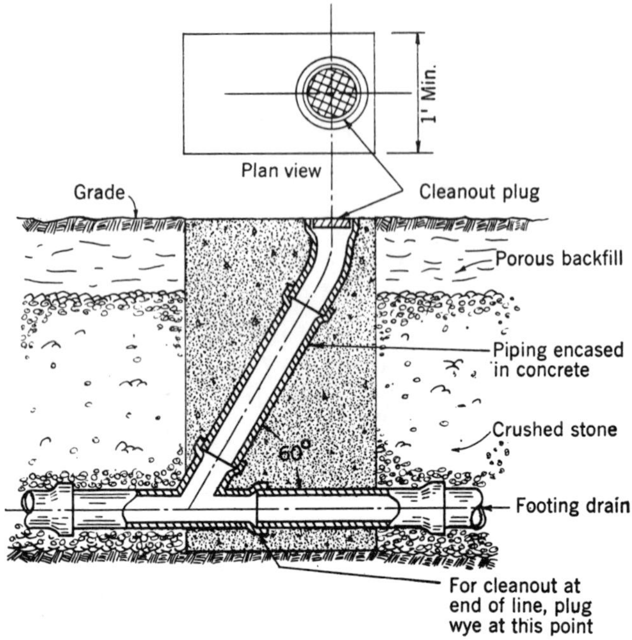

Fig. 3. Detail of a typical cleanout.

Although other types of cleanouts are often installed for this purpose, the cleanout shown has been chosen to illustrate this installation. As indicated, the top of the cleanout has been installed flush with the grade with a plug arranged for easy removal. The cleanout piping is encased in concrete to keep the cleanout in place and to prevent an unbalanced weight or strain on the drainage piping. During the normal operation of the system, small amounts of sand or earth will settle in the pipe, despite the preventative steps taken. The accumulation of such material will eventually impede the flow of water through the pipe. To clear the pipe of this material, it is only necessary to remove the cleanout plug and insert a water hose. By applying a gentle flow of water to the system, the sand and earth will be flushed through the system to the sand interceptor. The cleanout is also useful in locating obstructions in the system.

Part of the system may become blocked to the extent that the water will not drain or is slow in draining, due to severe settling of part of the piping, or a heavy concentration of earth material, or even a growth of tree roots in a part of the system. By running water into one cleanout at a time, and by checking the flow into the sand interceptor, the faulty part of the system can be quickly located and dug up if necessary. Bushes or shrubs that develop deep and fine roots should be avoided directly over or near the drainage pipe.

Sand Interceptor

The sand interceptor is a necessary component of the system because, as its name implies, it intercepts and retains the sand or earth carried in suspension by the water passing through the system. It prevents any sand or earth from entering the outside storm water drainage line into which this system empties. Deposits of earth materials may be periodically removed through the top.

The sand interceptor, Fig. 4, is constructed of concrete, although it may also be of cast iron or steel of which several patented types are available. This particular type of sand interceptor was illustrated because it also serves as a cleanout manhole for the piping between the interceptor and the storm water line. Water containing small quantities of sand or earth in suspension flows into the interceptor. Since there is no turbulence problem in the interceptor, this material will settle rather quickly and only clear water will be discharged into the storm water line. Often baffles are placed in the interceptor to assure positive separation of the earth material and the water. However, the construction of the particular type of sand interceptor and the use of baffles are individual design problems that satisfy particular requirements of a proposed project.

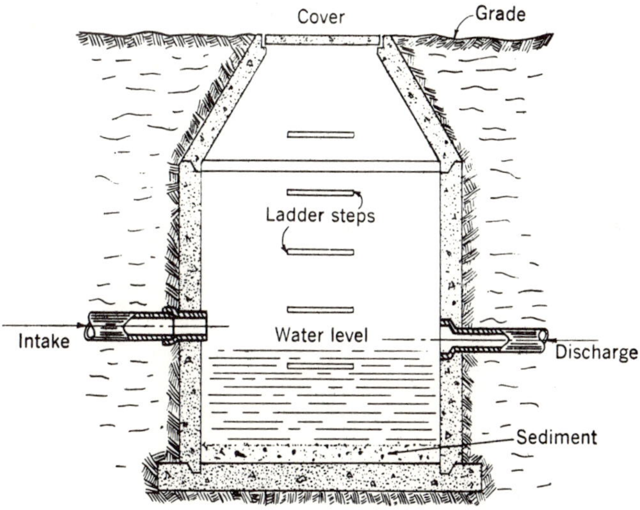

Fig. 4. Section through a typical sand interceptor.

Without an interceptor the flow of water from the footing drainage system, because of its low velocity, would leave deposits of sand or earth in the storm water system. Such deposits could cause line stoppage or sluggish action in this system, which would be troublesome during a storm. During a storm, storm water quickly passes through the storm water system whereas the water from the footing drains reaches this system much later depending on the percolation rate of the soil through which the surface water must seep before it reaches the footing drainage system.

If the storm water line is at a higher elevation than the footing drainage system, it is then necessary to conduct the water from the interceptor to a sump pump, through which it is raised to a sufficient elevation to drain by gravity into the storm water line. If a sump pump is required, it is usually located in the basement or cellar of the building, for easy access and protection of the motor and controls. Generally, when the sump pump is placed in the building, the interceptor is also placed in the building near the sump pit. The sand interceptor prevents the sump basin from acting as an interceptor and also avoids the damage that the sand or earth could cause to the pumping equipment. It is very important that the local authorities or code be checked for additional requirements relating to the piping hook-up between the interceptor and the point of discharge or between the sump pump and the point of discharge.

These suggestions apply to a complete system around a building or to a system serving just a part of a building. A well designed footing drainage system will provide for years of trouble-free operation and will function in the most efficient manner.

CHAPTER 14

How to Install Drainage Piping in Fill

Many new buildings now under construction, designed for varied occupancies, will not be equipped with basements or cellars. Instead, the first floor of the building will set on the grade level. This means that all the mechanical and electrical equipment, distribution piping and services will be installed within the area of the first floor. This form of construction is rather widely used and is very often found in the new single story, rambling type of school buildings and in single story industrial plants.

When buildings of this type are contemplated, the architect carefully studies the contours of the ground at the proposed site to select the location in which the building, when it is erected, will be prominent with regard to elevation and attractive appearance.

To achieve these desirable features, it is often necessary to either remove some of the existing grade or to add the fill necessary for the proposed elevation requirements of the building.

In this chapter, we are concerned with the situation that requires the addition of fill. When fill is involved, careful study must be given to the support of the sanitary and storm drainage piping that will be installed in the ground, under the first floor slab. Depth of the required fill often ranges from several inches to a good number of feet. If this fill had sufficient time to compact, there would be no problem relative to the installation of drainage piping. Since footings or piles are extended down to the existing hard, compacted earth, the construction work of the building itself is not dependent upon the stability of the fill. Grade beams are supported on these footings.

However, drainage lines must be installed before the floor slab is poured and very often at the time these lines are installed, the fill may not have had enough time to compact. As a result, there is the ever-present danger that the piping will settle and fracture so that leaks will occur. When sanitary or storm water lines leak under a

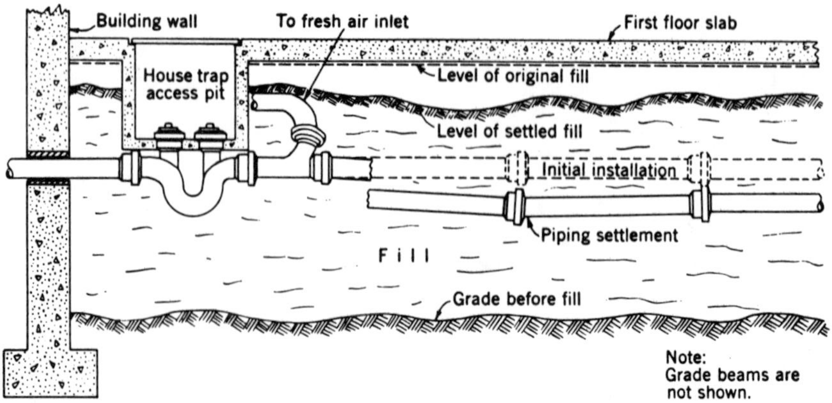

Fig. 1. What happens when piping is installed in loose fill. Dotted lines show initial installation.

building, the resulting damage can be considerable. Most unfortunate is the fact that after the slab is poured, the piping is inaccessible and therefore cannot be repaired or replaced without the expense of chopping up the slab. An illustration of what can happen when piping is installed in fill is shown in Fig. 1.

Compacted Soil

Where natural soil or compacted fill exists, the piping installation is made in a manner similar to that shown in Fig. 2.

In this detail it will be noted that the trench must be carefully graded and bell holes cut in the earth to allow for an even distribution of support for the piping. A thin layer of sand is placed under the piping because sand is easily compacted under the weight of the pipe. This sand furnishes a soft bed into which the pipe will so settle and it will be firmly supported for its entire length.

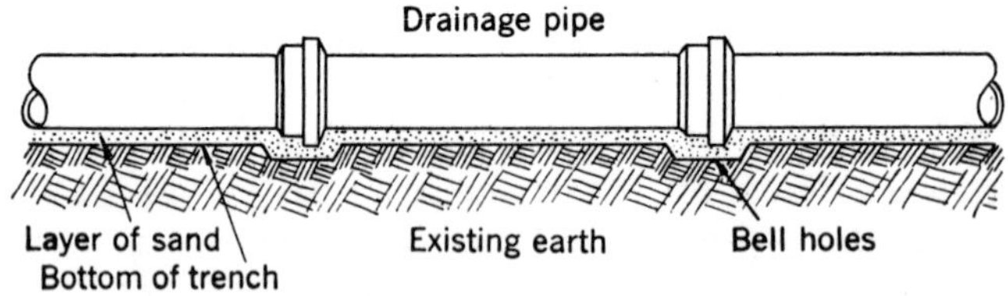

Fig. 2. Piping installed in existing earth or compacted fill.

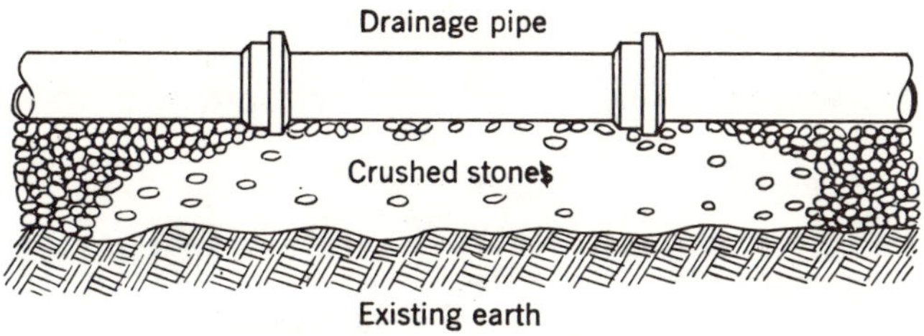

Fig. 3. Piping installed in shallow fill.

If the backfilling of the trench is carefully handled, the installation can be counted upon for long service and free of the leaks or breakage that follows pipe sagging.

Shallow Fill

Where shallow amounts of fill are encountered, piping may be installed as shown in Fig. 3.

Here the trench is excavated down to the existing earth and is backfilled to the proper elevation for the piping with crushed stone or similar material that is easily compacted. By these means a firm bed may be obtained for support of the pipe. Here again, if the final backfill is carefully handled, the piping system may be considered well installed.

Deep Fill

It takes time for fill to compact enough to support piping. Although this settling can be speeded by physical means, this is seldom done to any extent. The building construction often proceeds shortly after the fill has been applied so that there is not enough time for compacting the fill. Therefore, to protect the drainage piping, special supporting means must be considered. Although there are numerous methods, a few that adequately serve the purpose will be described. One such method is shown in Fig. 4.

Piping Support

With this method, the pipe is supported from the slab itself. Piping is installed in the trench after the grade beams have been formed, poured and set. A pipe clamp is fastened to the pipe, with the hanger rod and plate set in a vertical position. The hanger rods are wired to the reinforcing rods that have been set in place for the pouring of the floor slab. The plate is so arranged that it will be flush with the top of the rough slab to be later concealed by the

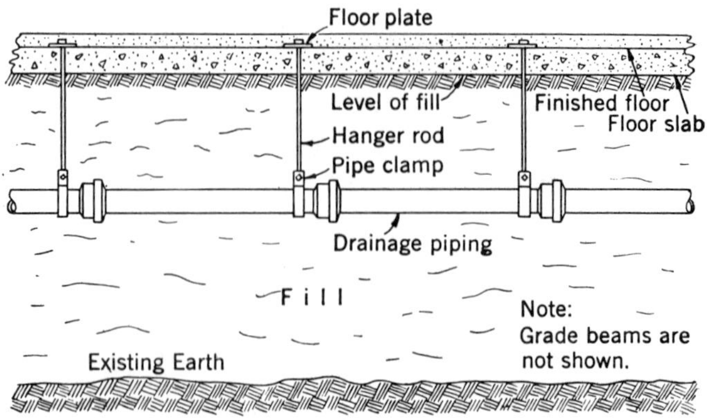

Fig. 4. A method of supporting pipe from slab by means of hangers.

finished layer of cement, as indicated. If a finished cement layer is not contemplated, the plate is then placed as a part of the slab itself. However, this method of support should first be discussed with the structural engineer to be certain that the slab can carry the weight load of the piping.

Another practical method is to support the piping from the grade beams as illustrated by Fig. 5.

Inserts are set in the forms used for the grade beams, before the concrete is poured. Later when the piping is installed, it is an easy matter to fasten the hanger rods to the angle supports as shown. However, the grade beams are seldom spaced close enough to provide sufficient supports for the cast iron piping. As a remedy, inter-

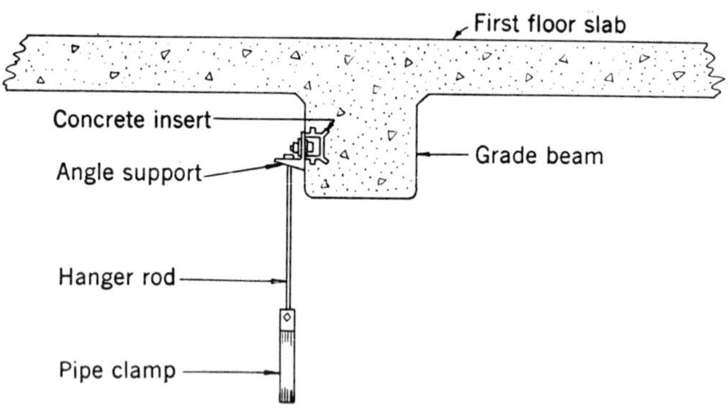

Fig. 5. Typical arrangement for the support of piping from grade beams.

mediate piping or structural supports are placed between the grade beams, to which the piping may be attached.

If very large diameter pipe is to be supported, piles may be utilized, in a manner shown in Fig. 6.

Piles are driven down into the hard natural earth to assure a sturdy support for the piping. Piers of concrete or cement blocks set on footings may also be used for the support of the piping.

Screwed Piping

Drawings show the methods that can be used to support pipe in fill. Cast iron pipe in 5-foot lengths was shown in these illustrations because it is the material most generally used for sanitary and storm water lines installed in the ground under a building. However, at this point it is recommended that when fill is encountered, cast iron pipe in long lengths with screwed or regular ends be used and material other than cast iron also be considered. A suggested material is galvanized wrought iron pipe or pipe with the same durability. The advantage in using galvanized wrought iron piping is that the pipe comes in long lengths with piping connections of threaded galvanized cast iron and recessed type drainage fittings. The engineer, however, must decide the proper material for his specific purpose.

These long length pipes provide a system that is more rigid than one made up of 5 ft. lengths of bell and spigot cast iron pipe and fewer hangers or supports are required. The wrought iron pipe can be supported every 10 to 12 ft. rather than every 5 ft. as required

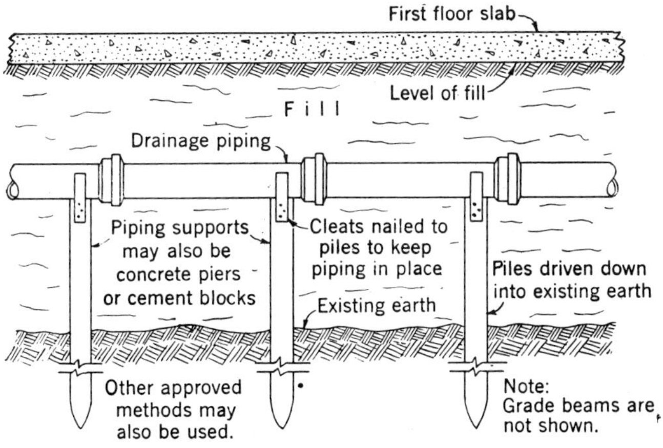

Fig. 6. Piping supported on piles or piers.

for cast iron. In addition, the wrought iron pipe will withstand the action of the fill as readily as the cast iron pipe. If desired the pipe manufacturer can be consulted to determine the corrosive properties of the fill so that he can specify the protective coating that should be applied to the pipe exterior to assure long and durable service. Moreover, wrought iron pipe can be supported in the fill by the same means as detailed in these illustrations, except that fewer hangers or supports will be required.

Engineer's Concern

If fill is contemplated for a proposed project, then the engineer must decide how the piping is to be supported, and whether this method be detailed by drawings or covered in the specifications.

If the fill has had, say six months to compact then piping supports are not necessary. If the fill has been firmly compacted by approved physical methods, then again supports may not be required. However, when physical means are employed, the compactness of the fill should be guaranteed by the contractor.

If it is known that the fill will not have had sufficient time to compact by the time the drainage piping work proceeds, then details showing the proper supporting methods to be used should be included in the contract drawings.

And should the project be of a nature that, during construction, there will be doubts concerning the compactness of the fill or the amounts of fill to be added, then the engineer should cover the subject of special supports for the pipe in his specifications. There can also be a provision in the specifications requesting the contractor to submit drawings for approval, which show methods he intends to use for the piping support, if fill is involved.

In any event when fill is encountered, every possible precaution should be taken relative to the proper installation of the drainage piping to assure a permanent system.

CHAPTER 15

Sub-surface Water Conditions

Numerous buildings are being erected throughout the country, and it is reasonable to expect that some of the structures will be built in areas where sub-surface or ground water conditions exist. Space is at such a premium in many of our cities that in order to erect a building within a certain area of the city, it may be necessary to utilize a plot of ground where ground water conditions exist. When this situation is encountered, careful attention must be given to the installation of an underground plumbing drainage system.

Let us assume that a building of say 15 stories is to be erected in a section where ground water is prevalent. In this case we shall concern ourselves only with the refinements in design that must be considered for that portion of the plumbing system and fixtures located at the lowest floor level in the building. Figure 1 shows a

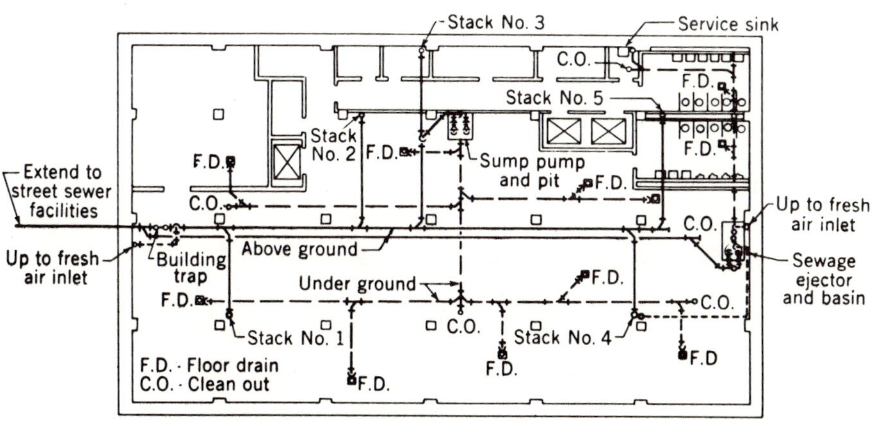

Fig. 1. Arrangement of soil and waste piping that are required at this floor level in a multi-story building.

typical plumbing system as may be found at the low level in the building which we shall call the sub-cellar.

In Fig. 1, certain toilet room fixtures, as well as a system of floor drains, are located at the sub-cellar level. A typical arrangement of the underground drainage piping in addition to the gravity drainage system, is illustrated. The sump pump and sewage ejector are also spotted on the plan. That part of the drainage system that is installed above the floor is represented in Fig. 1 by the heavy solid lines. The broken lines signify that part of the drainage system installed underground. Using this as a key plan, components of the underground piping and equipment will be described along with the manner in which this part of the system is installed.

Arrangement of Piping, Fixtures and Equipment

Figure 2 shows how the piping of Fig. 1 appears in elevation.

Floor drains at the sub-cellar level, receiving waste from mechanical equipment and floor surface, discharge into the sump pit while the waste from the plumbing fixtures is discharged into the sewage ejector. The sump pump lifts the waste and allows it to discharge into the building drain. The sewage ejector lifts the waste and conducts it to the street side of the building trap on the building drain line.

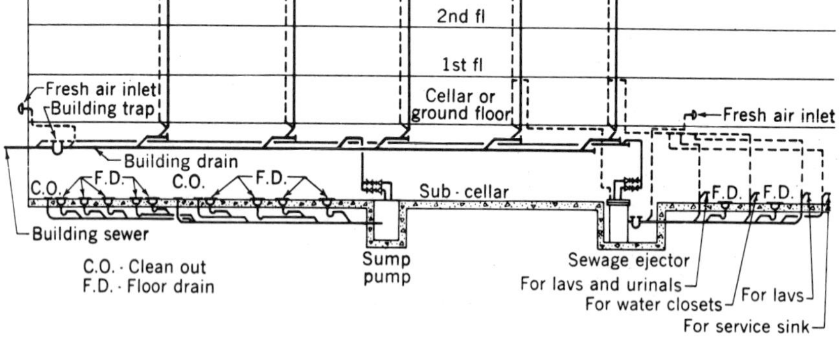

Fig. 2. Diagrammatic arrangement of soil and waste piping.

Both a fresh air inlet and a vent connection are installed at the sewage ejector to satisfy maximum code requirements. In this regard, it is always good practice to check with local plumbing authorities in a specific area for their requirements. The five plumbing stacks illustrated in Fig. 2 are included for the purpose of example only. For the sake of clarity, other plumbing systems normally required in a building of this type are not shown.

Waterproofing

Since ground water conditions, in the form of a high ground water table, springs or any other circumstances that create water problems, exist at the site, measures must be taken to render watertight both the floor slab and building walls in the sub-cellar. In this case, floor slab and foundation walls must be waterproofed.

There are a number of methods that can be employed to accomplish this. For example, in this case, we shall assume that a single floor slab of considerable thickness will be used. A finish coat of waterproofing will be applied to both the floor slab and the walls to ensure dry and watertight conditions. What then are the phases of plumbing design that should be given adequate attention? The following are some of the refinements to be considered.

Underground Drainage Piping

The installation of underground drainage piping presents somewhat of a problem where ground water conditions exist. This piping must be carefully installed since, once it is in place, it is not accessible. Therefore every precaution must be taken to insure proper installation and support. Assuming that a floor slab of approximately 18 in. thickness is used, the underground piping can be supported in one of several ways as illustrated in Fig. 3.

Detail *A*, in Fig. 3, shows the application of crushed stone. The crushed stone is placed under the piping as a support. It effectively distributes the weight of the piping and will in general prevent the piping from sagging regardless of the poor stability of the soil due to water conditions.

Detail *B* shows how the underground piping can be supported from the slab itself. The piping is installed with the pipe clamp attached and the hanger rod extending in an upright position. After the floor slab is poured, steel plates are placed over the hanger rod and rigidly secured in place. The layer of waterproof finish will conceal the steel plates.

Detail *C* illustrates how the piping is installed in a trench that

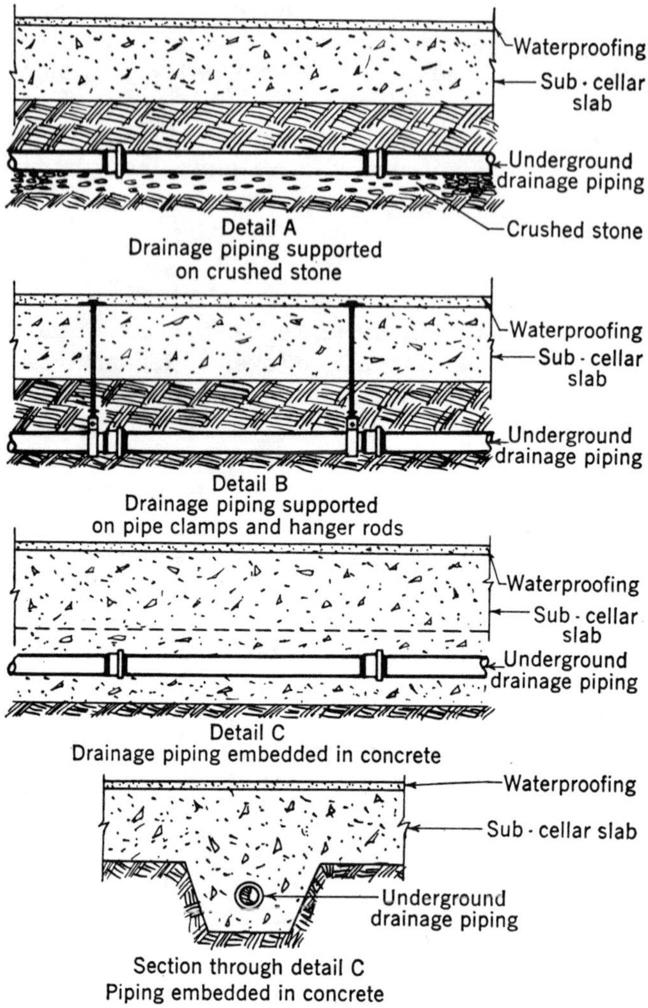

Fig. 3. Several methods for installing underground drainage piping.

is to be entirely encased in concrete when the floor slab is poured. The section through Detail *C* shows the trench and how the piping is encased in the concrete. Reinforcing rods hold the entire concrete mass together. The decision of just which method will be most suitable, rests with the engineer and the foundation consultant. In this chapter, it is assumed that the method of support shown in Detail *C* is being used.

Floor and funnel drains are necessary components of the drainage system. The floor drains are required to intercept any water that may accumulate on the floor while the funnel drains are needed for the various mechanical equipment from which waste or cooling water is discharged. Both types of drains can be installed in a watertight manner as shown in Fig. 4.

Floor drains can be installed in a watertight manner with the method suggested by Detail *A*, Fig. 4. In this instance, copper or

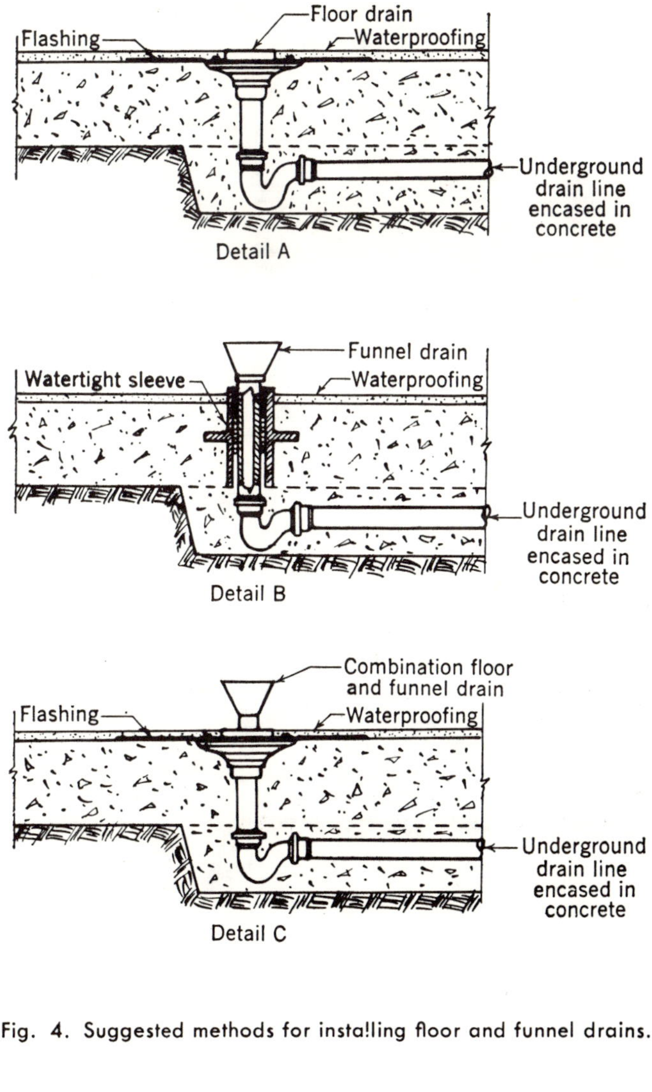

Fig. 4. Suggested methods for installing floor and funnel drains.

lead flashing is clamped into the drain as shown to prevent any seepage of water out from around the perimeter of the drain itself.

Funnel drains can be installed as indicated by Detail *B*. Here, a watertight sleeve is employed. The sleeve is then packed with oakum and lead or mastic compound. The sleeve is easily fabricated on the site and can be counted upon to prevent water seepage.

A very popular drain is the combination floor and funnel drain. This unit not only serves to receive the waste from equipment but also functions as a floor drain. Detail *C* shows how a drain of this type can be installed in a waterproof way, with copper or lead flashing.

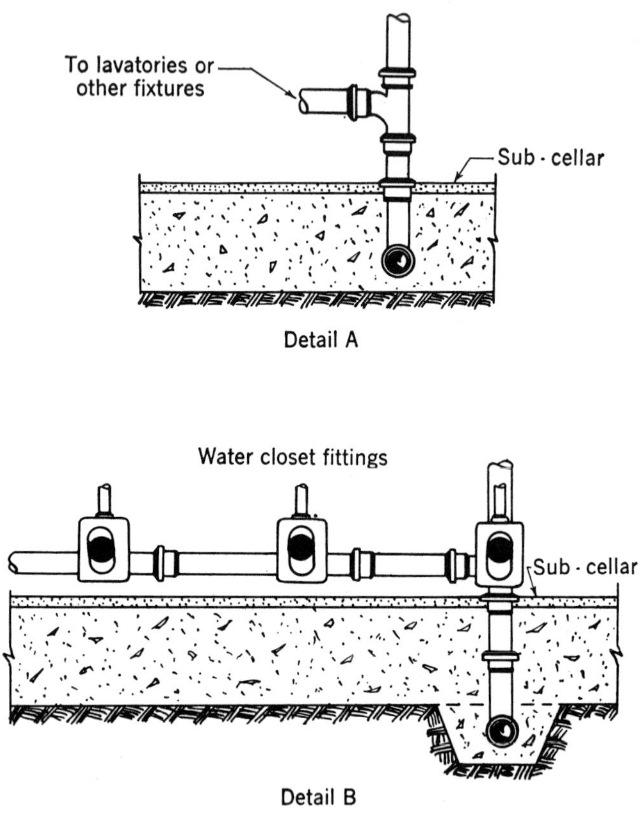

Fig. 5. Typical installation of soil and waste lines for sub-cellar fixtures.

Soil and Waste Pipe Penetrations

Soil and waste lines serving the plumbing fixtures should also be given some attention at the point where the lines penetrate the floor slab. Figure 5 represents typical soil and waste line penetration.

Detail *A* of Fig. 5 shows the installation of a waste line while Detail *B* portrays the installation of a soil line. Both installations, in this particular case, were accomplished without the use of watertight sleeves or flashing fittings. The reason is that, when the underground drainage line is included within the slab, or is encased in concrete as shown in both details, then it often is not necessary to provide any special devices. This is especially true when waterproofing liquid is mixed with the concrete used for the floor slab.

If, however, the waterproofing liquid is not used, and it is desired to have the positive protection of watertight devices, then either sleeves or flashing fittings can be used with the lines penetrating the floor slab.

Sewage Ejectors

Since a certain number of plumbing fixtures are required in the sub-cellar floor level, a sewage ejector is needed. The waste matter flows into the ejector basin from which it is pumped to a higher elevation and into the building drain on the street side of the building trap. From here it discharges by gravity to the outside street sewer facilities. A typical ejector installation might be somewhat similar to that shown in Fig. 6, where a pit has been provided to receive the ejector basin.

The pit interior is waterproofed as is the surface of the floor slab. Watertight sleeves are installed for the drainage piping that penetrates the side of the pit. Sand is placed around the basin to keep vibration of the basin to a minimum when the pump operates. The ejector basin is used for two reasons:

(1) It simplifies installation since the basin can be set in place after the concrete work and the waterproofing are done.

(2) Since the basin affords tighter connections than just the pit and cover, there is less possibility of odors.

The installation of piping and equipment, as illustrated in Fig. 6, reflects maximum code requirements. Local plumbing authorities should always be consulted for their particular regulation governing the installation of a sewage ejector. The checkered plate cover over

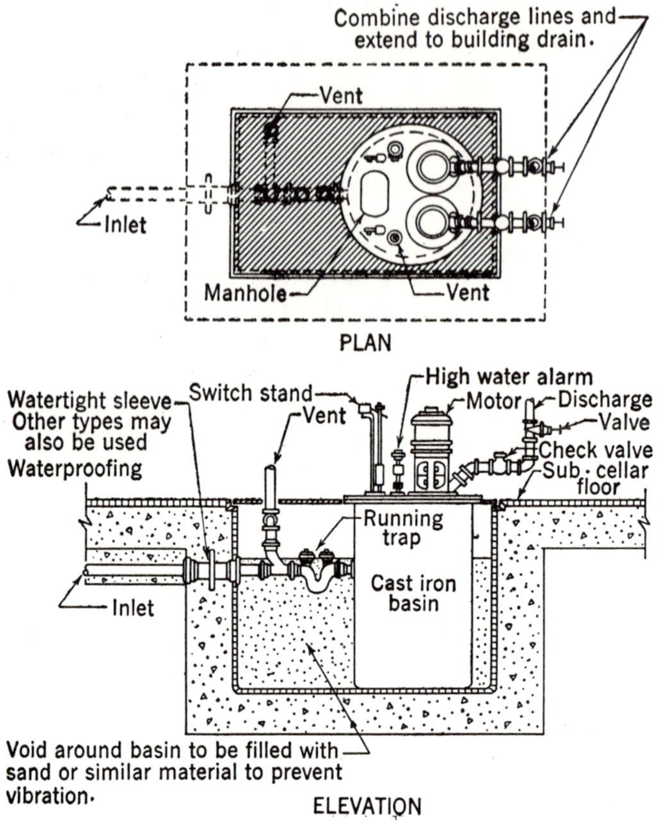

Fig. 6. Installation of a sewage injector.

the pit itself should be thick enough to withstand the weight loads imposed. In addition it should be braced sufficiently to form a snug fit around the basin as well as at the sides of the pit.

Sump Pump

A sump pit is provided to receive the waste water from the floor and funnel drains. A typical installation of the sump pit and pump is shown in Fig. 7.

As noted in Fig. 7, a concrete pit is constructed to receive the flow of waste water. Waterproofing is applied to the interior of the pit as well as to the surface of the floor slab. The sump pump lifts the waste water to an elevation high enough for it to drain by gravity into the building drain. In this instance a cast iron basin is not

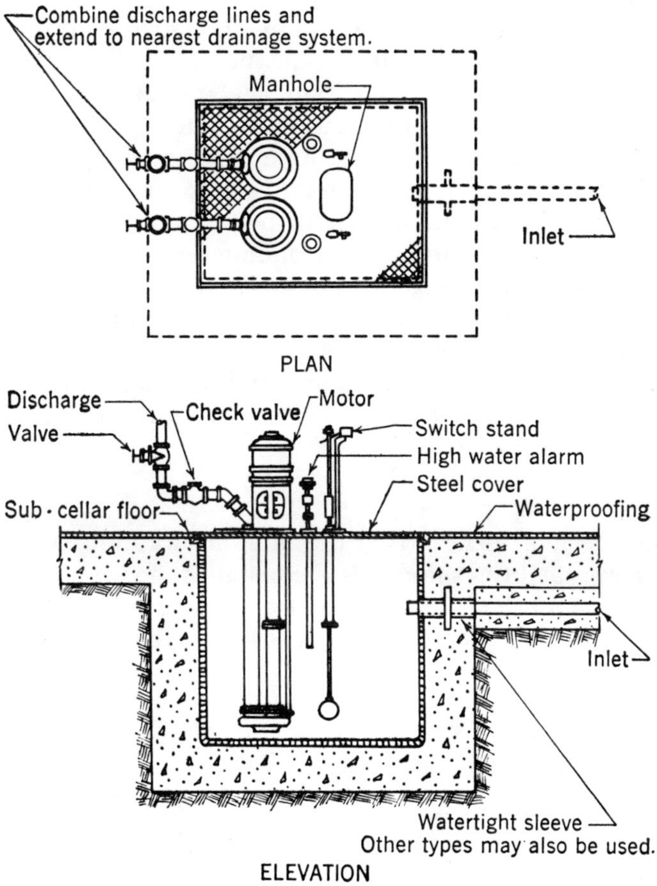

Fig. 7. Arrangement of sump pump and sump pit.

needed because the waste water is relatively clean and does not present the odor problem as does the soil line waste.

On occasion, the foundation engineer may wish to have the sump pit receive the ground water when it exceeds a certain level. This practice will reduce the possibility of the water exerting pressure on the underside of the slab. To do this, slots are furnished through one side of the pit. Both the pit and pumps must be sized to a capacity that is adequate to receive the quantity of water expected. The foundation engineer should be able to supply the necessary data concerning the rate of flow.

When these slots are included, it is well to remember that some emergency means of auxiliary power must be considered in the event of an electrical power failure. Otherwise, if the electrical power should fail, the water would not only flood the sub-cellar but would also damage much of the mechanical and electrical equipment contained at the sub-cellar floor level. Another method entails the placement of pipes in the side of the pit. Valves with operating stems that extend through the cover of the pit are placed on the pipes. When the danger of power failure occurs, it is easy to close the valves until the impending emergency is past.

Of course, when emergency power is included in the building for lighting as well as the operation of certain equipment, then steps should be taken to include the sump pumps in this case on the emergency circuit. The sump pump and its component piping must be installed in a manner that conforms with the requirements of the local plumbing authorities.

Watertight Sleeves

Drainage lines penetrating the building wall on its route to the street sewer facilities should also be installed in a watertight manner. This is accomplished through the use of watertight sleeves. Figure 8 illustrates two of several types of sleeves for this purpose.

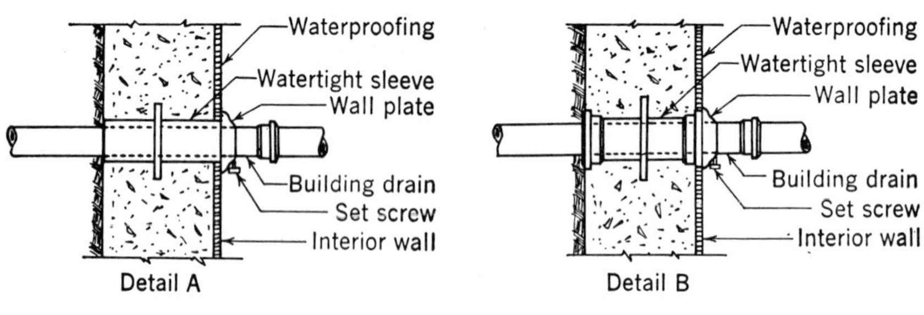

Fig. 8. Watertight sleeves for drainage lines penetrating a building wall.

Detail *A* shows the type of a sleeve that can be fabricated on the site. It consists of a piece of pipe of sufficient diameter to allow the drainage line to pass through it. A baffle is welded to this pipe to prevent any seepage of water around the sleeve itself. The void between the inside of the pipe sleeve and the outside of the drainage line is packed with oakum and lead or mastic compound to complete the watertight connection. The wall plate shown on the interior of the building wall is for the purpose of appearance only.

The sleeve shown in Detail *B* is of a type manufactured expressly for this purpose. It consists of a piece of pipe with hub joints at either end. After the installation is made, the joints are packed with oakum and lead or mastic compound. The wall plate shown is for the purpose of appearance only.

Other types of sleeves are also available for this use, and it is up to the engineer to select the type of sleeve he feels will best suit his needs. When confronted with installations that must operate at sites where ground water conditions exist, it is hoped that the suggestions outlined will be of some help.

CHAPTER 16

Vent Stack Terminal

A vent stack terminal is that part of the plumbing piping system which extends up through the roof of a building to the atmosphere. This vent pipe, although small by comparison to the overall piping, is nevertheless a vital part of a plumbing system for it is through this open pipe that the air at atmospheric pressure enters the system to hold in balance the water seal contained in each fixture trap. Without this balance of air pressure, the plumbing system, consisting of drainage and vent piping, would not function properly. In addition air movement in the system is necessary to keep the lines fresh and free of odors.

Factors that the designer should consider regarding the installation of the vent stack terminal will now be discussed. Figure 1 shows a portion of a typical riser diagram in which are illustrated soil and vent stacks, and the manner in which the stacks are projected up through the roof. In the circles above each stack, the number shown is the stack number and *P* is an abbreviation for "Plumbing."

The diameter of a vent stack terminal should never be less than four inches for two very important reasons:

1. A 4-inch diameter line allows an adequate volume of air to enter the plumbing system, and its effective opening is not as easily blocked by foreign matter as would be with a smaller diameter pipe.
2. In climates subject to freezing weather, the warm condensation vapors rising in the line tend to form frost at the pipe opening when contact is made with cold air. This frost coating can block or partially block the effective opening of a pipe smaller than 4 inches in diameter; this condition is remote with a 4-inch diameter pipe. Therefore, when smaller diameter stacks for fixtures other than water closets are involved, the vent stack is usually increased to 4-inch diameter below the

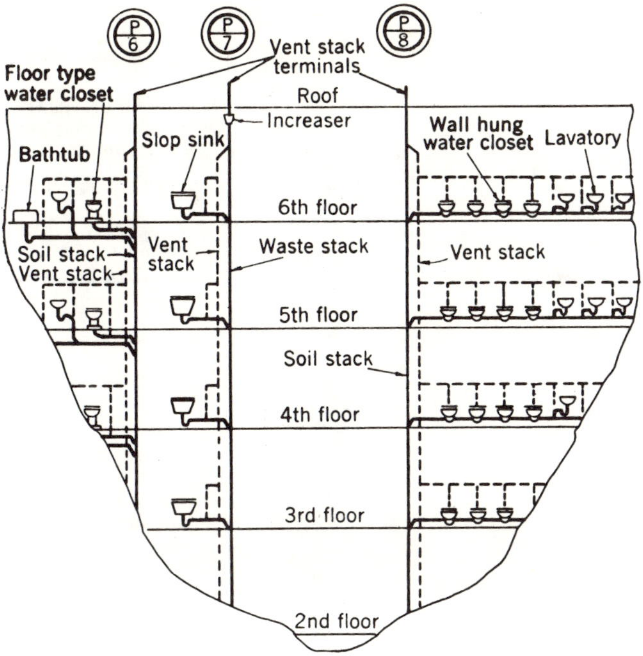

Fig. 1. Partial riser diagram showing vent stack terminals.

roof level. This larger diameter pipe is then extended up through the roof to serve as the roof stack terminal. A typical detail of this connection is shown in Fig. 2.

Projection Above Roof

The height to which a vent stack terminal should be projected above the roof is normally determined by local code requirements. This height generally varies from one foot to several feet. The code also has requirements pertaining to the location of the effective vent opening with regard to windows, doors or roof construction.

Roof Penetration

Since the vent stack will penetrate the roof, adequate steps must be taken to provide a watertight installation. There are a number of methods that may be used, and Fig. 3 indicates three of the more common methods used to furnish the necessary watertight connections.

Detail *A* of Fig. 3 indicates a copper vent stack terminal. The

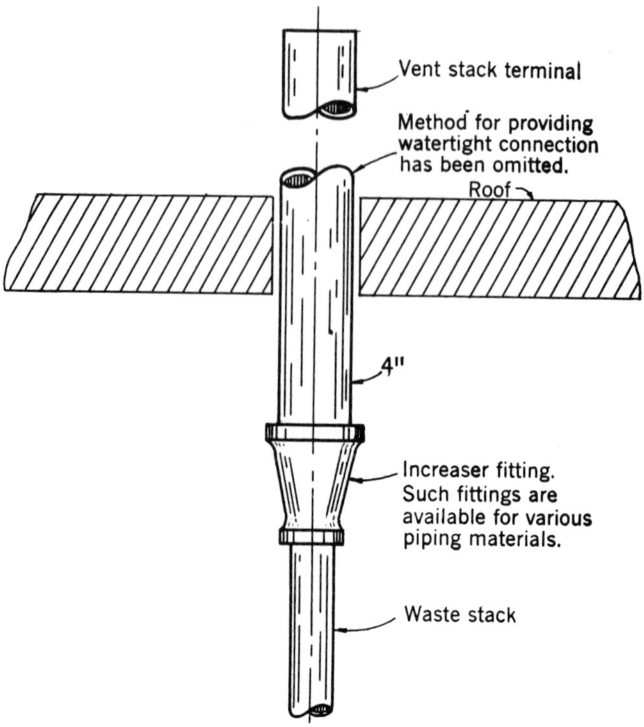

Fig. 2. Detail of an increaser fitting.

copper or lead flashing is soldered to the pipe as shown, thereby assuring a watertight connection. This method may be used when copper piping or tubing is the material specified for the entire drainage and venting system. Often, copper pipe or tubing is installed only for the vent stack terminals, with the rest of the system of other material.

Detail *B* shows a vent stack terminal consisting of cast iron pipe with the terminal itself of galvanized wrought iron pipe. The copper or sheet lead flashing is turned into the hub of the cast iron pipe as illustrated, and caulked in place with the wrought iron pipe. This method of installation also provides for a durable and watertight connection.

Detail *C* pictures a design using a galvanized wrought iron vent stack terminal and the application of a patented roof coupling, of which there are numerous types. The copper or lead flashing is slipped over the vent stack protruding through the roof. Next the roof coupling is slipped over the top of the flashing and is screwed to the piping. The upper portion of the piping is then screwed in place and the installation is complete with a watertight connection.

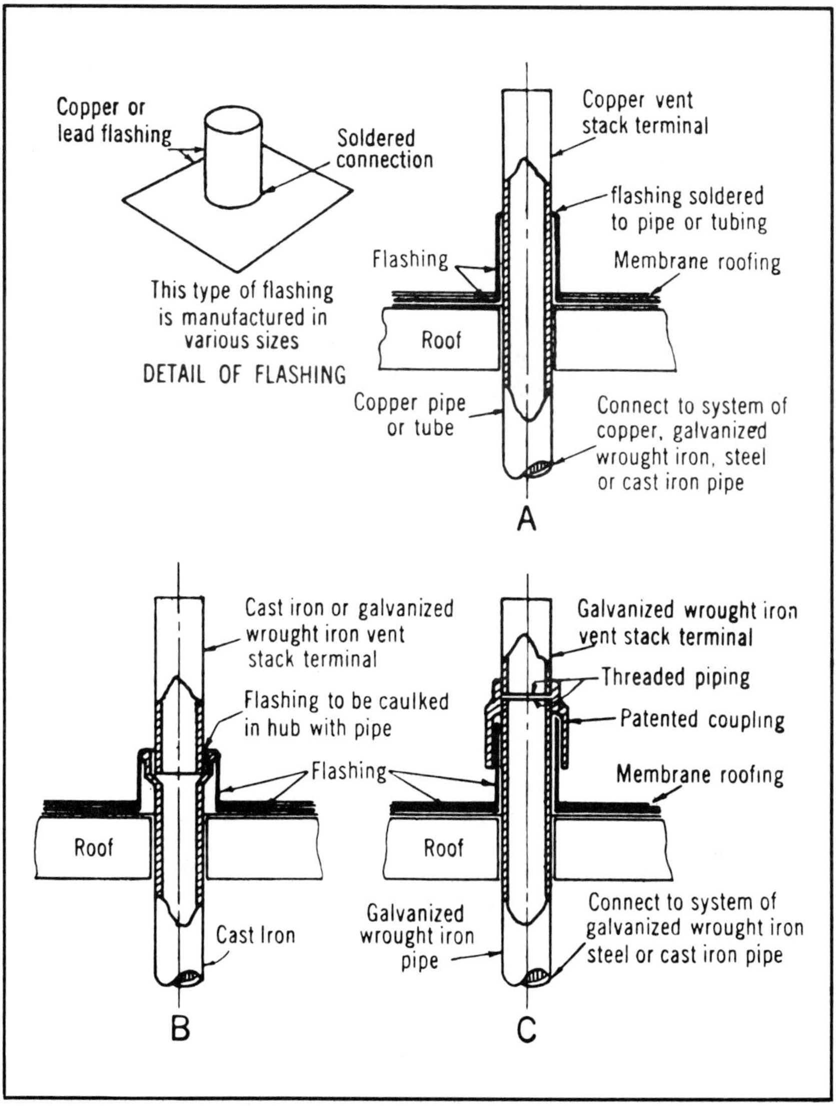

Fig. 3. Typical vent stack terminals with a detail of flashing.

Of the three methods shown, preference is given to the method using the patented roof coupling because it is easy to install, offers a neat appearance, and provides a watertight connection having durable qualities.

Effects of Wind

Winds of sufficient force can affect the function of the drainage and venting system. In certain areas where: (1) strong winds are prevalent, (2) when the vent stack terminal is located on the top of a tall building or (3) at a high elevation where it will be exposed to severe wind conditions, provisions should be considered to shield the vent from such winds.

A strong wind blowing across the effective opening of the vent stack terminal can create unbalanced air pressures within the piping system, thereby causing the fluctuation or loss of water seal in the fixture traps. This can result in the discharge of odors into rooms in which there are plumbing fixtures.

To protect the effective opening in the pipe from the direct blast of the wind, a device called a cowl is employed. The cowl is installed to deflect the wind away from the vent opening thereby eliminating the possibility of air pressure fluctuation in the piping system. There are many types of cowls available for this purpose, some stationary and some that revolve. Figure 4 illustrates two types of cowl frequently used for this purpose.

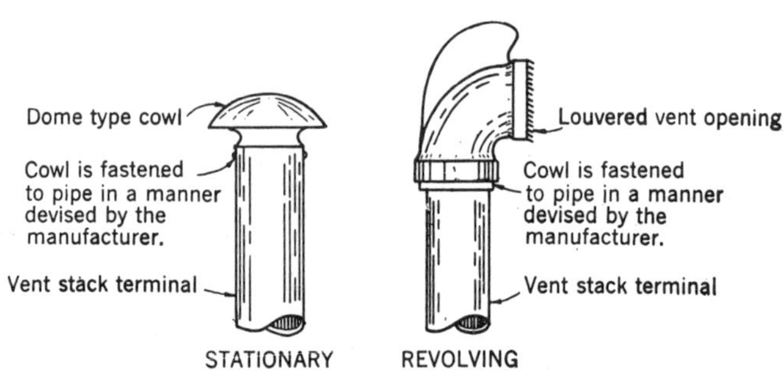

Fig. 4. Detail of typical cowls.

The stationary cowl of Fig. 4 is similar in shape to a dome. This dome will deflect the wind from the pipe opening and yet allow the required volume of air to enter the piping system. The dome will not hamper the flow of warm air generated in the piping system to the atmosphere. This cowl may be fastened to the pipe with bolts or set screws as provided by the manufacturer.

The revolving cowl of Fig. 4 is a patented type of cowl. It is designed to turn or rotate with the direction of the wind. By this means, the effective vent opening in the cowl will always be turned away from the force of the wind so that venting will not be affected by wind conditions. This type of cowl may also be fastened to the pipe in a manner devised by the manufacturer. Now for a note of caution concerning the use of the revolving type cowl. Unless the cowl is periodically inspected and maintained, the unit will stick and defeat its purpose.

Foreign Objects

Often vent stack terminals are installed in locations where they are within the reach of the children as in the case of apartment house roofs. If the vent opening is not protected, rubber balls, sticks or other objects may be dropped into the pipe. These objects can cause severe stoppages and many times present a difficult removal problem. Objects such as sticks can lodge in the piping and although they might not cause complete stoppage, they can block the line sufficiently to interfere with the proper venting of the system.

To prevent these occurrences, a vent hood is employed. This device is placed over the vent opening and is securely fastened to the pipe. Although there are numerous types manufactured for this purpose, one typical device is shown in Fig. 5.

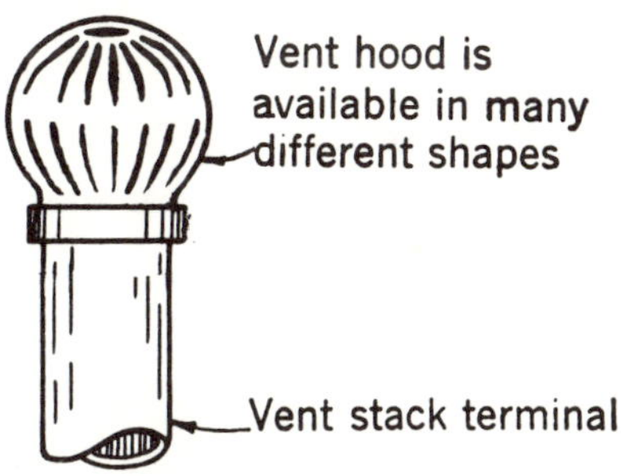

Fig. 5. Vent hood placed on top of a stack terminal.

From Fig. 5 it is apparent that the hood will prevent large objects from being dropped into the vent stack terminal. When cowls are used, this device is not necessary since the cowls are equipped with louvers or screens. In the case of the dome type cowl, there is not enough space between cowl and pipe opening to insert large objects.

The same provisions may be applied to sloping roofs. For this condition there are, similarly, numerous devices available for furnishing watertight connections with the various types of roofing encountered. Cowls similar to those shown may be used to shield the vent opening from strong winds.

When careful consideration is given to the proposed installation of the roof stack terminal, a properly vented plumbing system will be assured and a number of problems will be avoided.

CHAPTER 17

Preventing Sewage Backflow in Buildings

In the design of a building's plumbing drainage system, it is of prime importance to provide an adequate system of soil, waste and vent piping. If this system is sized adequately and the proper material is used, years of trouble-free service can be expected. Furthermore, if the building drain and the building sewer are installed with the proper slope, the discharge from the fixtures will be effectively conducted to the outside utilities. When the same consideration is given the storm drainage system, similar results can be expected.

With proper design and ideal site conditions, it is always possible to produce a sanitary or storm system that will ensure favorable results. Sometimes, however, there are certain conditions prevalent at the site which require extra consideration. One such condition is that of sewage blackflow which usually originates in the outside sewer lines and is caused by a number of reasons, a few of which follow:

1. Old and overloaded system of sewer piping.
2. Undersized, forced or pumped sewer lines.
3. Sewers, subject to flooding at point of discharge, due to tides, high water and similar reasons.
4. Peak periods of year as during summer, when larger quantities of sewage are handled.
5. The high velocity of flow in sewers caused by tall buildings.
6. Sewer lines that have sagged or settled and retard normal flow.
7. Stoppages in sewer lines.

Separation of Sanitary and Storm Systems

Where the danger of sewage backflow exists, the first important step to be taken by the engineer, is to design the sanitary and storm systems so that they will be completely independent of each other.

This means that the piping of both systems must be installed with separate connections to the outside sewer system or street sewers. If individual sanitary and storm sewers are available in the street, the installation will be simplified. However, if a combined sewer exists, the storm system should be connected to the downstream side of the sanitary connection to this sewer. Figure 1 illustrates the underground piping of the two systems with connections to the sanitary and storm sewers or the combined sewer.

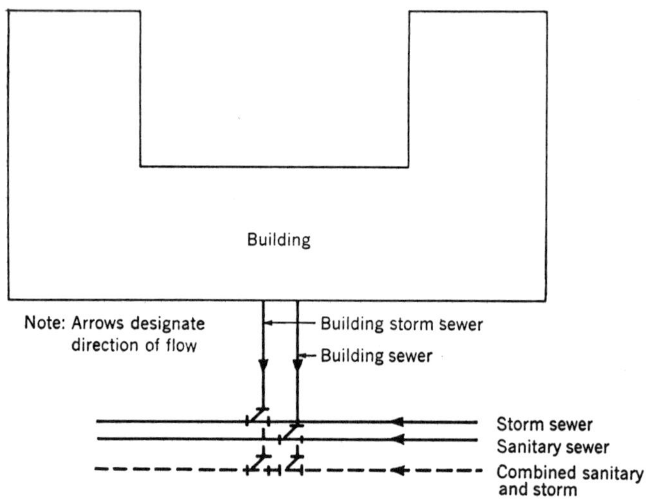

Fig. 1. Typical installation of separate sanitary and storm sewer.

It will be noted in Fig. 1, that both the sanitary and the storm piping are installed as separate systems and connect independently with the outside sewer lines. In the case of the combined sewer, the reason for the downstream connection of the storm system is to prevent the storm water, which is under pressure during a storm by virtue of its total head, from short circuiting into the sanitary system.

Piping Systems for Buildings with Ground Floors

There are several conditions that require protection from backflow. The first of these conditions is the case of a multi-story building with fixtures located on the first or ground floor level. A typical diagram of this arrangement is shown in Fig. 2 where the sanitary

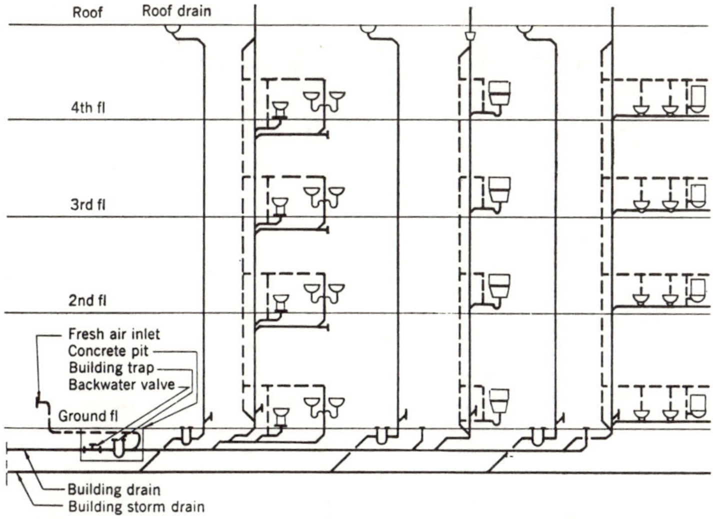

Fig. 2. An arrangement of sanitary and storm drain system within a building.

and storm systems are separate installations. The storm water system receives the roof water, drops below the ground floor slab and is run underground and out of the building to connect with the storm or combined sewer.

The sanitary system receives the discharge from the plumbing fixtures at the various floor levels and drops under the ground floor slab to receive the discharge from the fixtures at this level. Then it runs underground to connect with the outside sanitary or combined sewer. A building trap and a backwater valve are installed on the building drain in a concrete pit as shown. The concrete pit offers ease of access to the building trap and backwater valve for the purpose of inspection and maintenance. The installation of the building trap, fresh air inlet and backwater valve must be in accord with local code requirements.

The purpose of the backwater valve is to prevent the backflow of sewage into the building where it can flood the ground floor area. The backwater valve consists of a flapper that opens with the normal flow to the street but closes when the flow is reversed.

It will be noted that the sanitary and storm water lines are completely separate systems and do not connect at any point within the building. The reason for this is that any stoppage in the sanitary system would permit the storm water to flood the ground floor, in the event of a rainstorm or backflow conditions.

Piping Systems for Buildings with Basement

The other condition for which backflow protection is required is the case of a multi-story building with a basement or cellar in which fixtures are proposed. A typical diagram of this arrangement is shown in Fig. 3.

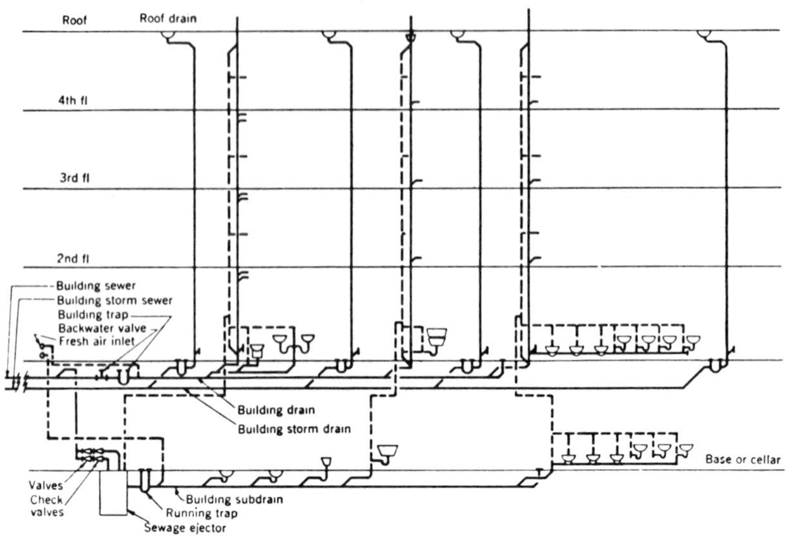

Fig. 3. A typical arrangement of sanitary and storm system within a building which has a basement.

Here again, the sanitary and storm piping are separate systems and connect independently with the street sewers. As shown, the sanitary and storm water lines are installed at the ceiling of the basement or cellar and discharge by gravity to the street sewer. A building trap and backwater valve are installed in the building drain as indicated in the illustration.

Fixtures in the basement or cellar discharge into the drainage line installed under the slab, connecting with the sewage ejector. The sewage ejector lifts the sewage up to the ceiling level where it ties in with the building drain. The check valves shown on the discharge lines at the ejector, prevent the backflow of sewage through the fixtures in the basement or cellar.

The particular hook-up shown for the ejector portrays maximum code requirements for this type of equipment. The unit with included piping should be installed in accordance with specific local code requirements for the area in which the building is to be erected.

Other Installations

The same recommendations previously illustrated and defined may be applied to similar installations, where backflow conditions are a regular or frequent occurrence.

Care of the Backwater Valve

When the building drain is laid underground, it is important that the backwater valve be installed in the building drain at a location that provides ease of maintenance. A concrete pit offers the necessary access. Figure 4 shows a typical arrangement of the building trap and backwater valve as installed in a concrete pit.

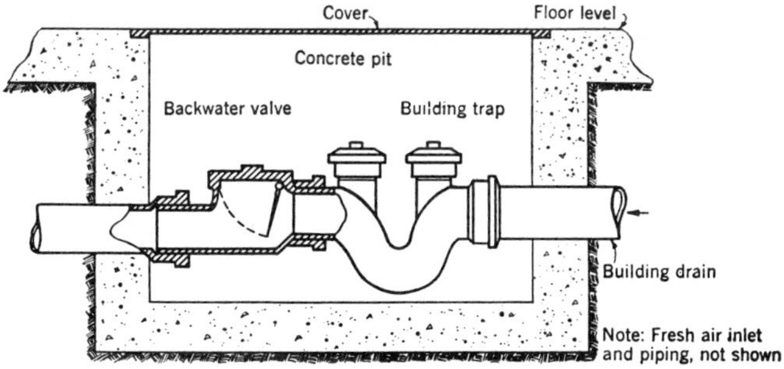

Fig. 4. Installation of a backwater valve.

The backwater valve requires a certain amount of attention since sediment will accumulate on the flapper valve seat thereby preventing the seat from closing tight during the event of backflow. Therefore, the cover should be removed periodically, the sediment removed, and the seat rubbed clean. This practice will ensure the proper function of the valve. When the building drain, including the backwater valve, is installed in an exposed location at the ceiling as indicated in Fig. 3, inspection and maintenance are simplified. However, the same care and attention should be given this valve seat as mentioned. Backwater valves of varied type and manufacture are available for this purpose. Although, all of these valves will serve their function well, they will only continue to work properly if they are kept clean and free of sediment.

Extra Preventative Measures

Often, in order to provide a more positive protection against sewage backflow, a quick closing valve is installed in the building drain. The valve is similar in operation to a gate valve with the exception that it is equipped with a lever handle for quick closure. The valve is generally located on the street side of the backwater valve. In the event of sewage backflow, the valve can be quickly closed, thereby eliminating the damage that can result from flooding. The danger of a quick closing valve is that although it prevents sewage backflow, it also stops normal flow to the street. As a result, when the valve is closed, steps should be taken in the building to curtail the operation of the plumbing fixtures, until the valve can be safely opened again.

Parts of System

The component parts of the sanitary and storm system were labeled in accordance with definitions given in the National Plumbing Code, ASA A40.8-1955. Following are the definitions of some terms used as given in the National Plumbing Code.

Building drain. The building (house) drain is that part of the lowest piping of a drainage system which receives the discharge from soil, waste and other drainage pipes inside the building and conveys it to the building (house) sewer beginning 3 ft. outside the building wall.

Building sewer. The building (house) sewer is that part of the horizontal piping of a drainage system which extends from the end of the building drain and which receives the discharge of the building drain and conveys it to a public sewer, private sewer, individual sewage disposal system, or other point of disposal.

Building storm drain. A building (house) storm drain is a building drain used for conveying rain water, surface water, ground water, subsurface water, condensate, cooling water, or other similar discharge to a building storm sewer or a combined building sewer, extending to a point not less than 3 ft. outside the building wall.

Building storm sewer. A building (house) storm sewer is the extension from the building storm drain to the public storm sewer, combined sewer, or other point of disposal.

Building subdrain. A building (house) subdrain is that portion of a drainage system which cannot drain by gravity into the building sewer.

Building trap. A building (house) trap is a device or fitting or assembly of fittings installed in the building drain to prevent circulation of air between the drainage system of the building to be served and the building sewer.

When conditions of sewage backflow exist or occur, the design considerations outlined will go a long way in preventing the sewage from entering the building. In addition, if adequate inspection and maintenance are provided at the site, the equipment can be depended upon to function properly. Such preventative measures are the means to stop sewage backflow in a positive manner.

CHAPTER 18

Importance of Cleanouts in a Building

After the plumbing work in a new building has been completed and the building is occupied, problems connected with maintenance begin. One phase of the maintenance work concerns the clearing of stoppages that may occur in a part of the plumbing drainage system due to any of a number of reasons. A few of the more common causes are:

1. Foreign objects lodged in the drainage line.
2. Accumulation of hair or other matter.
3. Deposits of grease fat or other substance.

Such objects can completely block the discharge through the drainage lines or at least reduce the effective flow. Blockage of the lines can, in the case of a complete stoppage, cause considerable water damage to the lower floors and contained equipment when the waste water backs up. Partial stoppages can create sluggish discharge conditions in the drainage lines and this will interfere with the proper venting of the drainage system. It will result in the discharge of foul odors in the rooms or areas. This is always a dangerous condition for when the capacity of the venting is unbalanced, siphoning of the trap seals will result.

Therefore, to avoid these troublesome conditions, certain considerations must be given to the cleanouts during the design stages to ensure complete access to all parts of the drainage system. Well planned points of access will permit the clearing of stoppages after the building is erected. Various cleanouts are illustrated in Fig. 1, and the purpose of each cleanout will be described.

The location of the cleanout governs the type cleanout that is installed.

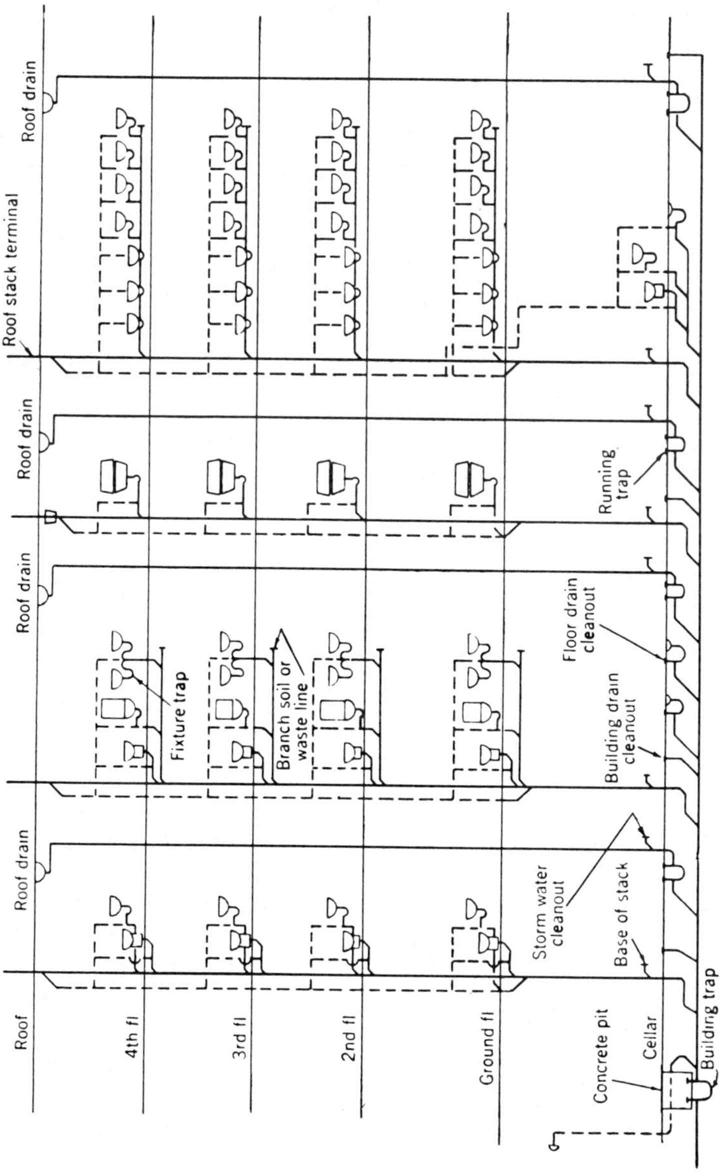

Fig. 1. A typical plumbing riser diagram to show the use of various cleanouts.

Cleanouts

Roof stack terminal. This is often used for cleaning a stoppage that may be lodged in the vertical soil or waste stack. Sticks, stones or even rubber balls are sometimes dropped down the vent stack terminal at roof level by children. Such objects can lodge in a section of the stack, between floors and presents a stoppage that is difficult to clear. Rags, rubber toys and similar objects provide the same problem when discharged through a water closet into a soil stack. These objects can be cleared or dislodged so they fall to the base of the stack where the cleanout at this point offers easier access to the object. To dislodge the object, a steel snake of sufficient length is inserted down the opening of the roof terminal until the object is reached and worked loose. This type of access is only possible when the building is not too tall, for otherwise stack cleanouts at intermittent floors are necessary.

Fixture traps. A means of clearing the waste line between the fixture and its waste connection to the vertical waste stack or horizontal waste branch is provided by fixture traps. Hair pins, safety pins or similar objects as well as accumulation of hair, soap fat or grease can cause stoppages in these lines. The fixture trap is equipped with a cleanout plug that may be removed for access purposes. By use of a snake or wire, the object can be removed or in the case of soap fat or grease accumulation, the substance can be worked loose by the motion of the wire. After replacing the plug, the line is flushed clean with hot water.

Branch soil or waste lines. Such lines are usually provided with a cleanout at the end of the branch. This cleanout assures access to stoppages that may develop in branch lines. Here again, the stoppages may consist of rags, rubber toys or just accumulation of soap fat or grease, especially if a sink is connected to the line. By removing the cleanout plug, the stoppages are easily cleared with the help of a steel snake. If the branch waste is installed above the floor slab, access may be provided by offsetting the cleanout pipe so that the cleanout plug will be flush with the wall. If the branch waste is installed below the floor slab, again the cleanout piping should be so arranged that the cleanout plug will be flush with the wall near the ceiling. For either installation, proper sized access doors might well be used to reach the cleanout plug.

Base of stack. A cleanout at the base of stack is quite important since foreign objects discharged or dropped into the vertical soil and waste stacks are generally trapped at this point. Such objects can cause severe stoppages resulting in considerable water damage to the lower floors. The discharge from the upper floor fixtures will back up and overflow through the plumbing fixtures on the lowest floor above the stoppage. Therefore, a cleanout at this point will offer the necessary access. Various types of manufactured cleanouts and access doors are available for use at this location.

Building drain cleanout. This type cleanout is installed either exposed on the building drain located on the ceiling of cellar or basement, or the floor level type consisting of fittings and pipe installed on the underground building drain and extending up to the floor level of the floor slab. Here again, various types of cleanout plugs and access plates are manufactured to afford the necessary access to the drain lines. Such cleanouts are usually placed between 40 and 50 ft. apart, at which distance a steel snake is more easily handled to clear stoppages in the building drain.

Building (house) trap. This type trap has a dual purpose for (1) it contains the water seal necessary to prevent the sewer gas from entering the building drainage system and (2) it serves as a cleanout for the building sewer, located between the building and the street sewer. Objects that cause stoppages in other parts of the drainage system in the building can also cause stoppages in this line. Occasionally, tree roots that enter the building sewer, constitute a more serious threat. To clear stoppages in this line, a steel snake is used. In the case of tree roots, patented devices are inserted into the line through the trap and the roots cut out and flushed out.

When the building trap is exposed, the cleanout plug is easily removed. However, when the building drain is underground, availability is offered by extending the cleanouts up to the floor level. More frequently, a concrete pit with a steel cover is constructed. This pit provides ease of access to the cleanout plugs. The concrete pit should be of a size that is proportionate to the depth of the underground building drain. This consideration will allow a man sufficient room in which to work.

Floor drain cleanouts. Such cleanouts are quite necessary where the floor drains are remotely located away from a branch waste that is normally equipped with cleanouts. The cleanout is an integral part of the floor drain, having a removable plug, thereby providing the necessary access. Many styles of floor drains with this feature are available, depending upon the particular function of the floor drain.

If a drain of this type is not used, then a separate cleanout should be installed at the end of the waste line.

Storm Water System

Cleanouts are also very much required for the storm water system. Stoppages can be caused in the storm water piping by leaves, gravel, grit, stones or even small sticks dropped by children through the openings in the dome of the roof drain. These objects that can cause stoppages, usually settle at the base of the storm water stack. A cleanout in the base of the stack at this point will provide the necessary access for clearing stoppages.

If a running trap is installed at the base of the stack as shown on the drawings, it will intercept any gravel or grit that drops in the line. Such matter would otherwise accumulate in the underground line where it can cause stoppages, especially during prolonged dry spells. The installation of the running trap should be in accord with the requirements of the local code.

Floor level cleanouts should also be provided every 40 to 50 ft. apart, whether or not the building storm drain is installed exposed at the ceiling of the cellar or basement, or if it is installed underground. The purpose of these cleanouts is the same as outlined for the plumbing drainage system. In general, the same careful consideration should be given the storm water system as is given the sanitary system.

CHAPTER 19

Providing Access for Building Traps

A building trap is regarded by many of the plumbing codes as a very important component of the plumbing drainage system for it has a threefold purpose: (1) it acts as a barrier against the gases in the street sewer that would otherwise tend to enter the interior plumbing system; (2) the trap offers a means of clearing stoppages that could develop in the building sewer between the building trap and the outside sewer facilities; (3) it is used in conjunction with the fresh air inlet to serve its function by permitting air to enter the sanitary system at its lowest point and circulate through the piping system to the vent stack terminal at the roof, thereby keeping the entire system fresh and clean.

Depending upon the location of the building drain, the building trap will be installed either above or below the ground. For example, if the building drain is located at the basement or cellar ceiling, the building trap will also be set in an exposed position. If the building drain is located underground, the building trap will also be installed underground.

This discussion will deal mostly with building traps located in the underground position. The terminology used for the various parts of the drainage system conforms with that given in the National Plumbing Code.

Shallow Piping Installations

When the underground piping runs in a building are not too long, the trap will be located rather close to the surface of the floor slab. A typical example of this condition is illustrated in Detail *A* of Fig. 1. Since the piping run is not too extensive, the trap by virtue of the drainage slope in the piping will be located rather close to the surface of the floor slab. This will permit the top of the cleanout

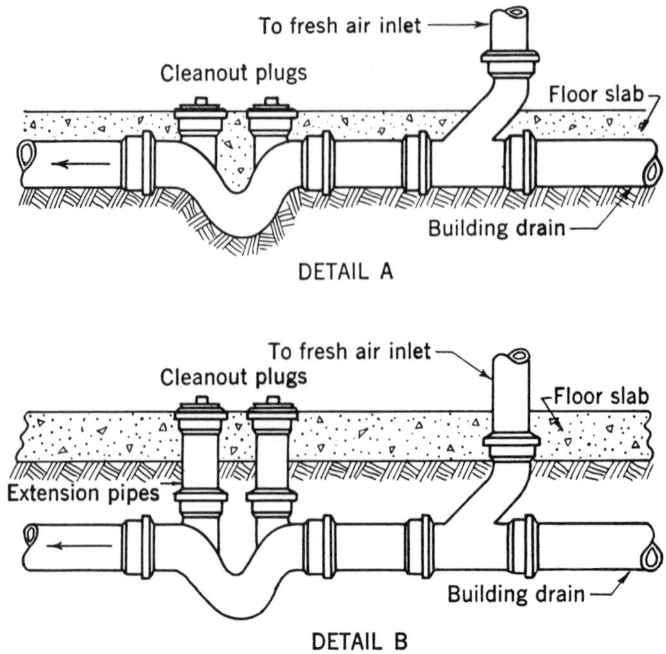

Fig. 1. Typical installation of building traps that are contained in a shallow bury.

hubs to be set flush with the floor slab where the cleanout plugs are easily removed.

Should the building trap be so located that it is impossible to terminate the top of the trap at the level of the floor slab, then extension pipes must be utilized. Detail *B* of Fig. 1, shows how this factor is achieved. The extension pipes are the most flexible means for arranging the installation, to enable the cleanout plugs to be flush with the floor.

A limit, however, should be placed on the length of the extension pipes used, before this arrangement becomes too awkward. Therefore, in order to offer the most practical limits of function, the extension pipes should not exceed 18 in. in length. For building traps that are buried deeper, concrete pits must be employed.

Concrete Pits

Concrete pits offer access to a building trap that is located at a deeper elevation in the ground. Figure 2 shows a typical location of this pit in a building.

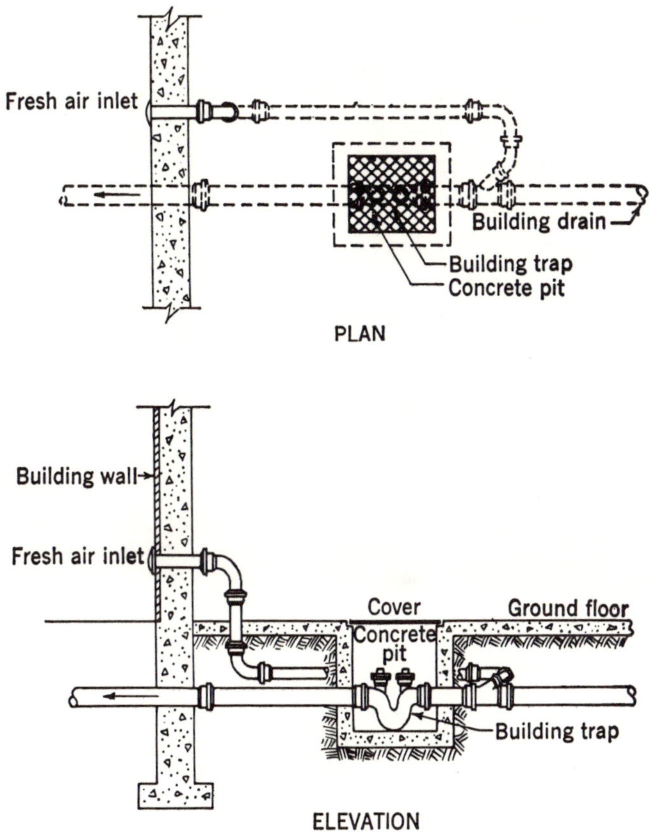

Fig. 2. Location of concrete pit and the arrangement of component piping.

Referring to the plan in Fig. 2, it will be noted that the pit is placed at a point just inside the building wall. Concrete pits of course, can be placed outside the building wall, but this article is concerned only with the interior installation. Care must be exercised to coordinate the location of the concrete pit with the general contractor, thereby avoiding interference with the underground footings for the building.

The elevation in Fig. 2 illustrates how the component parts of the drainage piping are arranged, including the hookup to the fresh air inlet. Many types of fresh air inlet plates are available for this purpose and each is connected to the piping at the face of the wall in the manner intended by the manufacturer.

Construction of Concrete Pits

As the concrete pits are generally furnished by the general contractor, he must be supplied with data on the required size of the pit. Figure 3 suggests several factors that should be included in the construction of the concrete pit to offer sufficient access space for a man to work.

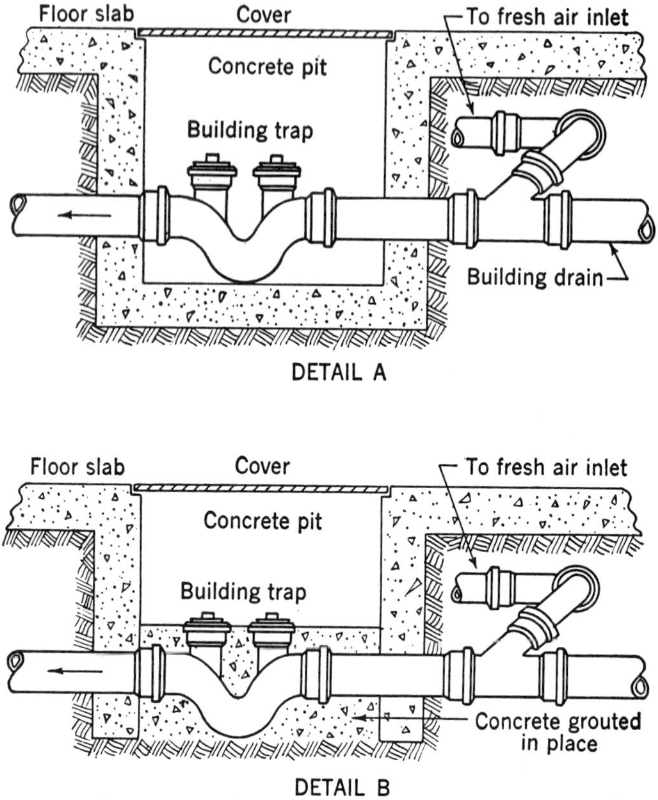

Fig. 3. Installation of building traps in concrete pits.

Detail *A* of Fig. 3 shows the type of pit normally provided in this instance while Detail *B* shows the type of pit provided when the bottom of the pit is left open for later grouting around the trap to a height that is level with the cleanout hub as shown.

Of first importance in the construction of a concrete pit is the physical size of the pit itself. The length of the pit should be sufficient to allow enough room around the hubs of the fittings to permit ease in pouring lead and the caulking of joints. In addition, the pit should be wide enough to offer a man complete freedom in performing the task of clearing stoppages. This provision is especially important when the pit is a fairly deep one because it means that a man will have to lower himself into the pit to accomplish the task.

A checkered steel cover is usually placed over the pit. The cover should be of a thickness that will best sustain the loads imposed on it. The cover will rest on a suitable frame set integral with the concrete itself. If the pit opening is rather large, then structural members may be required for the cover for additional strength. A recessed handle in the cover will allow for easy lifting of the cover.

This is the general type of pit that is provided for the building trap. Since the fresh air connection piping is not included in the pit, space will not have to be allotted for it. However, if it is desired to include this piping connection in the pit, then it is necessary only to lengthen the pit.

Backwater Valves

Where building drains are subjected to backwater conditions that originate in the street sewer facilities, suitable protection must be considered. Backwater valves, generally employed for a safeguard, prevent the sewer water, caused by storm, stoppages or other reasons from entering the interior piping system where it can cause flooding of the lower floor levels. When a backwater valve is installed, it should be included in the pit with the building trap. Figure 4 shows such an installation.

As detailed in Fig. 4, the backwater valve is included in the same pit with the building trap. Here again, the pit must be of sufficient length to permit the proper pouring and caulking of the cast iron joints. It is also vital in this instance that the pit be wide enough to allow ease in maintaining the backwater valve to keep the flapper and valve seat clean.

Watertight Installations

Another problem that frequently must be considered by the engineer is the installation of the building trap in a concrete pit that is to be waterproofed. When ground water conditions exist, special care must be taken to ensure a watertight slab. Likewise, the same consideration must be applied to the pit. The watertight sleeves are required for the passage of the piping through the walls of the pit. Figure 5 demonstrates how this provision can be accomplished.

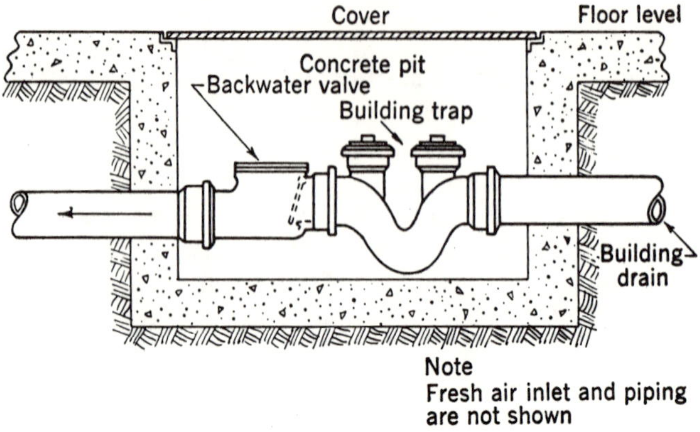

Fig. 4. Installation of building trap and backwater valve.

There are many types of manufactured sleeves available for this purpose. The sleeve shown in the illustration has been used for example only and does not indicate a preference. The drainage line passes through this sleeve. Oakum and lead or mastic compound is used to make watertight the void between the sleeve and the piping. When carefully installed, this method provides a dependable watertight connection.

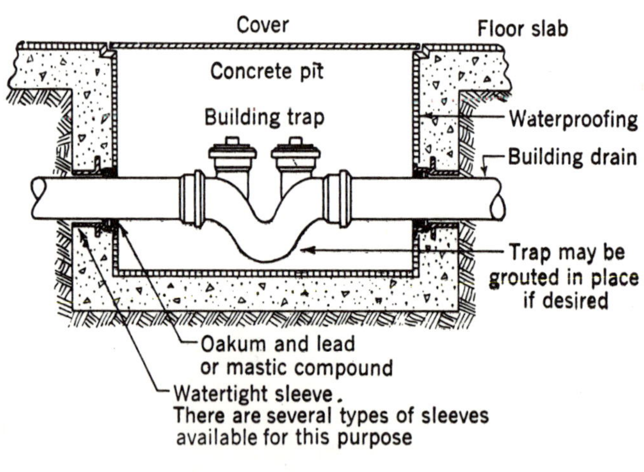

Fig. 5. Suggested construction of concrete pit in areas where ground water conditions exist.

Deep Concrete Pits

Very often, the drainage line because of its long run will be located at a considerable depth in the ground. When this occurs it is necessary to provide a deep pit, as shown in Fig. 6, constructed in the same manner as the others shown with the exception that the walls of the pit are of a heavier masonry construction. In addition, a maintenance work ladder consisting of iron rungs is embedded in a wall of the pit. It is of utmost importance, in this case, that the pit be of sufficient size to provide the needed space in which to work.

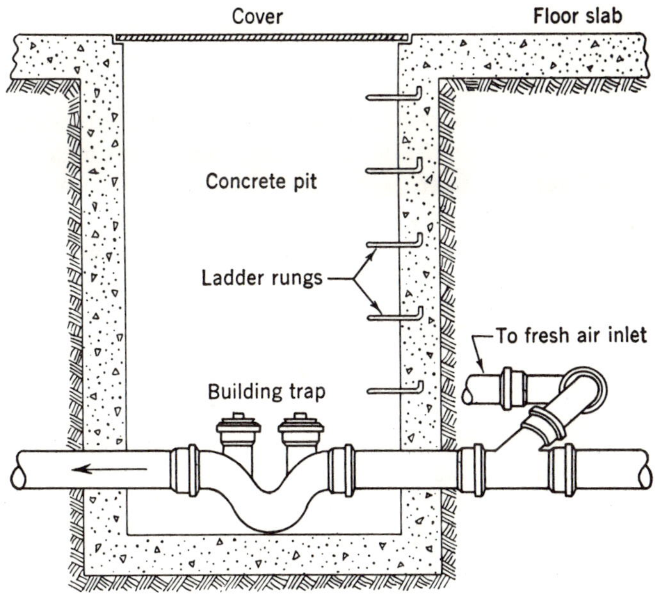

Fig. 6. Construction of concrete pit when underground drainage line is buried at a deep elevation.

Regardless of the conditions that may prevail at the site which have a direct effect on the building drain, proper access to the building trap can be provided by one of the methods shown. It is always good practice for the engineer to detail on his drawings the specific type of access he wishes to provide.

CHAPTER 20

Interception of Volatile Liquids

Multi-story garages serve a very definite need since they are the practical solution to the critical parking problems experienced in many of our congested cities.

These unique buildings are used for automobile parking, storage and repair. When erected above ground, such buildings range from one story to ten or more stories in height. When erected underground, the buildings may have from one to six or more levels. In either case, whenever automobiles are stored or parked in a building, they present a potential fire hazard. While fire protection systems will help to extinguish a fire once it has started, it is equally important that means be employed to prevent the outbreak of fire.

Floor Drainage Systems

A well designed floor drainage system is a necessary component of garage buildings. This type system is needed for many reasons, a few of which follow.

1. On wet or snowy days, automobiles, after they are parked in the building, will drip profusely. Floor drains are, therefore, required to receive the drippings and to keep the floor free from water accumulation.
2. The floor of the parking areas is regularly washed down as a cleaning routine. Floor drains are needed for this function.
3. Car washing facilities are often included in the building and also require drainage facilities.
4. In general, a prescribed number of floor drains must be included at each story level for miscellaneous needs as well as to receive the water discharged by the sprinkler system in the event of a fire.

The drainage system to which these floor drains are connected is always an independent system and does not connect to the sanitary system within the building. It does sometimes connect, when required, with the building drain at the building side of the building trap and fresh air inlet. Often, the system will discharge into the outside sanitary or storm sewers or other point of discharge as called for by the local code. This system will become a hazard when volatile liquids are introduced.

A few actual causes are listed to show how volatile liquids enter the drainage systems of garages.

1. Gasoline, kerosene or other volatile solvents normally associated with the cleaning of automobile parts, may carelessly be poured into a floor drain.
2. Discarded oil removed from a car is stored in drums, due to spillage or leakage, this oil will find its way to the floor drain.
3. Gasoline spillage will end up in a drain.
4. Since automobiles drip oil and grease when parked, the drippings will remain on the floor until the floor is cleaned. The oil slick and grease are washed into the drainage system to form a potentional and important fire hazard.

These are situations that permit volatile liquids to enter the drainage system. Such liquids in any form or concentration allow explosive conditions to develop both in the interior and exterior piping systems. Sufficient attention must, therefore, be given to the possible accumulation of the liquids and vapors and methods that can be employed to safely discharge the waste to a point where the combustible liquids can be separated from the waste water.

How Fires Start

Vapors liberated from volatile liquids are explosive and can be ignited by a spark provided by static electricity ignition or fire may start from burning cigarettes or matches. Discharge of combustible liquids into a street sewer presents such a potential hazard that a large number of larger cities require the installation of approved equipment to separate volatile liquids from the waste water.

Individual Drain Installations

There are several methods to separate combustible liquids from waste water. Although this subject is not a new one, greater attention is being focused on it to make this discussion a timely one.

Suitable drains are available for an individual installation, when one or just several drains are required in a specific area. Such drains will intercept and retain the volatile liquids and yet allow the waste water to pass through. A typical drain is illustrated in Fig. 1. Drains illustrated in Fig. 1 serve several purposes. When the waste water is discharged into the drain, the volatile liquids separate and float on the surface of the water. The gas cone prevents the escape of highly explosive vapors while the sediment bucket traps the sand and grit washed into the drain. The waste water will flow unimpeded through the drain.

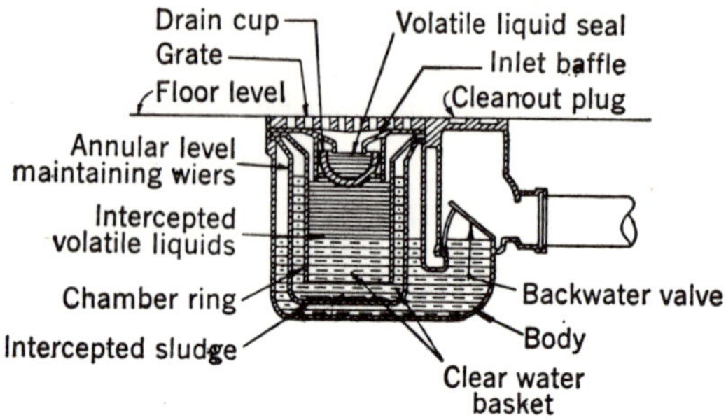

Detail A

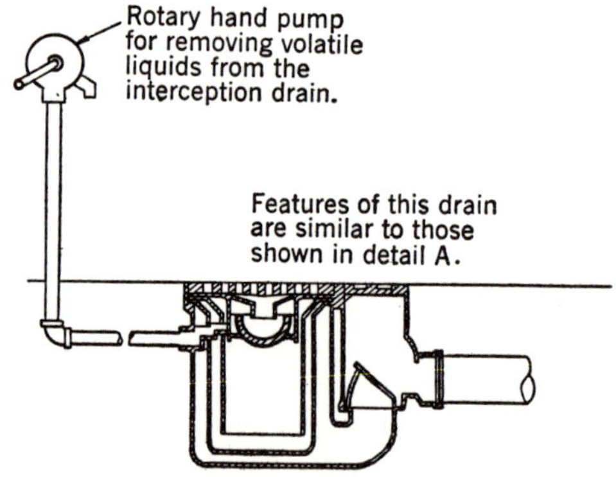

Detail B

Fig. 1. Recommended individual drain installations to intercept and retain volatile liquids. Detail B has a rotary hand pump for removing volatile liquids.

When required, the drain can be equipped with a rotary hand pump for ease in removing the volatile liquids. Such an arrangement is pictured in Detail *B* of Fig. 1.

Single Story Installation

When a waste system is required for a single story of a building, a method similar to that shown in Fig. 2 may be employed.

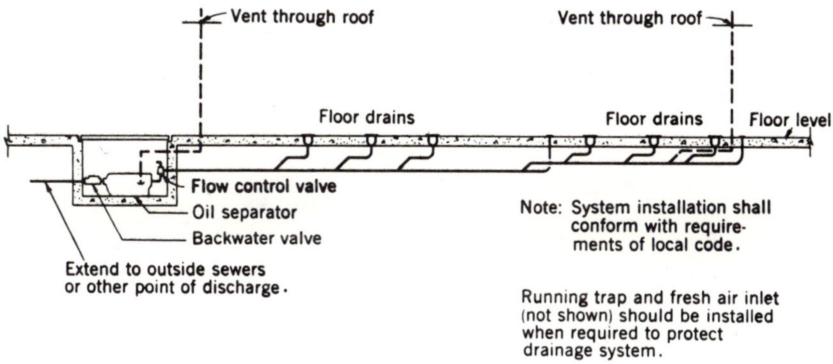

Fig. 2. Typical arrangement of equipment and drains for a one-story installation.

In the installation of Fig. 2, an oil separator is used. Waste water, along with any combustible liquids, is conducted by the drains and piping to the oil separator. Here, the volatile liquids are separated while the waste water passes through. The flow control valve regulates the velocity of the waste water through the unit. Periodic cleaning of the separator is required to remove the accumulation of combustible material from the unit. This can be done by lifting the cover and scooping out such material which floats on the surface of the water in the unit by the hook-up of a rotary hand pump.

This type system represents a safe means of safeguarding the building against the dangers of a fire or explosion.

Multi-Story Installations

For buildings that include several floors for automobile parking or storage, a system similar to that shown in Fig. 3 is suggested.

Figure 3 indicates that the drains on the several floor levels are

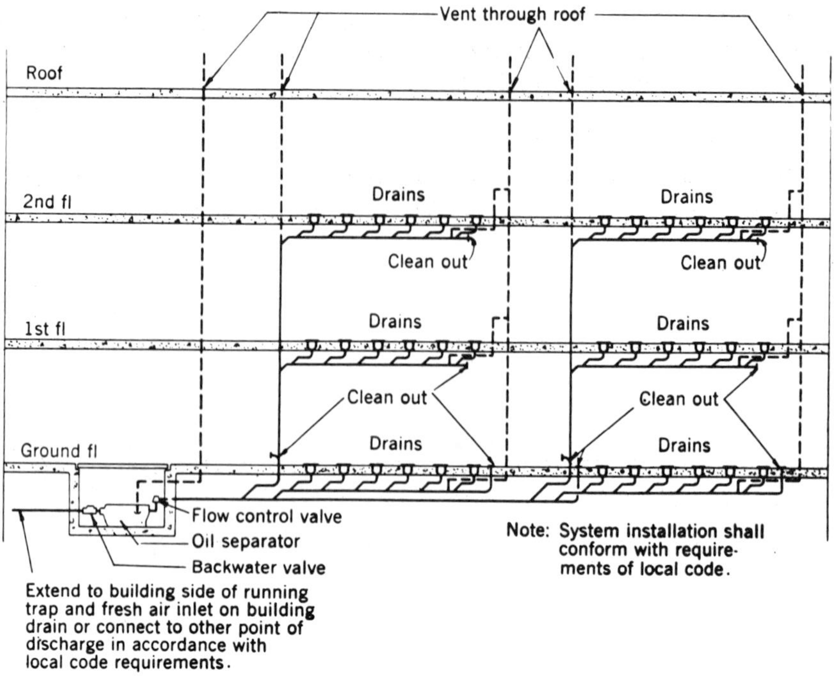

Fig. 3. Oil separator system for a multi-story garage building.

connected to a common waste system. This system of piping terminates at the oil separator system. Here again, the unit separates and retains the volatile liquids while the waste water continues in its passage to the street sewer facilities. The flow control valve regulates the velocity of the flow passing through the separator. Proper venting assures the displacement of accumulated explosive vapors. The separator, in this instance, also requires periodic cleaning for the removal of volatile liquids.

Multi-story Installations with Roof Parking

For buildings that will include several floors for automobile parking or storage as well as a roof area on which cars are parked or stored, the system becomes a bit more complex. Figure 4 illustrates such an automobile parking arrangement.

The system shown in Fig. 4 is similar to that illustrated in Fig. 3 with the exception that drains have been included on the roof area. These drains serve as parking area drains as well as roof drains.

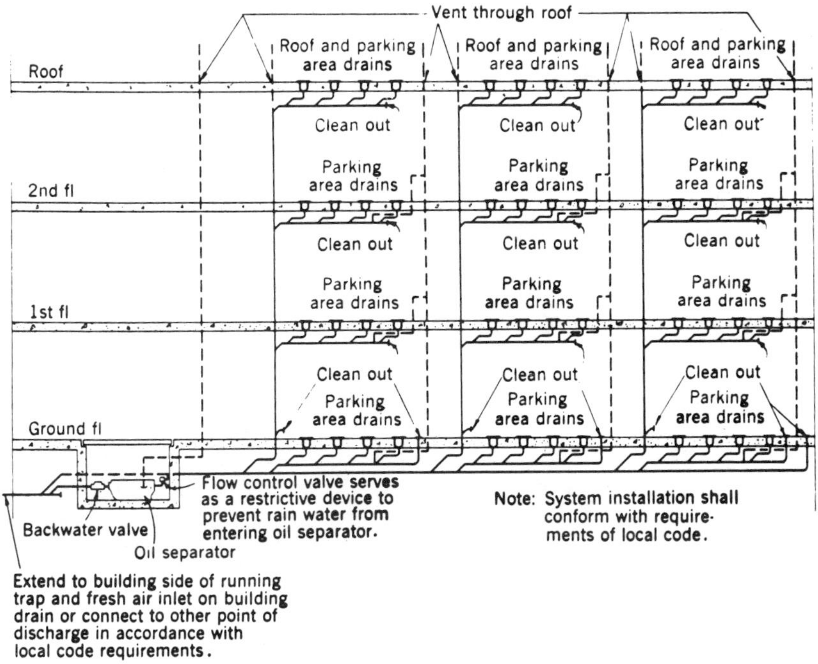

Fig. 4. Arrangement of an oil separator system for a multi-story garage building including roof and parking area drains.

There are three methods that can be employed to handle the storm and waste water the drains will receive. One method is to install the roof drainage as an independent system and allow the water to run directly to the outside storm sewers.

Another method is to provide the type of a roof drain that actually serves a duel purpose. This drain is equipped with an orifice at the bottom of the drain that permits the normal waste water discharged on the roof to drain into the oil separator system of piping. In addition, the drain is also equipped with an outlet at the side of the drain that will receive the rain water and conduct it to the storm system. When it rains, the restrictive orifice prevents all but a small quantity of water from entering the piping connected to the oil separator system. Thus the rainwater is conducted by the storm water piping to the outside storm water system or other point of discharge.

The final method is a system that utilizes the roof drain for both the waste water and the storm water encountered on the roof. This system is illustrated in Fig. 4. Such a system functions as follows:

Waste water, a result of washing the roof will flow into the drain to be conducted by the piping system to the point of the oil separator. Here because of the small quantity of water involved and the low velocity of its flow rate, the water will drain into the vertical pipe connection to the oil separator. It will pass through the flow control valve into the separator where the volatile liquids are separated and retained. The water itself will pass through and out.

However, when it rains, the water will flow into the same drain and system of piping to the point of the oil separator. The restrictive action of the flow control valve will permit only a very small amount of water to pass into the oil separator. The rain water by virtue of its quantity and rate of flow will bypass the oil separator and will discharge directly into the street sewer. Since water will follow the path of least resistance, the bypass action is possible.

The backwater valve located at the discharge side of the oil separator prevents sewage backflow from entering the oil separator.

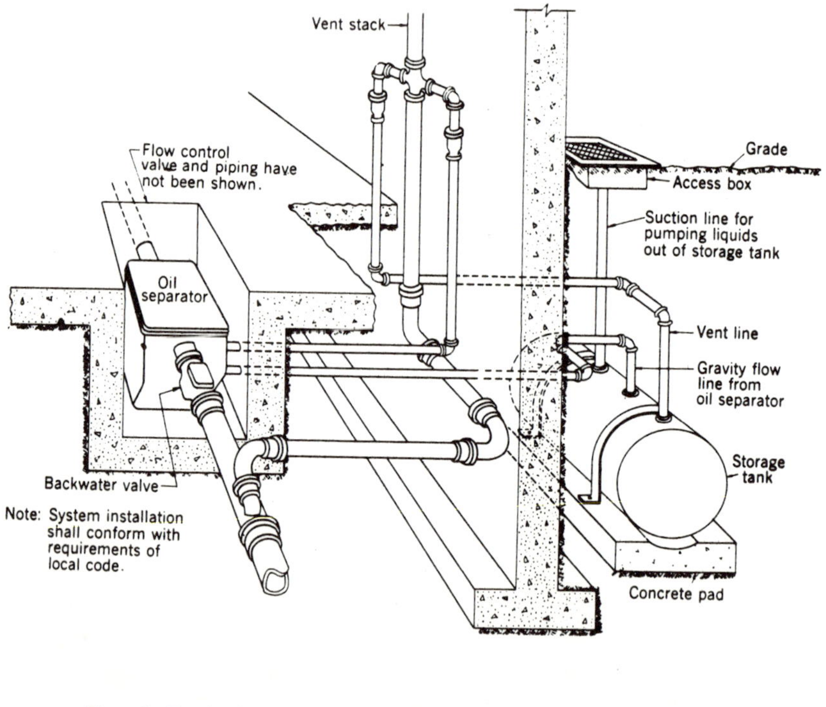

Fig. 5. Typical arrangement of oil storage tank, oil separator and component piping. When required, the vent lines should be extended up through roof and terminate as separate vent lines.

Storing Volatile Liquids

For systems that will experience large concentrations of volatile liquids, sufficient thought must be given to a proper means for receiving and storing the liquids. For this condition, a storage tank is usually employed and Fig. 5 shows a typical installation.

In Fig. 5, the storage tank is located outside the building. Its size and capacity are determined by the engineer for the particular requirements of the system. The tank must meet the requirements of the National Board of Fire Underwriters as well as the Code for Unfired Pressure Vessels and it is placed on a concrete slab or mat and securely anchored in place. A suction line, that is also used for testing the capacity of the storage tank, is extended to grade with proper provisions as shown.

The draw-off line from the oil separator is connected to the tank. An adjustable line in the separator is set at an elevation just above the water level and permits the volatile liquids to run into the storage tank by gravity. Adequate venting must be provided to remove any accumulation of explosive vapors. When desired, two vents may be connected to the separator to ensure a more positive circulation of air. If two vents are installed, one vent should be connected to the vent stack at a higher elevation than the other to induce circulation by an unbalanced air pressure.

Systems illustrated in this chapter represent a concept in the handling of the combustible liquids which is a definite phase of plumbing design. This subject is a broad one and extends to many other types of buildings including aircraft hangers where volatile liquids are stored, handled or encountered.

CHAPTER 21

Storm Drainage System for Flat Roofs

An important design consideration for every new project is a storm water system that will receive the rain water at roof level in an adequate manner and conduct it to an outside, independent storm sewer, combined sanitary and storm sewer, dry wells, a stream, the curb or other point of discharge. Such systems might be regarded as a simple phase of design, requiring but a minimum of thought and effort. Actually, storm drainage systems require considerable thought and design effort to produce not only the best system for the purpose intended but also one that is economical. Unfortunately, there exists a general tendency among engineers to treat this subject rather lightly.

Attention is directed in this chapter to some of the factors that should be considered when the design of a storm drainage system is contemplated. Only flat roofs will be covered at this time.

Rainfall Intensity

The very first factor to be studied in connection with storm drainage systems is the amount and rate of rainfall expected in the vicinity of the proposed project. This information can usually be obtained from the local plumbing authorities or is available in tables on Rainfall Intensity-Frequency Data, as issued by the United States Department of Agriculture. These tables supply rainfall data for areas in all parts of the country.

Once the frequency and rate of rainfall has been determined, it is easily applied to the square feet of roof surface to enable the engineer to calculate the diameter of the storm water piping needed. Various tables have been worked out to assist the engineer in his design work. An example of suggested tables than can be used for this purpose are contained in the National Plumbing Code. Local plumbing codes also generally contain helpful tables.

Spacing of Roof Drains

Where should the roof drains be placed on a flat roof? The rainfall tables indicate the diameter of vertical leaders required for a prescribed square foot area of roof surface, based on the prevalent rate of rainfall in inches. A number of drains are needed when small diameter roof drains and leader piping are used. Fewer roof drains are needed as the diameter of the drains and leader piping are increased. Figure 1 represents a typical roof surface and suggests a method of arrangement in roof drains.

Referring to Fig. 1, for which 4 in. diameter roof drains have been used, it will be noted the drains are spaced evenly apart. This particular method prevents the accumulation of puddles on the roof after a storm. Even though the roof is flat, there will always be uneven areas on the surface where the water will remain for a number of days until it is dried up by the process of evaporation. With the greater number of small diameter drains the occurrence of puddles is greatly reduced, and the entire surface is drained rather easily. A lesser number of larger diameter drains will not drain the roof area as thoroughly. The decision on the number and diameter of roof drains to be used rests with the engineer and is based on the specific needs of the project.

Type of Roof Drains

After the number and spacing of roof drains has been accomplished, the next step is to select the appropriate type of roof drain. For flat roofs there are numerous types of drains from which to make a selection. Figure 2 illustrates several of the types commonly used.

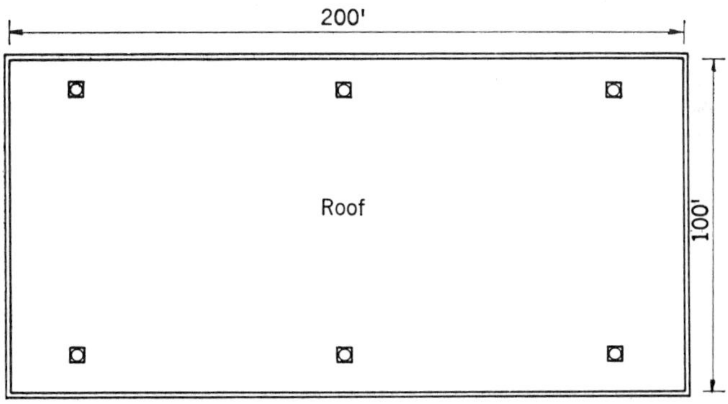

Fig. 1. Typical arrangement of roof drains to afford complete drainage of roof.

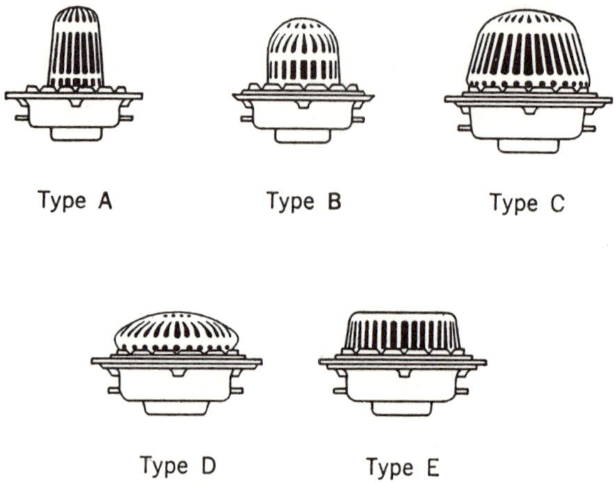

Fig. 2. Several of the many types of roof drains available for flat roofs.

Although each of the roof drains shown in Fig. 2 are similar in body construction, the shape of the dome strainer is different. Each of these dome strainers was designed to fit a certain particular need in storm drainage control. We will not attempt to outline the specific advantage to be gained with each type of strainer but will only advise that all of them are suitable for flat roof surfaces.

Regardless of what the special requirements may be, roof drain manufacturers have drains that will satisfy even the most unusual needs. In addition, roof drains can be furnished in one of several materials or material compositions to comply with the wishes of the owner or architect. When necessary, a material can be used to withstand the corrosive action of the impurities found in the air at certain locations.

Retarded Flow Drains

For buildings like industrial plants, warehouses, automobile plants and similar buildings which are rather spread out and occupy acres of ground, the installation of the storm water system can become an expensive construction item. A great deal of thought has been expended on the subject of cost and as a result, great strides have been made in lowering the cost.

One method that has been used with a good deal of success consists of installing roof drains that are undersized for the rain load imposed. The storm water piping is then also undersized in the same proportionate manner. This causes the roof to act as a reservoir for the water during a severe storm. The result is that flow of rain water is retarded and although it will take longer for the roof to drain, a great savings is realized by furnishing small diameter pipe rather than the larger diameter pipe otherwise required. Tremendous savings are accomplished in this manner. Moreover, by prolonging the time needed for the roof to drain, the outside storm system in the street will also benefit from the reduced concentration of rain water flow. This is an important consideration since many of the present outside storm systems are undersized and are not adequate for the many new buildings being erected.

Often, it is not desired to limit the capacity of the building storm water system but instead provide a system that is of sufficient size for the prevalent rainfall. If a retarded fall is still required for any reason, it is something relatively easy to furnish. In this method, roof drains of the correct size are used. However, the roof drain will be equipped as shown in Fig. 3.

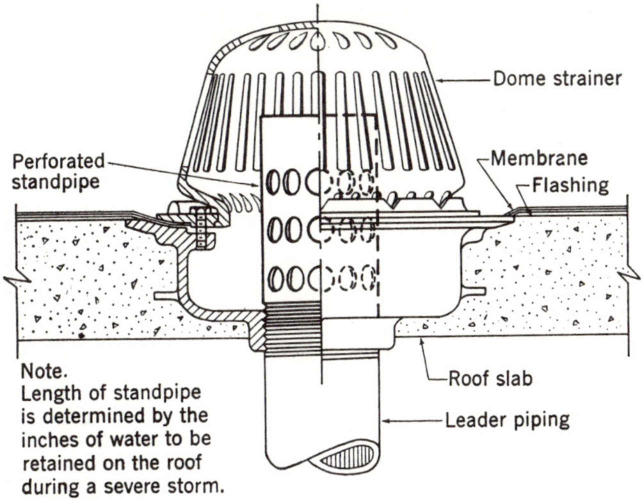

Fig. 3. Modification to a roof drain to provide a retarded flow.

As indicated in Fig. 3, the roof drain is of the regular type generally used for flat roofs, with the exception that a perforated standpipe is inserted into the drain. The purpose of the standpipe is to retard the flow of rainwater discharged into the piping system. If a 6-inch drain is required for the particular area of roof to be drained, then perforations equal to the effective open area of a 4- or

5-inch pipe can be furnished in the standpipe. Depending on the number and size of the perforations, any desired rate of flow can be accomplished.

The roof in this instance also acts as a reservoir for the rain water during a heavy storm. An advantage of the standpipe is that it governs the amount of water buildup on the roof. When the water exceeds the height of the standpipe, it will overflow into the standpipe and is then discharged at a rate for which the size of drain was intended.

If for any reason at a later date, the owner wants the roof drains to function at the rate for which they were sized, it is necessary only to remove the standpipe. This drain also has application on those roofs where it is desired to maintain several inches of water on the roof for cooling purposes.

Roof Construction Materials

Tremendous strides have been made in lightweight roof construction. Concrete slab, concrete planks, and preformed steel decks are some of the many materials available. The preformed steel decks offer the biggest problem as far as the installation of the roof drain is concerned. Figure 4 shows a typical roof drain installation in such construction.

The type of steel roof decking illustrated in Fig. 4, is known as

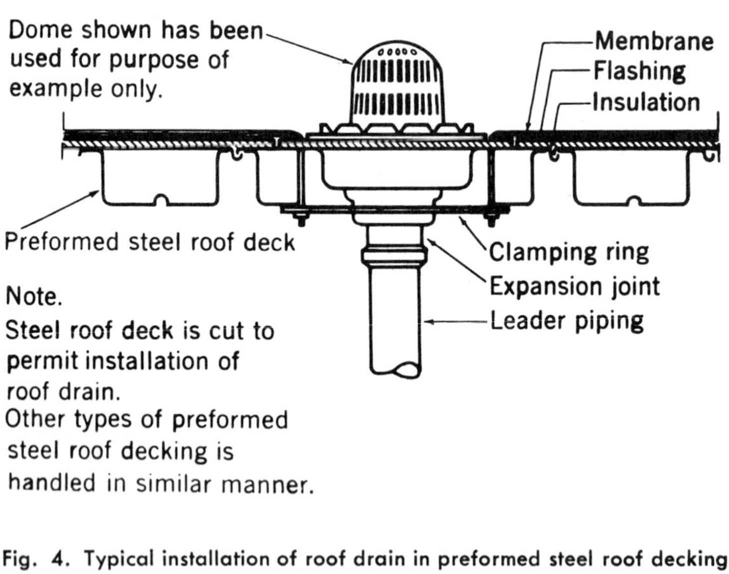

Fig. 4. Typical installation of roof drain in preformed steel roof decking.

Fenestra. It is used here only for the purpose of example. When preformed steel decking is used, a hole must be cut in it to allow for the installation of the drain. It is recommended that the hole be cut in a strong point of the decking. When in doubt as to where to provide the hole in any type of preformed steel roofing, it is suggested that you consult either the structural engineer designing the roofing or to check with the manufacturer of the particular material.

For the installation shown, you will note that a clamping ring has been attached to the drain. This ring is tightened against the underside of the formed steel sections to anchor the roof drain securely in place. This modification to the roof drain represents one of the many little refinements that can be added to the drain in order to satisfy the various types of roof construction. Roof drain manufacturers produce many such modifications to render the drain suitable for any condition of installation. Regardless of the type of roofing experienced, there is a roof drain available of the type that can be used for a particular design.

Connections to Roof Drains

Roof drains are equipped with either a caulked or threaded connection depending upon the material used for the piping system. The materials most generally employed for the piping system are cast iron, galvanized wrought iron, galvanized steel or copper tubing as determined by the engineer.

Expansion and Contraction

Expansion and contraction are factors that must be given some attention with storm drainage systems. Most expansion is generally caused by a roof that heaves when it is heated, and contracts when it cools off. When this condition becomes a problem on your project, one of several methods can be utilized to prevent rupture of the roof drain connection. Figure 5, shows two suggested methods.

In most cases, the movement can be taken care of by offsetting the piping from the vertical leader to the roof drain as shown in detail *B* of Fig. 5. This piece of horizontal piping is generally sufficient to compensate for any movement.

Another method is by means of an expansion joint that is an integral part of the roof drain as shown in detail *A* of Fig. 5. This method is used when it is not possible to offset the piping. The sleeve of the joint slides in the direction required to take up the motion. Thus, the danger of rupture to the connection is removed.

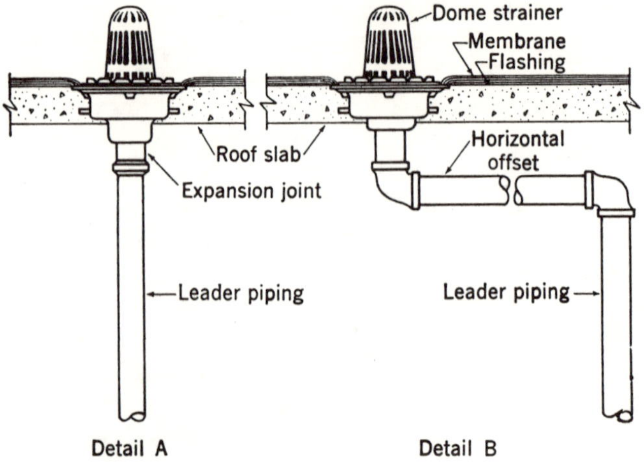

Fig. 5. Suggested methods of compensating for expansion and contraction.

Arrangement of Storm Water Piping

A storm water system is installed as an independent system or is combined with the sanitary system within the building. An independent system is a preferred one, because when the systems are combined and a stoppage occurs in the sanitary system there is always the danger of the rain water backing up and flooding the lower levels of the building.

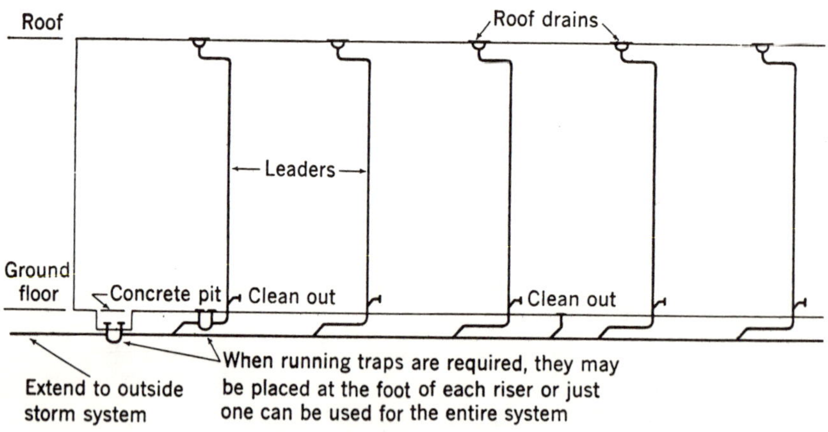

Fig. 6. Arrangement of storm water system for a single story building.

The system portrayed in Fig. 6, is one for a single story structure used as an industrial plant or for similar buildings. Cleanouts are placed at the base of the vertical risers to clear any obstructions that could cause stoppages at this point. The underground piping is carefully installed and connects with the outside storm system in the street or other point of discharge. Cleanouts are also installed every 40 to 50 ft. along the underground line to clear the debris or deposits of sand or grit that often accumulates.

If running traps are required, they are placed either at the foot of each riser or just one trap is installed in the line near the building wall as shown in Fig. 6 to serve the complete system.

CHAPTER 22

Draining Watertight Floor Slabs

Every now and then it is necessary to locate plumbing fixtures or equipment where water seepage through the floor slab on which the fixtures or equipment are located cannot be tolerated. The space below this slab may be an area where important electrical apparatus is installed, a finished lobby, recreation room, display area, or storage room. Normally, the architect will not locate rooms or areas which include water and waste facilities directly over the type of spaces previously mentioned, but sometimes it is unavoidable.

Where, for example, a rest room, or kitchen, or utility room, or heating or air conditioning equipment space is directly over critical areas, it is imperative that the floor slab be waterproofed. Many methods are available. The decision as to which method should be used rests with the architect and the structural engineer. Regardless of the specific method used, lead or copper flashing is used in most cases to assure a watertight condition. Sheet lead or copper is formed to the shape of the room or area and extends up at least 6 to 8 inches at the sides to prevent seepage (Fig. 1). When copper is used, the copper sheets are soldered

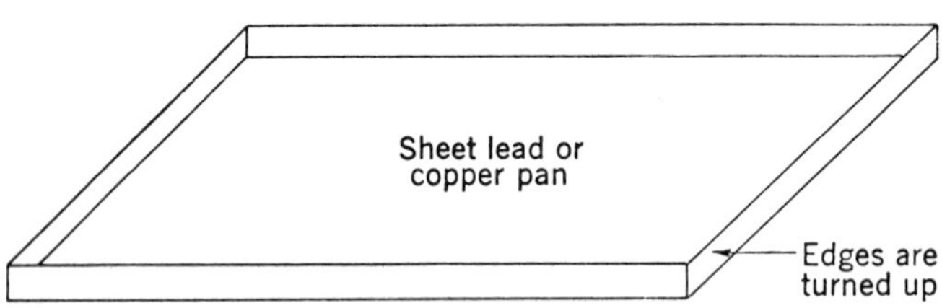

Fig. 1. Sheet lead or copper watertight pan fabricated at the job site.

together to form the required pan. When lead is used, strips of sheet lead are joined together to form the desired shape of pan. The weight of material is specified in ounces for copper and in pounds for sheet lead.

Typical Application

While the watertight pan will prevent seepage through the slab, provision must be made to intercept water that may be discharged on the floor. The rest room, Fig. 2, presents the typical case. Note that a floor drain has been included to intercept any water from an overflowing fixture. The floor drain can be placed in any spot that seems practical. It has been placed in this particular location so (1) it will not appear too obvious in the finished room and (2) it will be in line with, and close to, the vertical soil stack to which it is connected.

Typical Elevation of Rest Room

A typical elevation of the rest room is shown in Fig. 3. The watertight pan is shown in place. The watertight pan is fabricated on top of the rough floor slab. Above it is a layer of lightweight

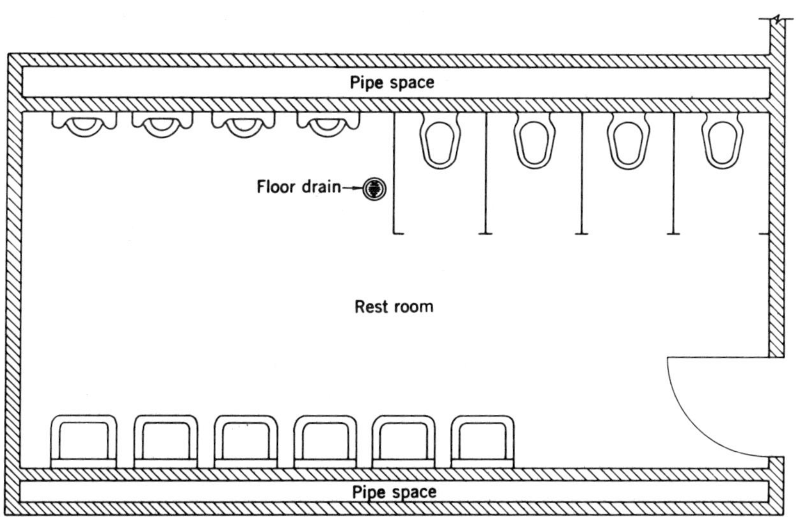

Fig. 2. Location of floor drain in a typical rest room.

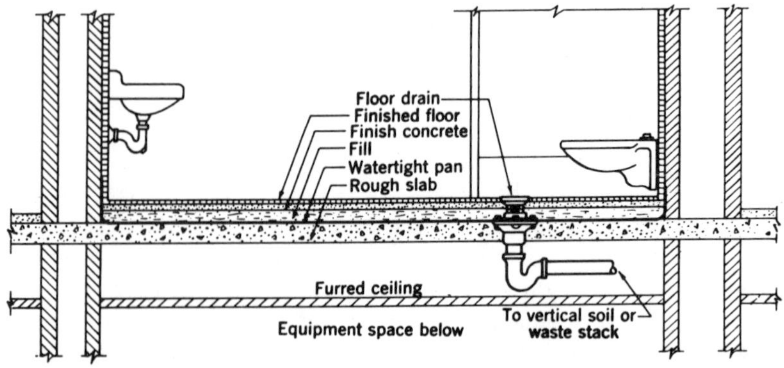

Fig. 3. Elevation of a rest room with floor drain.

concrete, cinders or other type of fill material. On top of the fill, a layer of concrete is placed for the setting of the tiled floor.

The plumbing fixtures, including water closets, urinals and lavatories, are the off-the-floor type to avoid any floor penetration by the roughing of the fixtures. There are no special provisions for the pipe spaces, since these generally extend down to the basement and therefore present no hazard. If the lavatories cannot be located as shown, then they should be located on the same wall with the water closets and urinals.

The floor drain serves a dual purpose, intercepting spillage at the floor surface and also seepage that may accumulate at the level of the watertight pan. The drain clamps to the watertight pan to offer a leakproof connection. Weep holes in the drain take care of the seepage at the level of the watertight pan.

The drain trap has been installed in the furred ceiling of the floor below. If entry to the trap is required for cleanout purposes, an access door must be provided in the ceiling of the floor below. If venting is required, it must be accomplished in the furred ceiling, in a manner prescribed by the local plumbing code.

Typical Floor Drain

Figure 4 is an enlarged detail of the floor drain shown in Fig. 3. The drain body is contained in the rough floor slab. A clamping arrangement secures the lead or copper pan to the drain. Weep holes in the clamping collar drain any seepage that may accumu-

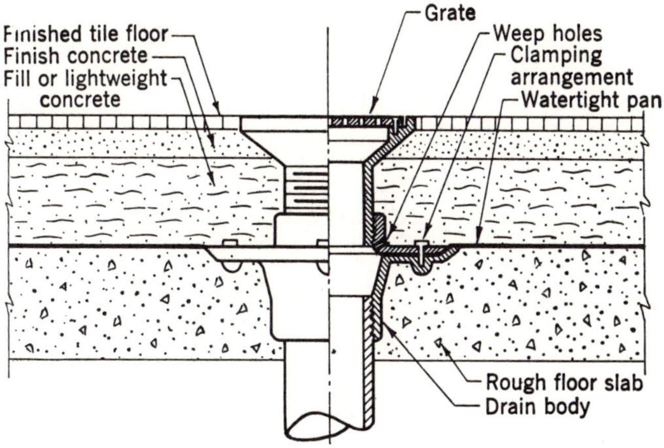

Fig. 4. Floor drain with provisions for intercepting seepage at the watertight pan.

late on the watertight pan. It is rather difficult for water to seep through the tiled floor. Nevertheless, drainage must be provided and the weep holes in the clamping collar serve this function in the event that water does accumulate in the pan. The strainer shank is adjusted so that the top of the strainer will be level with the surface of the floor and thus intercept any water that overflows from the fixtures. It is important to mention here that the floor surface should be sloped toward the floor drain for positive drainage.

Kitchens and Equipment Rooms

Figure 5 shows a type of drain that is recommended for kitchens or equipment rooms above critical areas where leaks can be destructive. It is the same type shown in Fig. 4, with the exception that a funnel has been incorporated in its grate. This funnel has a definite advantage in kitchens where much of the waste facilities is of the indirect type from some of the kitchen equipment. This type drain serves a triple purpose: as a funnel drain for the indirect waste lines, as a floor drain, since the floor surface water will be intercepted by the grate; and, since the clamping arrangement is equipped with weep holes, as an interceptor of water that has accumulated at the watertight pan.

This particular drain is also ideal for spaces that house heating

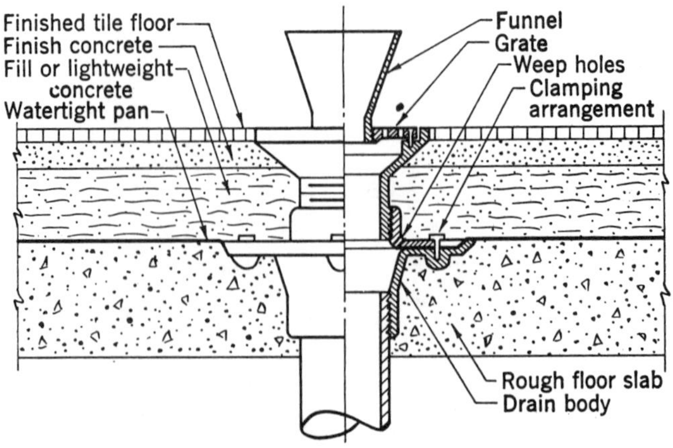

Fig. 5. Arrangement of a floor drain that also serves as a funnel drain.

or air conditioning equipment because the funnel serves as an indirect waste to handle the drip piping from the various items of equipment.

Above-Slab Installation

All the drain installations described have had the waste piping from the floor drain included in the furred ceiling space. Sometimes it becomes necessary to install the piping above the rough floor slab, as in Fig 6. Here, a drain with shallow physical dimensions is used to enable it to fit tight space requirements. The drain is placed on top of the watertight pan and the waste piping is run in the fill to the point where it connects with the vertical soil or waste stack.

The shallow body floor drain does not offer any provision for intercepting water that may accumulate on the watertight pan. Therefore, a separate telltale drain must be installed, as indicated. The small drain consists of a drain body, a clamping collar, and a small dome covered with copper or plastic mesh. This arrangement is used where a number of drains are required for equipment or apparatus in an area, and where the drain piping, by virtue of the nature of the construction, is to be accommodated above the floor slab.

Shallow body floor drains are equipped with a grate and a cleanout that is level with the floor area. The cleanout provides access to the waste piping in the event of a stoppage. The telltale

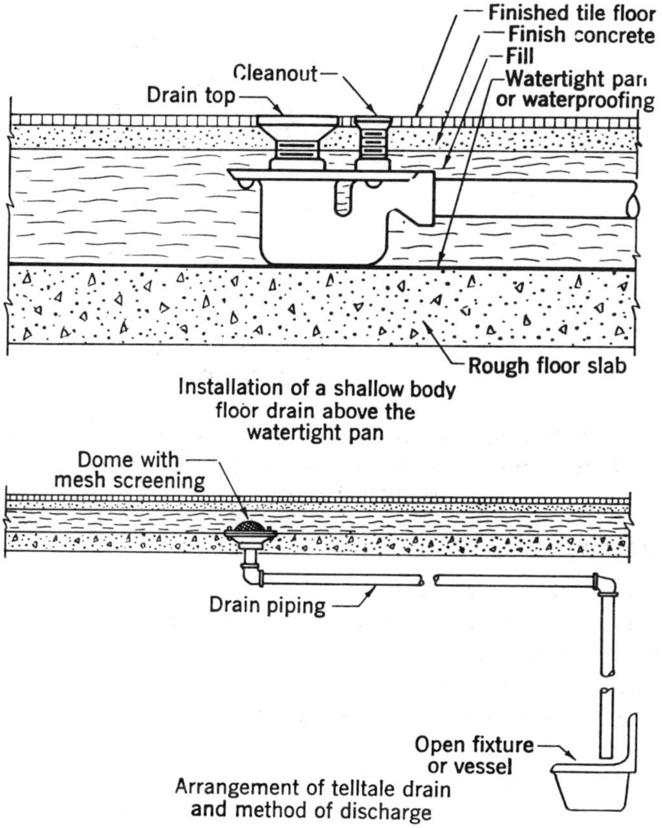

Fig. 6. Manner in which a shallow body floor drain can be installed.

drain is placed in an out-of-the-way location, at the ceiling below, so it will not be too obvious. Waste piping usually discharges into the nearest fixture, most generally a service sink on the floor below. The mesh screening around the dome of the drain prevents fill from entering.

In connecting with the vertical soil or waste stack, the waste piping from the floor drain must pass through the edge of the watertight pan. This penetration must be done carefully to avoid any possibility of leakage. To make a proper seal, bitumastic or other compound or cement is applied to the point of penetration around the piping.

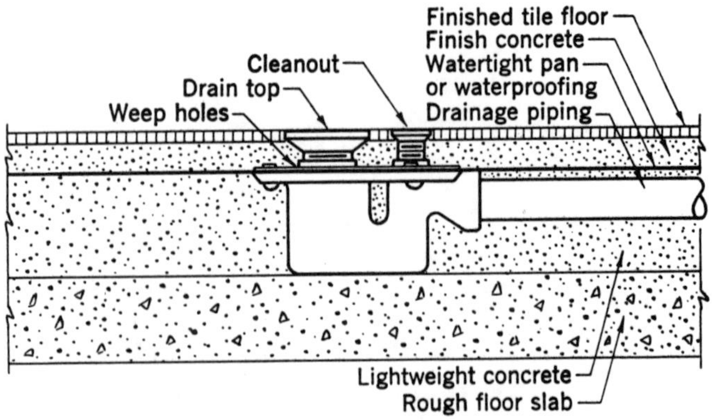

Fig. 7. How a shallow body floor drain can be installed with its clamping arrangement engaging a watertight pan or waterproofing material.

Alternate Method.

Another method for installing the shallow body floor drain is shown in Fig. 7. In this instance, the watertight pan or waterproofing is placed *above* the fill, as shown. The clamping collar arrangement engages the pan of waterproofing to provide a tight connection. Weep holes in the clamping collar intercept any seepage that may accumulate at the level of the pan or waterproofing. When required, a funnel is placed in the grate to serve indirect waste or drip lines from fixtures and equipment. When venting is required it must be installed in the fill in a manner that conforms to the applicable local code regulations.

CHAPTER 23

Cleanouts for Underground Drainage Lines

Whenever drainage lines are installed underground in a building, one must always anticipate that stoppages will occur in these lines. Stoppages can be caused by foreign objects such as bits of cloth, pieces of newspaper or accumulations of grit and grease. Therefore, some means of access must be provided in the piping system to permit the clearing of blocked lines.

Since concrete floor slabs, floor finishing and floor construction,

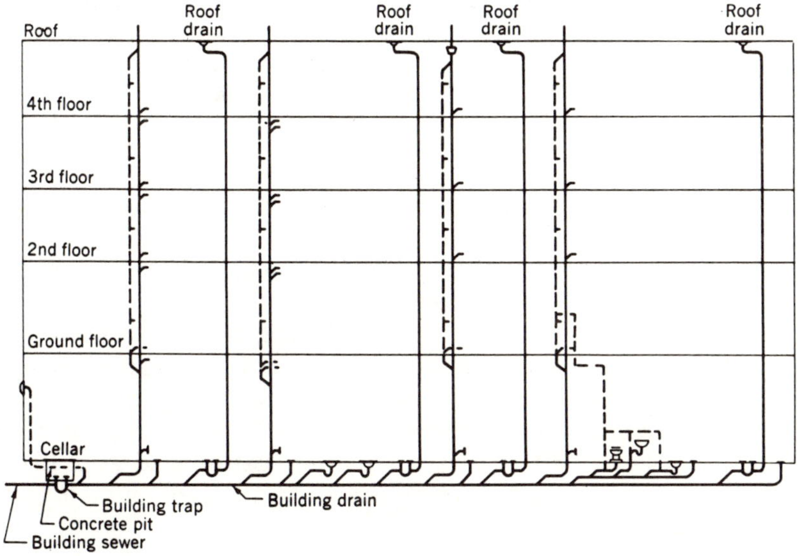

Fig. 1. Building riser diagram showing location of cleanouts for the lower floor level.

in general, vary considerably in each building, the type of device selected will be different for each condition.

Figure 1, a typical riser diagram, suggests locations where cleanouts may be placed at the lower floor level in the building. Soil, waste and storm water lines are included in the diagram. The cleanouts consist of the building trap and devices placed along the underground run and at the base of the soil stacks. Also included are the running traps placed at the base of the leader stacks.

The underground drainage line shown is a combination sanitary and storm drain line. When this combination system is used, it is vital that traps be placed at the base of the leader lines to intercept objects that find their way into the roof drains through children's pranks or other means.

It is recommended practice to have separate sanitary and storm drainage lines within a building to eliminate any danger of flooding the lower floor level as could occur when a stoppage develops in a combination line during a rain storm.

Cleanouts should be placed at least every 40 to 50 ft on a run of underground piping and should be placed at every change in direction of the underground piping. These are the minimum needs. The engineer must place the cleanouts at locations where he believes they can best offer access to all portions of the underground drainage line. It is always good practice to check with local codes for requirements in placing cleanouts.

Building Traps

The building trap is a cleanout as well as a device that prevents the intake of sewer gas into the interior piping system. The cleanout offers access to the building sewer line between the building and outside street sewer facilities.

Variations in the installation of building traps are shown in Fig. 2. Detail *A* illustrates a building drain with cleanout extensions to the finished floor level. Access is difficult when the extensions are longer than 12 inches, especially with regard to a building trap. Detail *B* shows a building trap placed at a lower elevation. A concrete pit permits access to the cleanout plugs.

Methods for Installing Cleanouts

Since a wire snake or specialized line clearing device must be inserted through the cleanout and into the drainage line, the

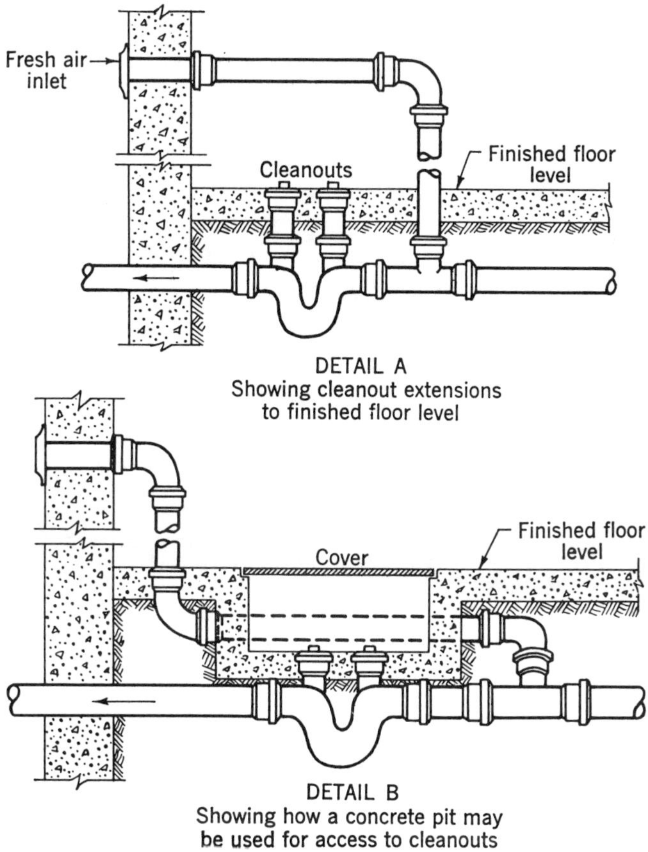

Fig. 2. How building traps may be installed for clearing stoppages in the house sewer.

cleanout piping must be arranged to facilitate operation of the apparatus. Figure 3 shows several different methods.

Method 1 shows the use of an 1/8-bend that terminates at the finished floor level. Method 2 shows an installation that utilizes an 1/8 bend with an extension between it and the TY fitting. Method 3 illustrates an end-of-the-line cleanout where an 1/8 bend may again be employed. Method 4 shows an end-of-the-line cleanout with a vertical piece of pipe and a long turn ell.

Other methods may be used, but the important point is that the piping should be so arranged that the wire or apparatus can be

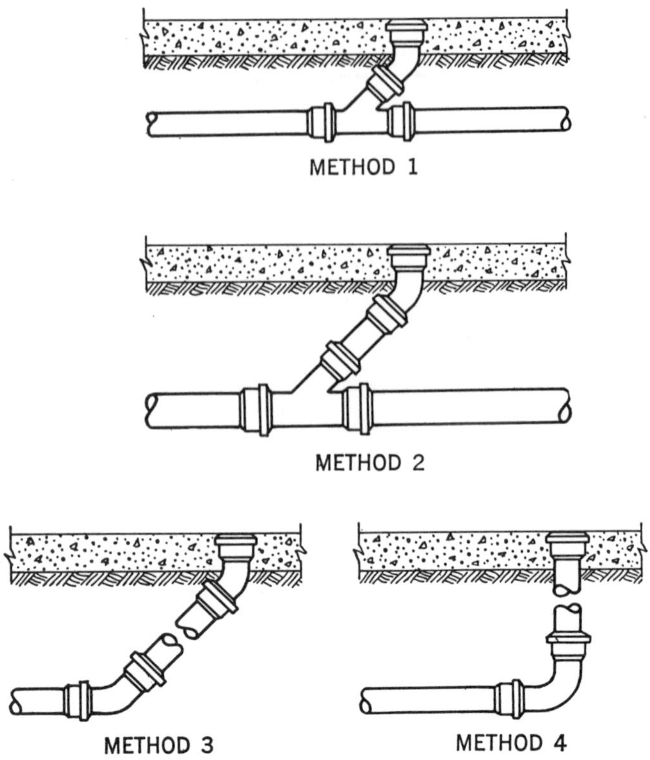

Fig. 3. Methods for extending cleanouts up to finished floor level.

inserted into the piping in the direction of the flow in the drainage line.

Although not shown, cleanouts with a recessed or countersunk socket have been used so that a projection will not be evident at the floor level.

Types of Cleanouts

Three of many different types of cleanouts available are shown in Fig. 4.

The first type is the exposed cleanout which can be used in areas where appearance is not a factor. Either a raised head or countersunk type plug has application in this location.

The access box offers a neat appearance and is ideal where floor finishes are applied. Furthermore, it affords a durable surface

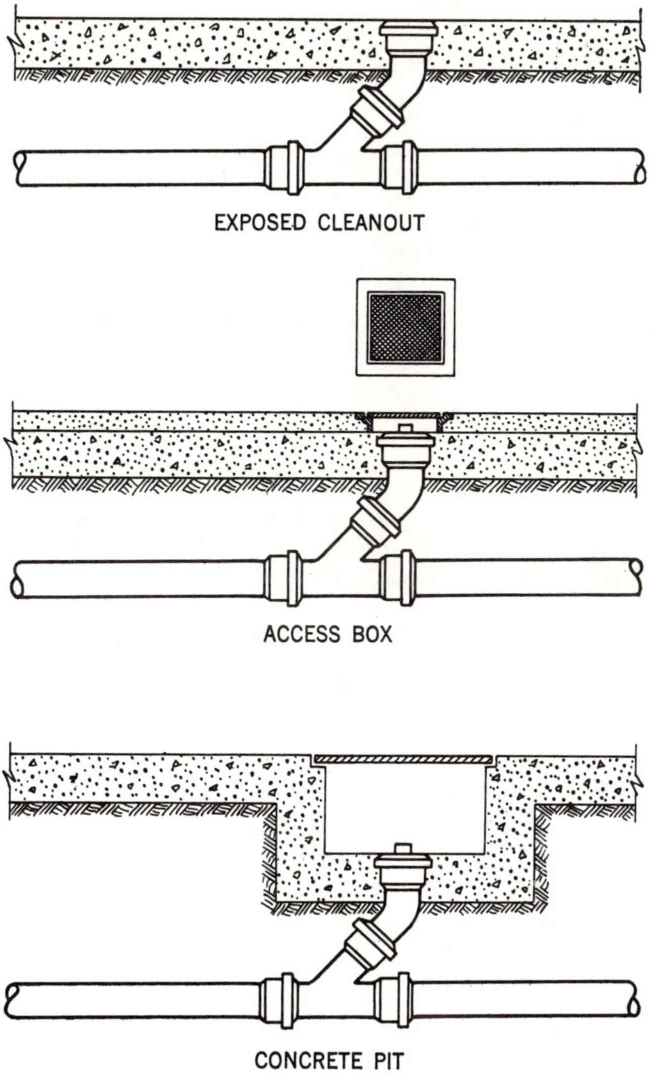

Fig. 4. Methods for providing cleanout access to underground drainage lines.

over which heavy traffic loads may pass without imposing a strain on the cleanout itself. The access box is easily set in place over the cleanout before the finished flooring is applied. Once the finish is added, the box is anchored securely in place.

While the concrete pit is desirable for certain applications, it is often more costly to provide the pit and cover than to install one of the manufactured devices.

Finished Flooring

Sometimes the drainage piping will be located below a corridor at the ground floor level in a building. When the floor finish in the corridor is of such a nature that cleanouts would be objectionable, the cleanout can then be installed as indicated in Fig. 5. Here, the piping is installed so the cleanout is set in the wall above floor level.

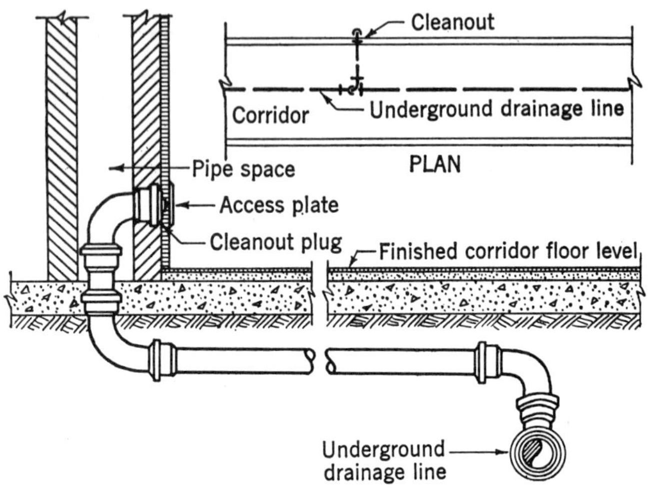

Fig. 5. How a cleanout may be offset and installed above the floor when it is desired to keep the floor surface clear of such devices.

Cleanout Plugs

The basic cleanout is a plug which, when installed in a ferrule or threaded connection, offers a watertight and gastight fit. This plug must be easy to remove to provide access to the drainage lines. Several different plugs as well as applications are shown in Fig. 6. There are the regular cleanout plugs of which there are two types. One is the raised head plug while the other is the countersunk plug. These may be constructed of brass, cast iron or other approved materials.

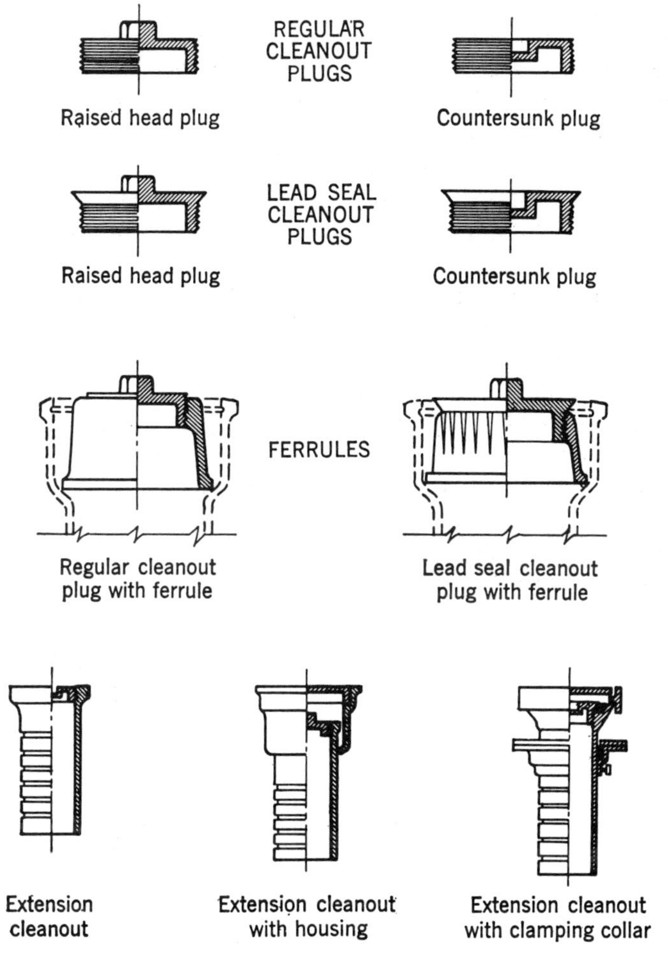

Fig. 6. Types of cleanouts that are available for various applications.

Lead seal cleanout plugs include the raised head and countersunk type plugs. This type of plug is equipped with a tapered flange that will seat against the lead poured in the hub of a length of soil pipe. The purpose of this innovation is to provide a connection that will not rust or freeze in place and thus can always be easily removed and replaced when access to the drainage lines is needed.

Ferrules are the components of cleanout plugs. Illustrated are the ferrules that may be used with the regular and lead seal type cleanouts. Soil pipe hubs are shown dotted to indicate how the installation will look when the plug and ferrule are added.

In addition to the cleanout plugs and ferrules, there are many variations in manufactured devices for cleanouts to fit any specific construction need. The extension cleanout is an example of this type device. It is very helpful because it fits into the hub of soil pipe and can be cut to any desired length to satisfy the prescribed limitations of construction and terminate at the floor level.

The extension cleanout with housing is a unique device because it allows the cleanout cover plate of the housing to be installed level with the finished floor. However, since the housing is free of the extension, a certain amount of flexibility is incorporated in the unit. Should the underground piping sag or settle slightly, the extension has freedom to move with the piping. This feature removes the danger of loose hub connections at the piping.

Where waterproofing is involved, the cleanout assembly can be furnished with a clamping collar to anchor to the waterproofing material and thus prevent seepage around the extension. The

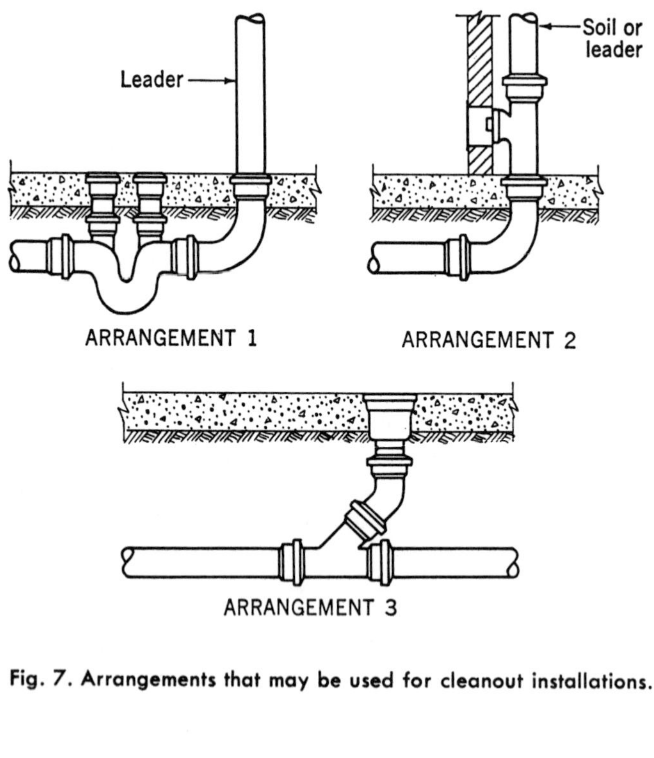

Fig. 7. Arrangements that may be used for cleanout installations.

clamping collar is a loose arrangement and can be placed at any elevation on the extension shank. Once it is located, it can be secured to the extension shank by means of the set screws and the collar caulked as shown by the cross section through the unit.

Cleanout extensions are equipped with cut-off rings for ease in cutting the extension to the prescribed length.

Other Cleanouts

Other types of cleanouts are required in the drainage system, and three are shown in Fig. 7.

Arrangement 1 shows a running trap often utilized at the base of a leader stack. This trap intercepts sticks, stones or other large foreign objects that are dropped into roof drains by children or through other means. Foreign objects will lodge in the trap where they are easily removed. This arrangement uses a running trap with extensions to the floor level.

Arrangement 2 shows a cleanout at the base of a stack, either a soil or leader stack. It is placed in a vertical position above the floor. An access door provides entry to the cleanout. This type of cleanout also permits a wire snake or other apparatus to be inserted into the drainage line underground.

Arrangement 3 illustrates the extension cleanout and housing, as indicated in Fig. 6 It can be applied to installations where there is a possibility of the underground line settling or sagging. The extension will move with the piping and will not impose any strain on the housing.

CHAPTER 24

Upper Floor Cleanouts

The last chapter covered floor type cleanouts and how they were utilized at the lower floor levels in the building to clear underground drainage lines of stoppages. This chapter deals with cleanout installations at the upper floor levels in a building. These cleanouts offer access to the vertical soil, waste and leader lines as well as the horizontal branches.

Stoppages in vertical or branch lines can be quite troublesome, since plumbing fixtures may overflow and cause considerable damage. Furthermore, stoppages can present a serious hazard to public health in the building. It is therefore extremely important that such stoppages be cleared as quickly as possible. This

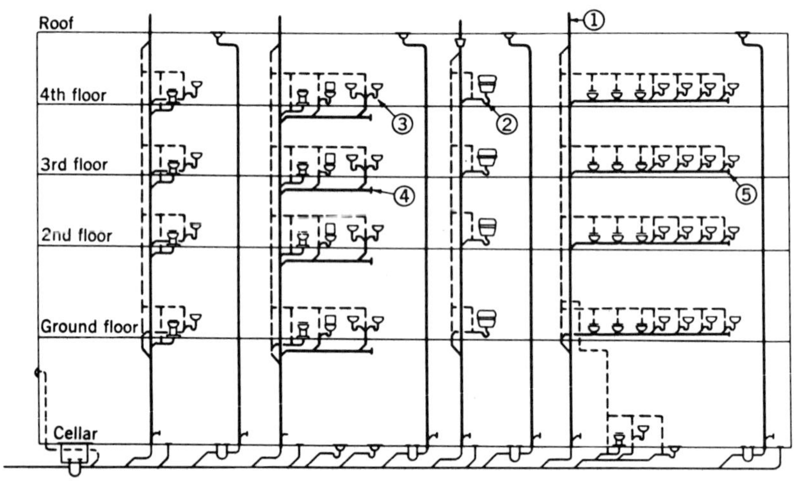

Fig. 1. Typical plumbing riser diagram showing placement of cleanouts. 1. Stack terminal. 2. P-trap at service sink. 3. P-trap at lavatory. 4. End-of-branch waste below floor. 5. End-of-branch waste above floor.

can be accomplished only by careful placement of cleanouts offering access to all parts of the system.

The general placement of the necessary cleanouts is illustrated in Fig. 1. Numbers indicate the locations where cleanouts have been placed in the system.

Waste Line Above Floor

A cleanout installed to protect a horizontal waste branch above the floor is illustrated in Fig. 2.

Here, a battery of three water closets and four lavatories is involved. The cleanout is installed at the end of the waste branch, as shown. The water closets are the wall mounted type and thus permit the horizontal waste branch to be installed above the floor.

Objects such as paper, cloth and bars of soap, discharged through the water closet, cause stoppages in the waste line. Once in drainage line, these objects will block the line and cause overflow at the fixtures. A reduced diameter branch has been used as the waste line from the lavatories. If desired, this line can also be of the same diameter as the line for the water closets. Factors that cause stoppages in lavatory waste lines are hair, hair pins and grease deposits that may be due to soap.

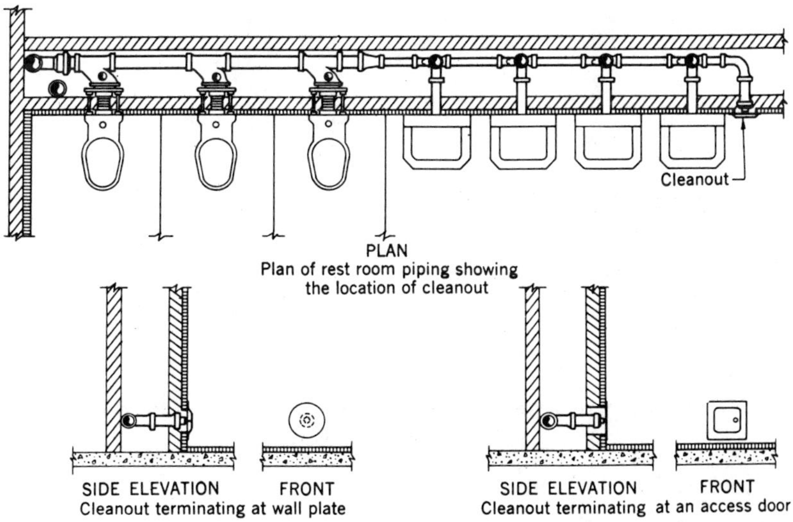

Fig. 2. How a cleanout may be installed on piping located above the floor slab.

In this instance, piping is of the threaded type. The cleanout consists of a drainage elbow and an extension cleanout with a wall plate covering an access to the cleanout plug.

In clearing a stoppage, the cleanout plug is removed and a wire snake inserted in the line to snag and pull out the foreign object or to push it into the vertical line. The snake will also dislodge grease deposits. Since the means of access to the cleanout plug is exposed in the room, it is important that a device is selected that will harmonize with the architectural treatment of the room. There are many type units available.

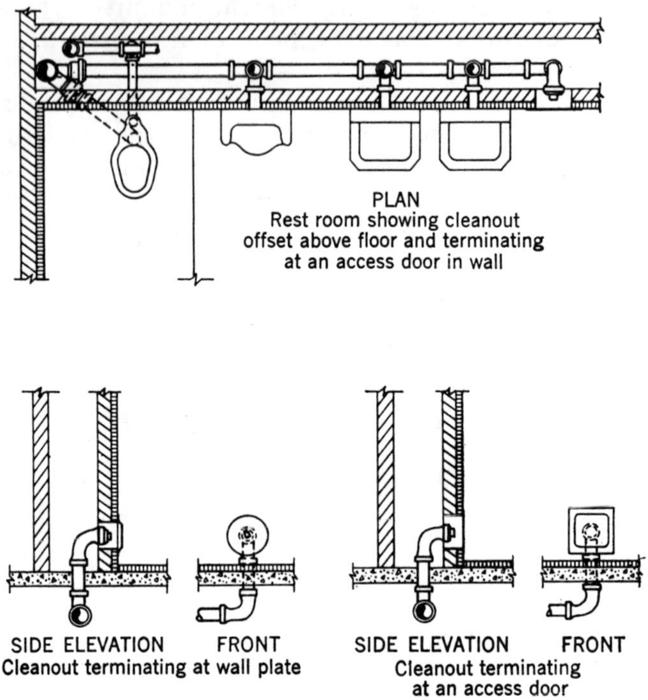

Fig. 3. How a cleanout may be installed on piping located below the floor slab.

Waste Line Below Floor

Figure 3 gives an example of how a cleanout may serve a waste installed below the floor slab. Here, the installation consists of a floor mounted water closet, a urinal and three lavatories. The cleanout is shown as being extended above the floor level and terminating at an access door set in the wall. This cleanout could just as easily be placed at the end of the branch under the floor, with its termination at the wall of the floor below. However, this method is not too practical, for in the event of a stoppage, the waste line will be filled with water, and this accumulation of waste water will discharge when the cleanout plug mounted on the wall of the floor below is removed.

There is less possibility of this occurrence when the cleanout

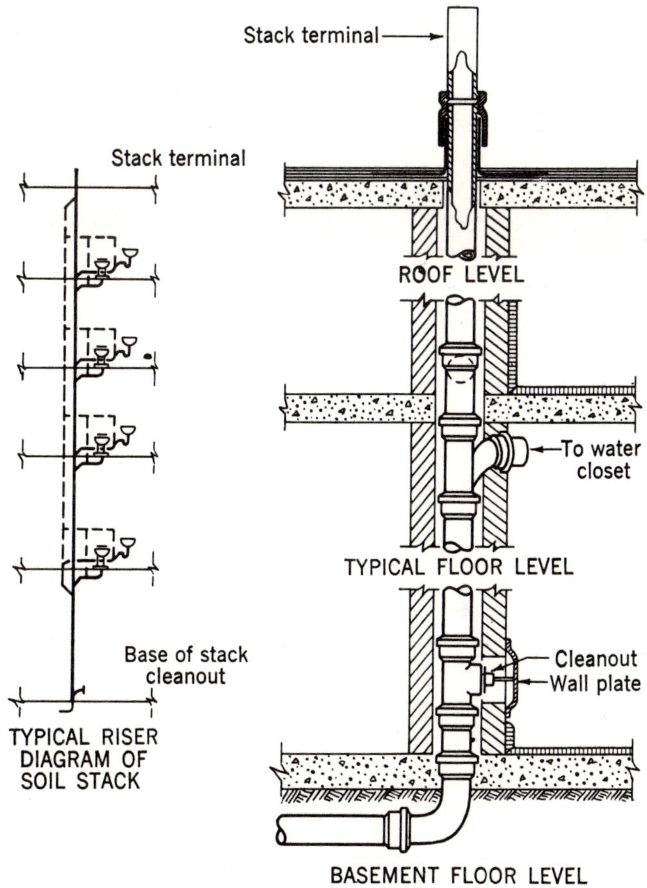

Fig. 4. How cleanout access may be provided for a vertical soil or waste stack.

plug is elevated above the horizontal waste line. In the event of a stoppage, the cleanout plug is removed and a wire snake inserted to clear the matter that is blocking the line.

The elevations included in Fig. 3 show how both a wall plate and an access door can be used to provide the means of entry to the cleanout plug. The piping arrangement is illustrative only; other arrangements can be used with equal success.

Vertical Stacks

Sometimes, stoppages occur in vertical stacks and must be cleared to prevent overflow at fixtures above the point of stoppage. Figure 4 shows a vertical stack and the locations where cleanouts provide access for clearing the stoppages. The typical riser diagram indicates the arrangement of the stack used for the example shown. Note the stack terminal at the roof, and the base of stack cleanout at the lowest floor level.

A stoppage in the vertical stack can be difficult to clear since large foreign objects or pieces of matter are involved. The objects may be sticks that are dropped in the stack by children at the roof level opening. These objects may lodge in the line or may drop to the bottom of the stack. If they are lodged in the line, it becomes necessary to insert a wire snake in the stack at roof level and attempt to snare or dislodge the object. If the object is dislodged, it will drop to the base of the stack where it can be reached by the cleanout there. Sometimes, a weighted ball at the end of a stout cord is inserted in the stack and manipulated to dislodge an object.

Since most objects or matter generally drop to the base of the stack, the cleanout at the base of the stack is most useful in clearing the stoppage at this point. The base of stack cleanout consists of a cleanout tee, plug and a wall access opening which is covered by either a wall plate or an access door. Note that the wall plate is held in place by a threaded rod which engages the cleanout plug.

Fixtures

Most plumbing fixtures come equipped with a cleanout plug. Figure 5 shows two plumbing fixtures and their cleanouts. Service sinks are supported on P-trap standards equipped with a cleanout plug, as shown. When the plug is removed, access is gained to the trap and the horizontal waste line to its point of connection with the vertical stack. Stoppages that occur in the waste

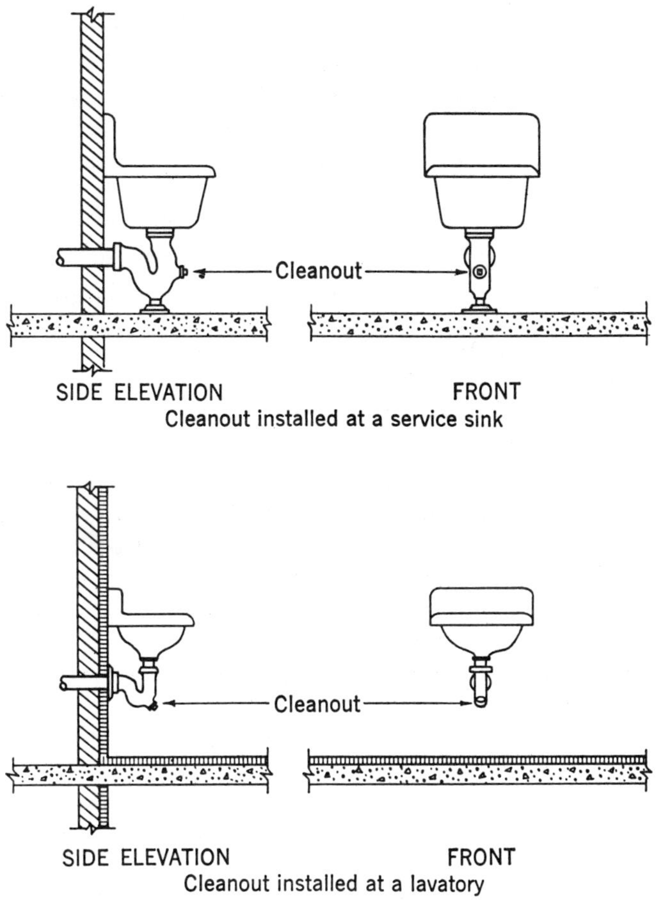

Fig. 5. Cleanouts installed at fixtures to keep waste lines clear of stoppages.

piping between the fixture and the vertical stack are easily cleared with a wire snake.

Water closets, by nature of their construction are equipped with an internal trap where foreign objects discharged in the bowl will cause stoppages. There are devices especially constructed for clearing such stoppages.

All other fixtures that are prone to stoppages are provided with a cleanout plug to offer access to the waste line.

Types of Access Devices

There are numerous devices that may be used for cleanout and access purposes. Several of these are shown in Fig. 6.

Type 1 is a cleanout tee installed in a vertical soil, waste or leader line. It is a caulked fitting with a threaded outlet into which a cleanout plug is placed. The cleanout plug may be of brass, cast iron or other approved material. This arrangement will serve, as shown, for exposed piping. When the piping is concealed, a wall access opening must be provided, and a wall plate or other means incorporated in the wall to facilitate access to the cleanout plug.

Type 2 is a threaded cleanout tee with a threaded outlet to receive the cleanout plug. Several types of tees are available. If a copper drainage system is used, there are copper cleanout tees available.

Type 3 is a caulked cleanout tee. It is equipped with a caulked outlet into which a ferrule is caulked. The ferrule is tapped to receive the cleanout plug. Lead seal type cleanout plugs may also be used with this combination. However, the lead seal plug and ferrule must be ordered as a unit. In addition, there are several types of extensions that may be used with this cleanout tee to satisfy the specific needs of the installation. Some means of wall access must be provided.

Four of the devices that can be used for ease of access to the concealed cleanout plugs are shown in Fig. 6.

Detail *A* is a wall plate that can be furnished in chrome plated brass, nickel bronze, stainless steel or other desired materials to satisfy the decorative wall finish desired. The wall plate is secured in place by means of the threaded rod that engages the cleanout plug. There are a number of different sizes and shapes of wall plates available to offer a broad selection.

Detail *B* consists of a frame that is anchored to the wall construction. The cover plate is threaded and is removed with a spanner wrench. This offers some vandal proof protection. The unit is generally constructed of chrome plated brass or nickel bronze. Different sizes and shapes are available to satisfy any specific need.

Detail *C* shows a square access frame and cover. The frame is anchored to the wall construction and the cover is secured with four screws. Here again, several different sizes and shapes and materials are available. When required, vandal proof screws can be provided.

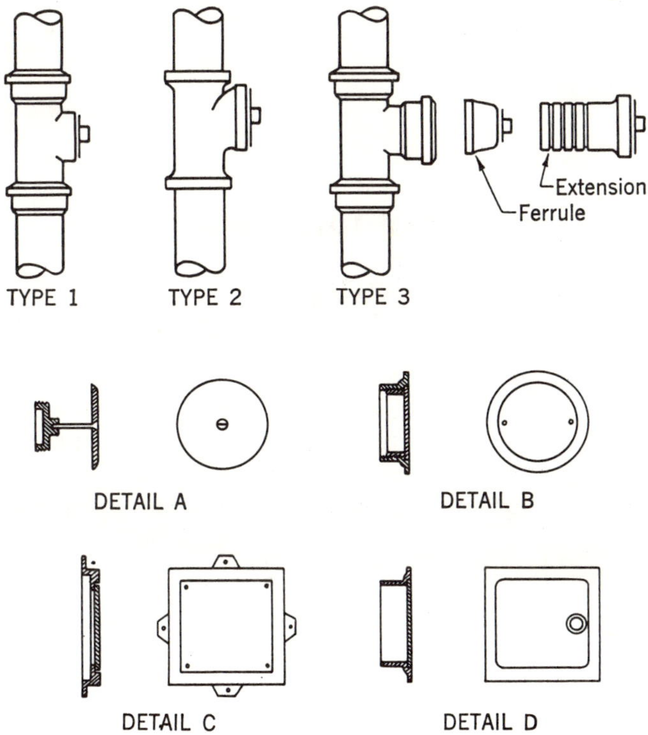

Fig. 6. Various devices that may be used for cleanout purposes.

Detail *D* illustrates access cover and frame, a very popular means for permitting entry to the cleanout plug. The unit consists of a frame and hinged cover. The frame is anchored to the wall construction and the door is opened by either a screw driver latch or a lock and key. When this unit is equipped with a cylinder lock, it insures a tamper-proof means of access. Once again, several different sizes and shapes are available, constructed of steel, stainless steel or brass. They can be finished to any degree of lustre or color to harmonize with the wall finish.

The access box Detail *D* is also very suitable for installation in ceilings to offer access to cleanout plugs placed at the end of a run or at offsets in piping that are installed in furred ceiling spaces. In this instance it is advisable to select a unit with a snug fitting cover and thus prevent plaster or construction grit from passing through cover voids and settling on the floor or a room or corridor below.

CHAPTER 25

Water Accumulation Around Buildings

Many schools, manufacturing plants, research laboratories and other types of buildings are being erected on land in suburban areas where the property terrain often has a considerable slope and where, by virtue of rock formation and other soil conditions, it is impractical or uneconomical to grade the property.

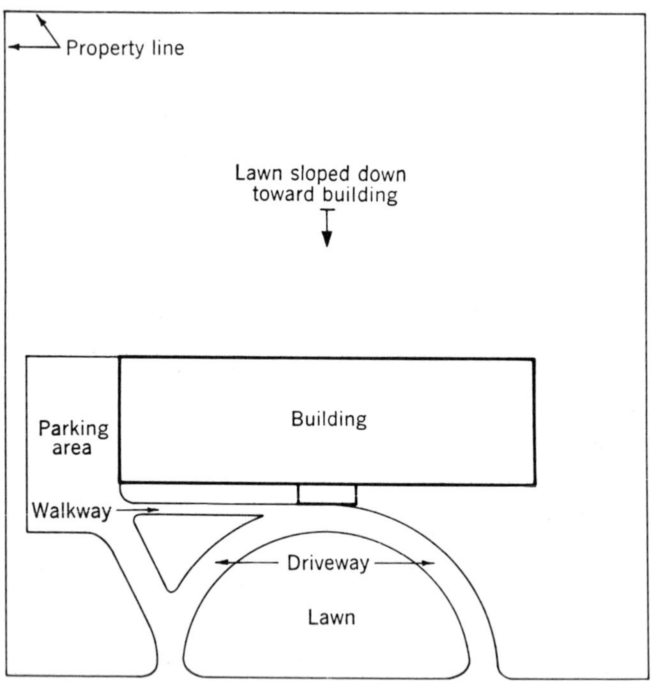

Fig. 1. Site plan showing the arrangement of building on sloping ground.

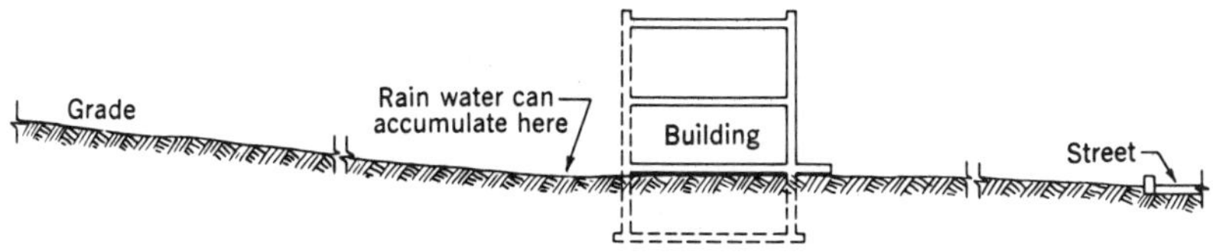

Fig. 2. How grade slopes toward building to form a pocket for rain to accumulate at the rear of building.

Let us suppose that the property (Fig. 1) has a rather sloping terrain from the rear of the property to the front. Since the building is rather long, water accumulation at the rear of the building is a potential danger unless measures are taken to keep the area dry. Accumulation of water can cause the building to settle and create a damp condition at the rear of the building detrimental to the foundation.

Figure 2 is an elevation of the building on the site. Note the pocket at the rear of the building where rain water can accumulate. After a rain, the water may remain in this pocket for days until it is finally absorbed by the earth. There are several ways to drain the water pocket, related to the specific nature of the ground conditions.

Exterior Drains

When the slope directly behind the building can be graded to a certain degree, exterior drains can be employed. This method is shown in Fig. 3, showing five exterior drains installed a short distance away from the rear of the building and properly spaced. The grade on either side of the drains is sloped in the direction of the drains. The grade is planted in grass to offer more rapid drainage and to prevent the water from washing earth into the exterior drains. The water is conducted by underground piping to the building interior where it connects with the system of piping that terminates at the exterior storm sewer facilities. Otherwise, it is conveyed to the interior sump pump arrangement and is pumped out to the exterior storm sewer facilities.

The exterior drains are of the large capacity type and are equipped with slotted grates to permit the rapid drainage of water accumulation.

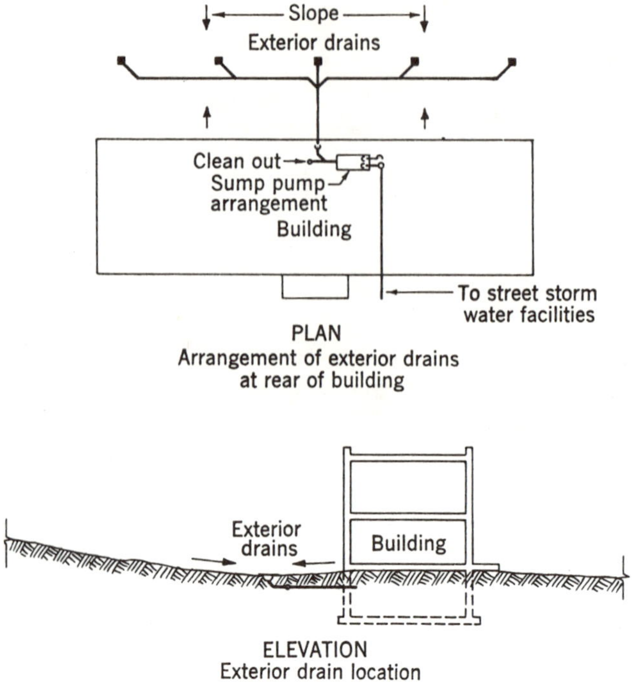

Fig. 3. Location of exterior drains. Grade is sloped towards the drains.

Underground piping may be of cast iron, tile, concrete or other approved material. Should stoppage occur in the underground lines, it is easily cleared by introducing a wire snake into the drain outlet to the point of stoppage. Generally, stoppages consist of sand, earth or grit deposits. This system of drainage is very dependable but it is important that the grading be sloped toward the drains to ensure the adequate removal of the accumulated water.

Drains and underground drainage lines are sized in accordance with the rainfall intensity in the area and the absorption rate of the soil. Rainfall intensity data can be obtained from the local plumbing authorities or from rainfall intensity tables.

Dry Wells

Dry wells may be used when ground conditions are favorable. The soil must have good drainage characteristics.

Dry wells, detailed in Fig. 4, are equipped with a slotted grate at grade level. Water accumulation drains into the wells and thence into the earth. The wells are placed in the same relative locations as are the exterior drains shown in Fig. 3.

The grate is set in a cast iron frame that is secured to the concrete cone. The dry well itself is reinforced concrete, formed in sections. The particular type of dry well illustrated has superior load-bearing strength over the type of structure often erected at a site.

Crushed stone is placed around the dry well to facilitate rapid drainage of the water into the earth. The drainage slots in the unit let the water escape but are designed to prevent backfill from entering.

Dry wells, like drainage systems, are sized in accordance with the rainfall intensity in the area and soil absorption rate. When in doubt concerning the absorption rate, it may be desirable to make percolation tests. Dry wells will handle large amounts of water.

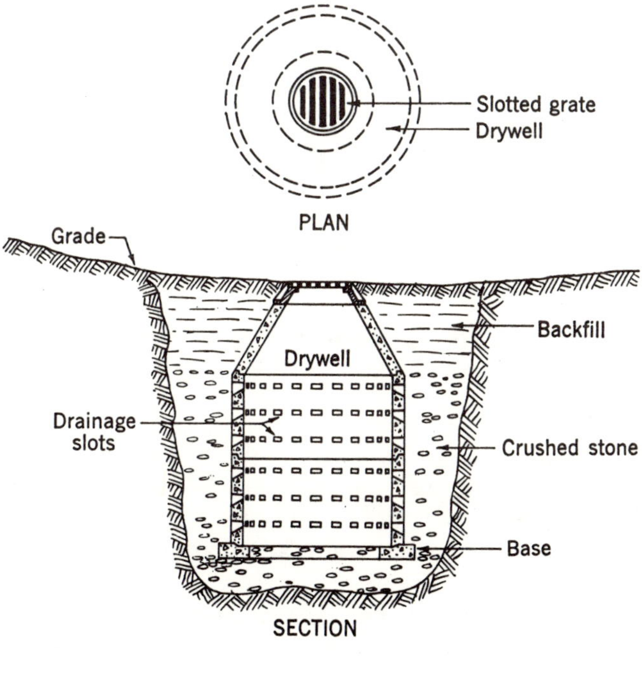

Fig. 4. Details of a dry well provided with a slotted grate.

Footing Drainage System

When, by nature of ground conditions, both surface *and* sub-surface waters are factors that must be considered, then a system of footing drains, Fig. 5, may be in order.

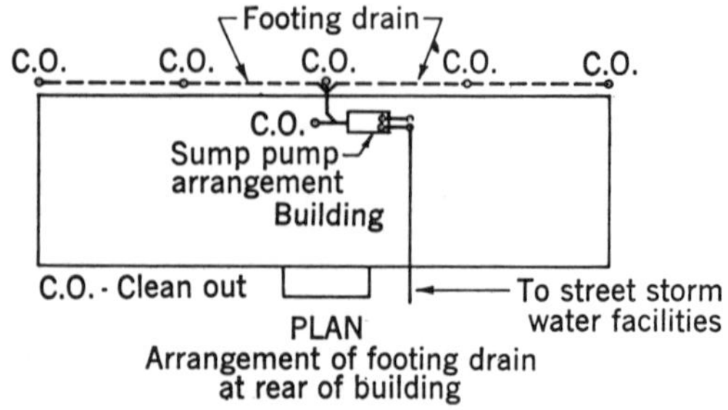

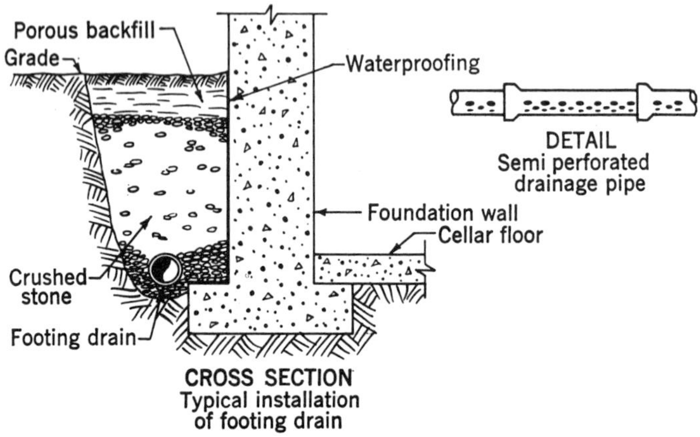

Fig. 5. Footing drainage system that can be used to drain any water accumulations at the rear of a building.

A footing drainage system will drain surface water that accumulates near the rear of the building. When, by nature of the earth strata, water will flow beneath the surface, the footing drains will intercept this sub-surface water and thus safeguard the foundation wall.

The plan shows the general arrangement of the footing drainage system at the rear of the building. The piping is semi-perforated clay piping. Water is intercepted by this piping and conducted to the sump pump arrangement in the building, where it is pumped out to the street storm water sewer facilities. Piping that enters the building from the footing drainage system must be cast iron.

The footing drainage system is equipped with cleanouts spaced at sufficient intervals to permit the ready clearing of stoppages in the line. Stoppages in this case are generally deposits of sand, earth or grit and they are cleared by inserting a hose into the cleanouts and gently flushing out the underground line with water.

The cross-section shows the specific arrangement of the footing drains near the foundation wall. The foundation wall below grade is coated with a heavy layer of waterproofing compound to prevent any moisture from penetrating the wall. The footing drains must be properly sloped to assure an adequate flow of water to the pumping arrangement. The excavation is filled with crushed stone to facilitate a rapid passage of water to the drainage piping. The upper portion of the excavation is filled with a porous backfill to allow the surface water to drain more readily. When carefully installed this system can safeguard the foundation wall and yet drain any surface water accumulations as well as any sub-surface water.

Concrete Trench

The concrete trench is a method used when the nature of the slope and ground conditions are such that the water will accumulate at the rear of the building but the soil (hard clay, for instance) will not readily absorb the water. Details are shown in Fig. 6. A concrete trench is installed for the length of the building, as shown. Grating is fitted into the recess atop the trench. Drains spaced at prescribed intervals in the trench receive the water flow. A system of underground piping conducts the water to the sump pump arrangement where it is pumped to the street storm water sewer facilities.

The cross-section detail shows the construction of the trench which is usually reinforced with steel rods. The trench is constructed next to the foundation wall and a concrete pad extends outward, away from the trench, to serve as a walkaway, if so desired.

The grating is of cast iron, steel or bronze, as selected for the purpose and is generally fabricated in sections 3 ft. or longer. The bottom of the trench is sloped toward the drains to afford

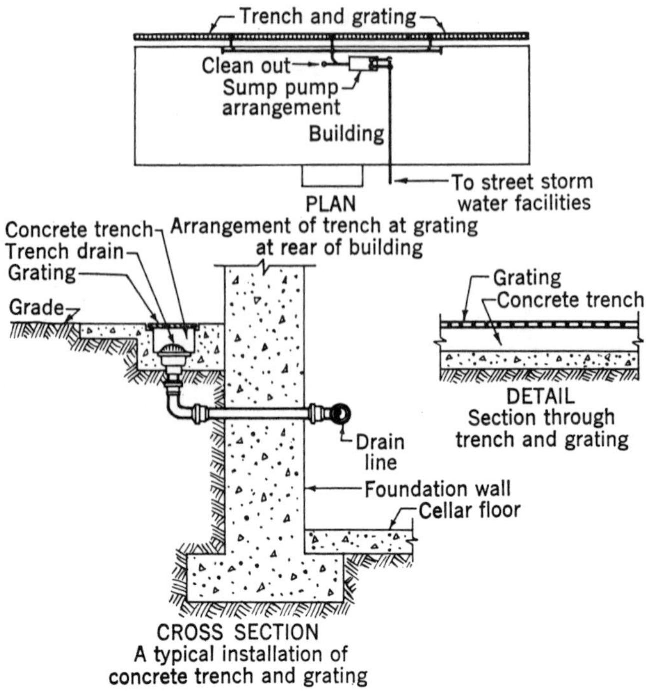

Fig. 6. How a concrete trench and grate can be used to drain any water accumulations at the rear of a building.

complete drainage. Trench drains with a dome strainer are used to receive the water flow. The dome strainer prevents clogging by bits of paper, leaves or cloth.

Inside the building, the piping is run along the foundation wall to a point where it drops below the floor slab to enter the sump pump arrangement. Suitable brackets are utilized to support the piping for its entire run along the foundation wall.

The small detail of Fig. 6 shows a cut through the concrete trench and indicates the manner in which the slotted grating is recessed in the trench.

Sump Pump Arrangement

All the methods shown, with the exception of the dry wells, conduct the water to the street storm sewer facilities either by gravity, or by a pump as in Fig. 7, illustrating how piping enters the foundation wall, drops through the concrete slab and then

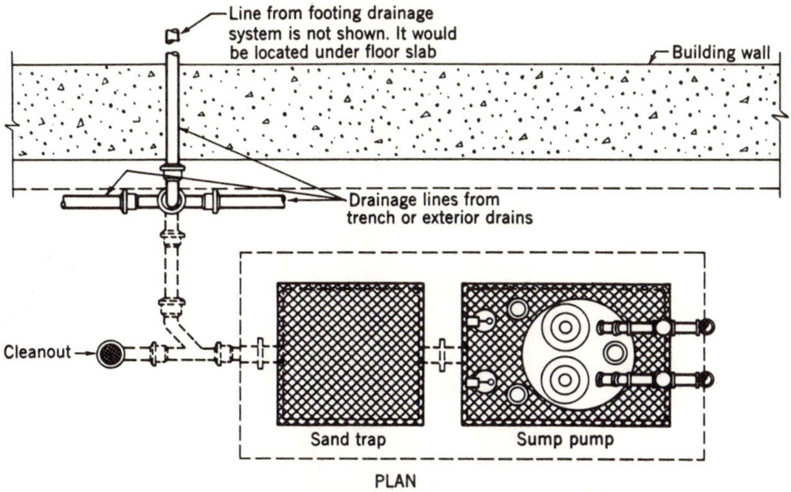

Fig. 7. Suggested arrangement in plan of a sand trap from an exterior footing drainage system, trench drainage and sump to handle discharge system, or exterior drains.

enters the sand trap and sump pump pit. The footing drainage piping will enter the foundation wall below the floor slab. Therefore, extra measures must be taken with regard to the construction of the foundation footing to compensate for the entry of the footing drain.

A cleanout has been installed at the offset of the underground piping to the sand trap. This is needed to clear any stoppages that might occur at this particular spot. Several different types of cleanouts are available for this purpose.

Steel covers are placed over the stand trap and sump pit to permit foot or heavier traffic and yet are easily removed for the proper maintenance of the pits. Careful attention must be afforded these covers, with concern to their rigid construction. The piping should be installed in a neat manner so that it will not waste floor space near the sump pump arrangement.

Section Through Pits

Figure 8 is a sectionalized detail of the sump pump arrangement. The piping first enters the stand trap. This trap intercepts any sand, earth or sediment that may be contained in the water entering the pit. The foreign material will settle at the bottom

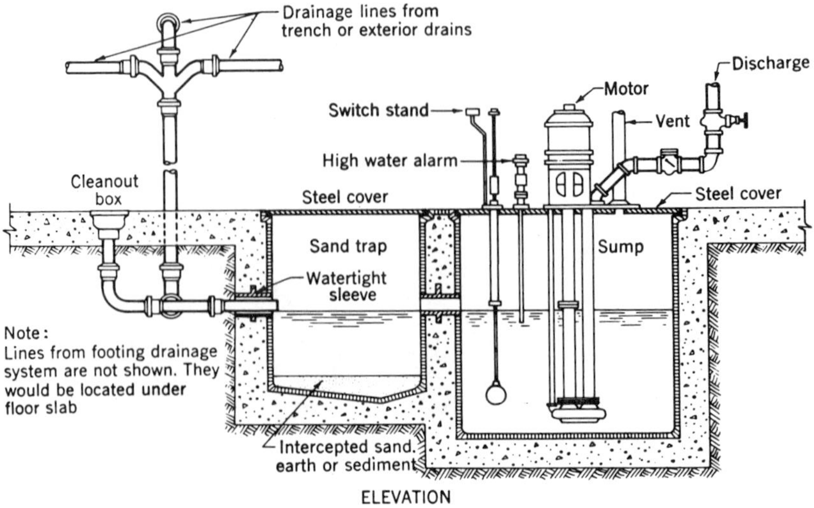

Fig. 8. Elevation of arrangement shown in plan in Fig. 7.

of the stand trap where it is available for easy removal. The sand trap prevents the material from entering the sump pit where it can cause damage to the pumping equipment.

The clear water from the sand trap then enters the sump pit where it will accumulate until a sufficient quantity activates the pumping equipment. Duplex sump pumps assure the positive discharge of water to the street storm sewer facilities. The sump pumps and pits must be sized in relation to the quantity of water that must be handled in each system. Here again, the rainfall intensity charts for the area offer the important data needed by the engineer to enable him to size his equipment properly.

The interior of the sand trap and sump pit is lined with a waterproofing finish to prevent the water contained from seeping through the concrete structure. The watertight sleeves offer a waterproof connection for the passage of piping or overflow of water from the stand trap to the sump pit.

Both the sand trap and sump pit must be carefully constructed to offer the durable qualities needed by such structures. Reinforcing rods are also usually included for the extra strength they add to the concrete structure.

CHAPTER 26

Piping Sleeves

One very important consideration for piping systems which seldom receives more than a mention in the specifications is piping sleeves. Although these sleeves have many applications in a building, their use is most often left to the judgment of the contractor.

Typical Installation

Each type sleeve has a specific application.

Figure 1 is a riser diagram of a building plumbing system comprising soil, waste, vent and storm water leader lines. The locations where sleeves are used are noted. For the leader lines, a sleeve is shown at each floor level. This is necessary because

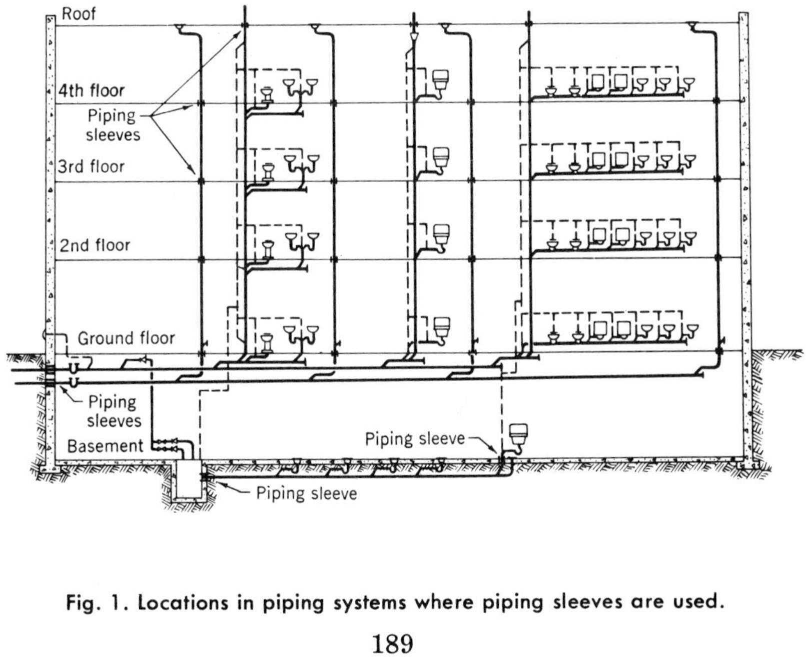

Fig. 1. Locations in piping systems where piping sleeves are used.

the leader lines are generally run independently of other services and therefore a hole must be provided in the floor slabs for their passage.

Sleeves are also used at the top of soil and waste stacks for the passage of the vent terminal through the roof. Sleeves are not provided at each floor level for the soil and waste stacks since a slot is usually provided in the floor slab for the passage of this as well as the water piping.

Piping sleeves on the different piping systems that pass through the foundation wall ensure a watertight installation.

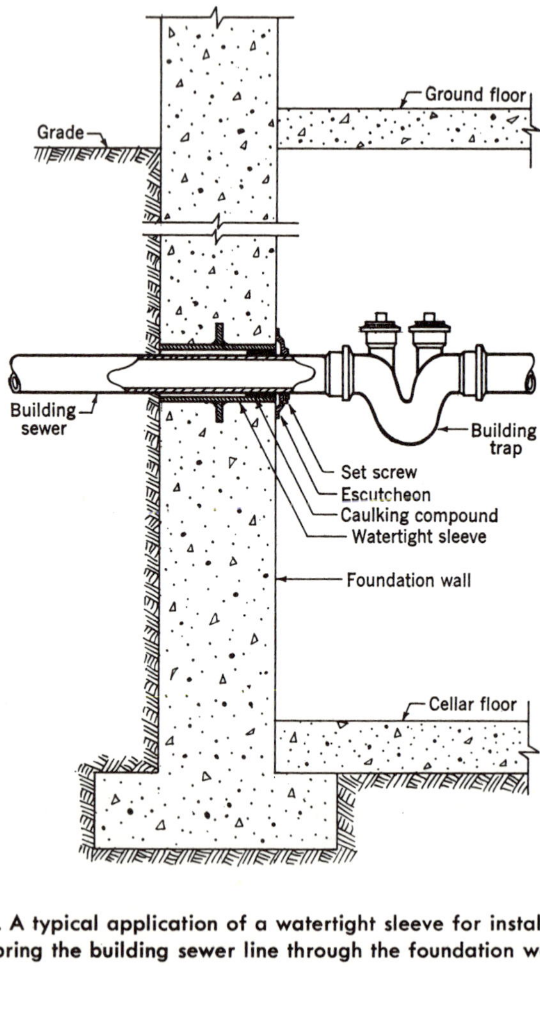

Fig. 2. A typical application of a watertight sleeve for installation to bring the building sewer line through the foundation wall.

Sleeves are also needed for various reasons at the lower floor level in the building, including the sump pump pit.

Foundation Wall

Building sewer, storm water, sprinkler, standpipe and gas lines are among the services that must pass through the foundation wall. Figure 2 shows a building sewer passing through this wall. The sleeve shown is fabricated and set in the forms for the pouring of the concrete foundation wall. The piping is then installed and the void between the piping and the sleeve filled with caulking compound to offer a watertight connection.

For any line passing through a foundation wall, there is always the danger of water forcing its way around the service or sewer line and eventually seeping into the building. This creates a nuisance extremely difficult to correct. Usually, it is necessary to chip away some of the concrete around the piping at the interior of the building and fill with cement plaster. In a short period of time, the water will again force entry and the sleeve must be repaired again.

Sleeve Types

Three types of watertight sleeves available are illustrated in Fig. 3.

Fabricated sleeves are made at the job site. The contractor cuts a piece of pipe to the proper length and of sufficient diameter to permit passage of the line or service with which it is to be used. Steel piping is generally used. A hole is burned in a disk of steel which is slipped over the sleeve and welded in place. This flange serves as an anchor for the sleeve in the concrete in addition to preventing moisture from forcing its way around the sleeve.

The sleeve is then placed in the forms for the pouring of the foundation wall. After the service or sewer line is installed, the void between the sleeve and the line is filled with caulking compound to ensure a tight seal. An escutcheon placed on the line in advance is then slid into position against the wall and its set screw tightened to secure it in place. Eventually, when the caulking fails and the water seeps through, it is an easy matter to recaulk the void.

Double-hub sleeves are manufactured devices, very practical

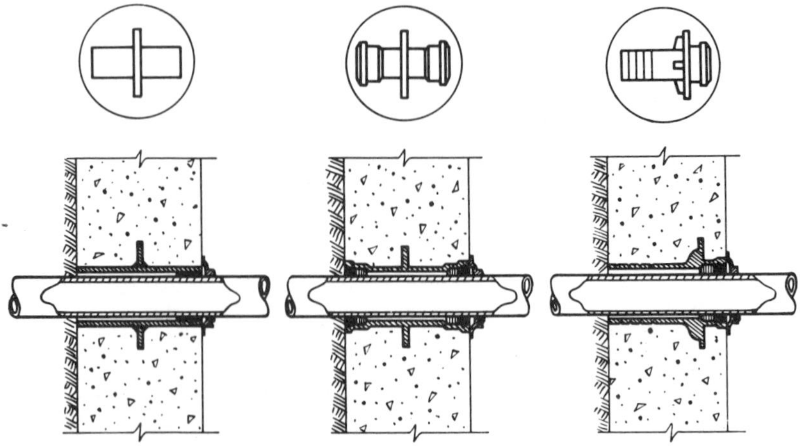

Fig. 3. Types of watertight sleeves that may be used for piping through a foundation wall.

for watertight applications, and made of cast iron in any prescribed length. There are hubs at each end and flanges in the middle.

The sleeve is set in the forms before the foundation wall is poured. After the service or drainage lines are installed, the hubs are packed with oakum and caulked with lead. An escutcheon at the interior wall offers a neat appearance. Should moisture seep through at the inner hub, a watertight seal can be accomplished by recaulking.

Single-hub sleeves are also manufactured cast iron devices. The sleeve is equipped with cut-off rings to allow the contractor to fit the unit to the proposed thickness of the foundation wall. The cut-off segments are easily removed with hammer and chisel to fit a desired length.

Underground

In sump pump installations, as in Fig. 4, in all underground pits, and especially in the case of sewage ejectors, the watertight sleeve prevents contamination or other damage that leaking water can do. When soil, waste or leader lines pass through lower floor slabs, adequate measure must be taken concerning sleeves.

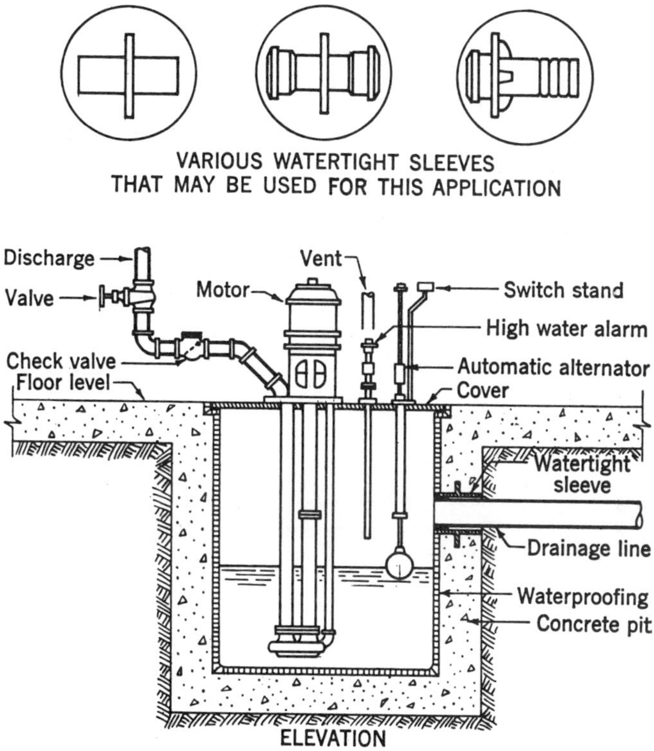

Fig. 4. How a watertight sleeve can be used at a sump pit for drainage and for passing a pipe line through the pit wall.

Two methods that are very practical are shown in Fig. 5. At *A* is a fabricated type sleeve that is slid over the piping during installation. Detail *B* shows a sleeve with a single hub.

Watertight Floor Slabs

When ground water conditions prevail at the site, the floor slab must be waterproofed, as in Fig. 6.

The sleeve is slipped over the piping during its installation and is then carefully positioned for the pouring of the rough floor slab. The flange of the sleeve is so placed that it will be level with the rough slab.

A waterproofing membrane is applied to the rough slab and is tightly compressed between the sleeve flange and the clamping collar which in turn is secured in place with screws. Next,

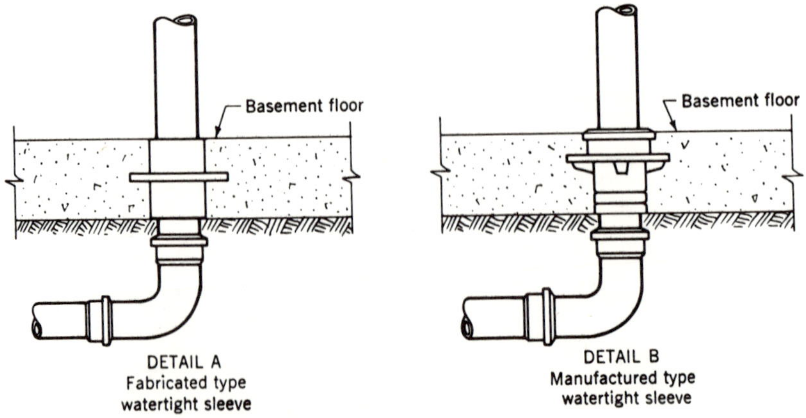

Fig. 5. Use of sleeves for making a watertight joint where pipe passes throgh a lower floor slab.

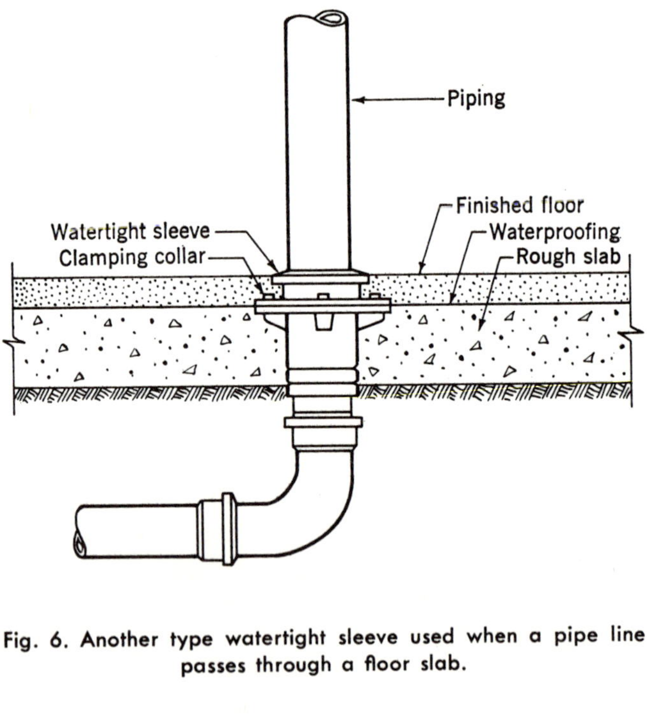

Fig. 6. Another type watertight sleeve used when a pipe line passes through a floor slab.

the floor finish is applied. After this is accomplished, the sleeve hub is packed with oakum and caulked with lead to offer positive assurance against any ground water seepage.

This type of sleeve is also very practical for upper floor watertight slabs. Here the sleeve will prevent seepage through the floor in the vicinity of the piping. When the floor area below is a finished area or contains equipment, moisture can be a damaging factor, especially if electrical equipment or apparatus is in that space.

Vertical Risers

Several applications of sleeves to risers are shown in Fig. 7.

Type *A* of Fig. 7 shows the application of a fabricated sleeve which, in this cas, is used for the passage of leader piping. The sleeve is set in the forms before the floor slab is poured. Later, the leader line, in this case hub and spigot piping, is so installed that the hub will rest on the sleeve. This practice thus provides a dependable support for the vertical piping. For this installation,

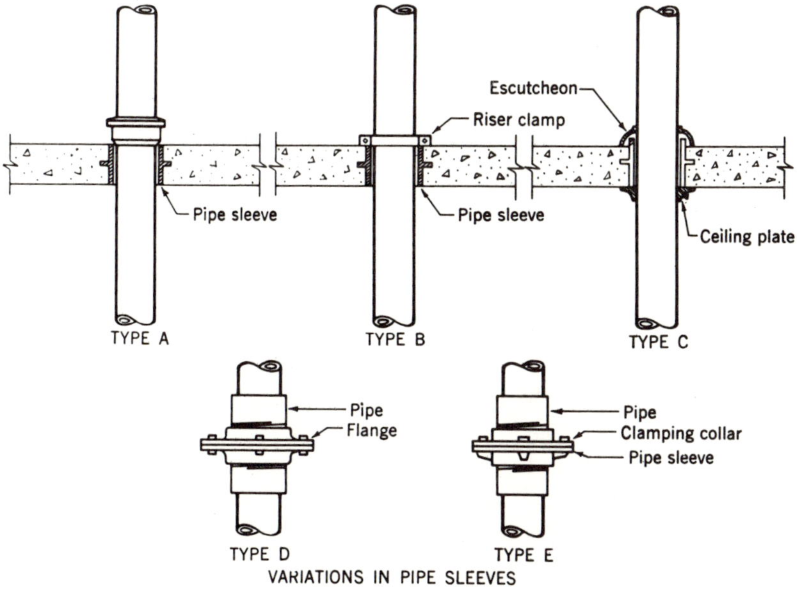

Fig. 7. How pipe sleeves may be used for vertical piping that passes through concrete floor slabs.

an ordinary piece of pipe with anchoring lugs or a flange welded in place will serve as an adequate sleeve.

Type *B* shows a method of providing a sleeve for all types of threaded piping systems. The sleeve is similar to that described in Type *A*. A riser clamp is used to support the piping. Several different types of riser clamps are available for this purpose, including the type of riser clamp that is of the offset type.

Type *C* shows a situation where the sleeve is extended above the floor, as may be required in exposed areas where floors are subject to washing with water. This arrangement prevents water from running down the piping to the floor below and thereby prevents staining of the piping. The escutcheon gives a neat appearance at the floor level while a ceiling plate is used to conceal the sleeve at the floor level below.

Type *D* shows a method of fabricating a pipe sleeve. It consists of a pipe flange into which prescribed lengths of pipe are threaded. Both halves of the flange are bolted together and if desired, gasket material can be inserted in the flange to offer a tight seal. The flange serves as an anchor in the concrete. The piping should be of a diameter that is sufficient for the passage of the line it will handle. This is a quick method of fabricating a sleeve on the job for a specific installation.

Type *E* illustrates a manufactured device into which pieces of pipe are threaded to serve as a sleeve. This unit can be equipped with a clamping collar to permit installation of the unit in a watertight floor slab. Both *D* and *E* sleeves were shown to indicate the type of devices that can be assembled at the job site to serve specific needs. Each contractor will have a type he favors when furnishing sleeves is his responsibility.

Vent Terminal

Piping sleeves are also required at roof level. Two are shown in Fig. 8. The first is a fabricated sleeve. It is placed in the forms for the pouring of the roof slab. Later the stack terminal piping is passed through the sleeve and terminates at a prescribed elevation above the roof level. A watertight condition is accomplished at roof level by means of the copper flashing and the adapter coupling. This method of waterproofing is clearly shown in the detail.

The other type of installation consists of a single hub sleeve that is set in the forms for the pouring of the roof slab. The

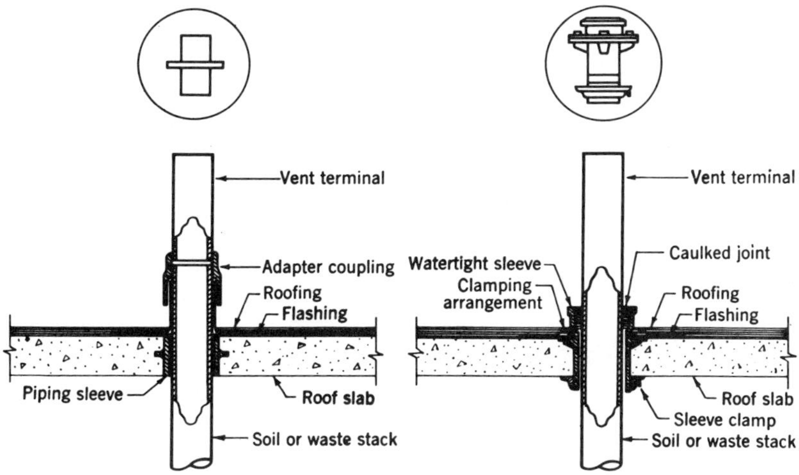

Fig. 8. Two types of piping sleeves that may be used for vent stacks that pass through the roof slab.

flange of the sleeve is set flush with the top of the roof slab. Lead flashing and roofing is then applied. The lead flashing is secured in place by means of the clamping collar. The final application of roofing is then added to completely cover the clamping collar. An under-deck clamp is employed below roof level to lock the sleeve unit in place. The under-deck clamp is secured with set screws.

In the meantime, the hub is packed with oakum and then caulked with lead to provide a dependable watertight seal. Both of these methods have been shown for example only, since there ar a number of ways in which the vent terminal can be installed.

Piping sleeves are an important component of well designed piping systems. Therefore, they should be carefully handled in the drawings and specifications to be certain that the desired piping sleeve devices are used in each case.

CHAPTER 27

Gasoline Station Drainage

The gasoline station service area generally is divided in two. One half includes a lift and is devoted to greasing or under-car repairs; the other half is used for car washing or repairs of a general nature. Larger stations will have more compartments.

For the sake of discussion here, a typical gas station is shown in Fig. 1. The station consists of an office space or sales area, toilet rooms, storage area, utility room and service area. Two floor drains in the service area connect with the sanitary drainage line.

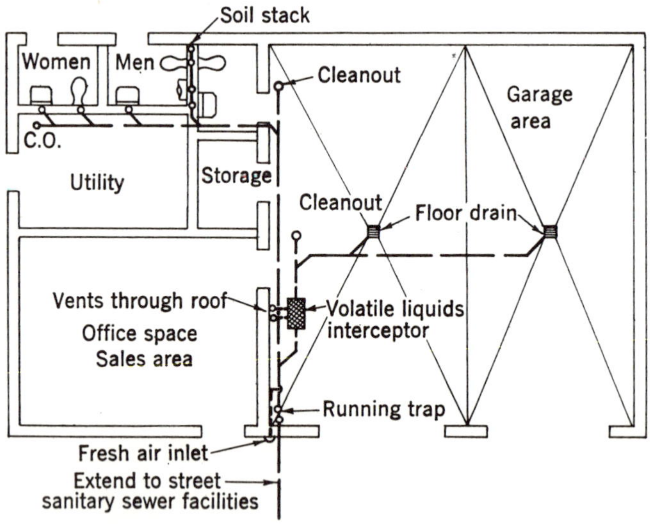

Fig. 1. Suggested arrangement of floor drains and interceptor for a typical gas station.

They are intended primarily for draining the water used in washing the floor area. One drain handles, in addition, water dispensed by the car washing apparatus.

Volatile Liquids

A potential hazard exists in the service area by virtue of the oil and grease that drips from cars and is washed into the drains when the floor of the service area is hosed down. A can of used motor oil might very well be discharged into one of the drains. Quite often, gasoline or other volatile liquids are used to clean parts of a car and this too, may be carelessly discharged into one of the drains.

Serious explosions and fires have resulted from accumulations of the volatile liquids. Many plumbing codes now require that some means must be afforded to intercept and retain the volatile liquids. One means of protection is the volatile liquids interceptor, shown in Fig. 1. The interceptor is installed in an out-of-the-way location where it can be easily maintained. When properly installed, this unit will intercept and retain for easy removal, any volatile liquids that are accidentally or carelessly discharged into the floor drain system, as explained later in this chapter.

Floor Drains

A floor drain for a gas station should:

1. Have a grate of sufficient strength to withstand the weight of a car passing over it,

2. Have drain and opening of adequate size to facilitate rapid drainage, and

3. Be equipped with the means to intercept grit and debris.

Three commonly used floor drains are illustrated in Fig. 2. The first includes a sediment bucket and is equipped with a P-trap. Slots in the side of the bucket and small openings in the bottom permit complete drainage of the waste water but retain coarse sediment and debris. The trap offers a water seal to prevent sewer gases from escaping. Venting the floor drains must be in accord with the local code.

A trapped floor drain is used when an interceptor is *not* placed in the line. The second drain is similar to the first with the exception that the trap has been omitted. In both types the grate cannot be set in position unless the bucket is in place. This

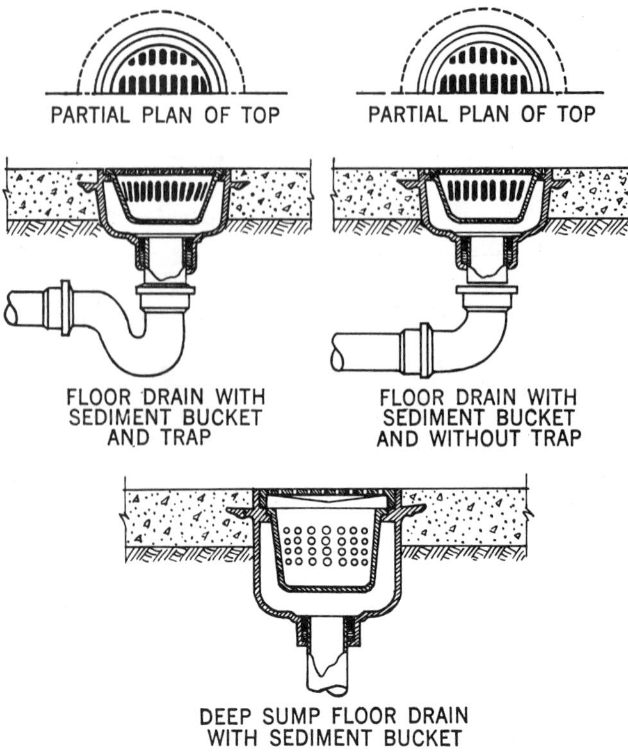

Fig. 2. Details of drains that can be used in gas stations.

feature avoids the possibility of neglecting to replace the bucket after a cleaning operation.

The third drain is a deep sump type. This drain, too, can be installed with or without a trap. The grate is of a heavy duty type to withstand the heavy loads to which it will be subjected. The bucket is of large capacity to intercept the large volume of sediment and debris, consisting mostly of the matter that adheres to the underside of cars, which enters the drain when the floor is washed.

Such debris can enter the drain and cause partial or complete stoppages in the underground drainage piping for it will accumulate in a portion of the piping and harden like concrete. With sufficient build-up, it will stop the flow of waste water.

There are many other styles and shapes of drains that may be utilized in the service area of a gas station with success equal to the ones illustrated.

Slope of Floor

Quite often, the proper slope of the floor in the service area of a gas station is overlooked. When this situation prevails, the water dispensed on the floor does not readily drain to the floor drains and the floor remains wet for a longer period of time. It is highly recommended that the floor be sloped in a positive manner toward the drains. For example, see Fig. 3.

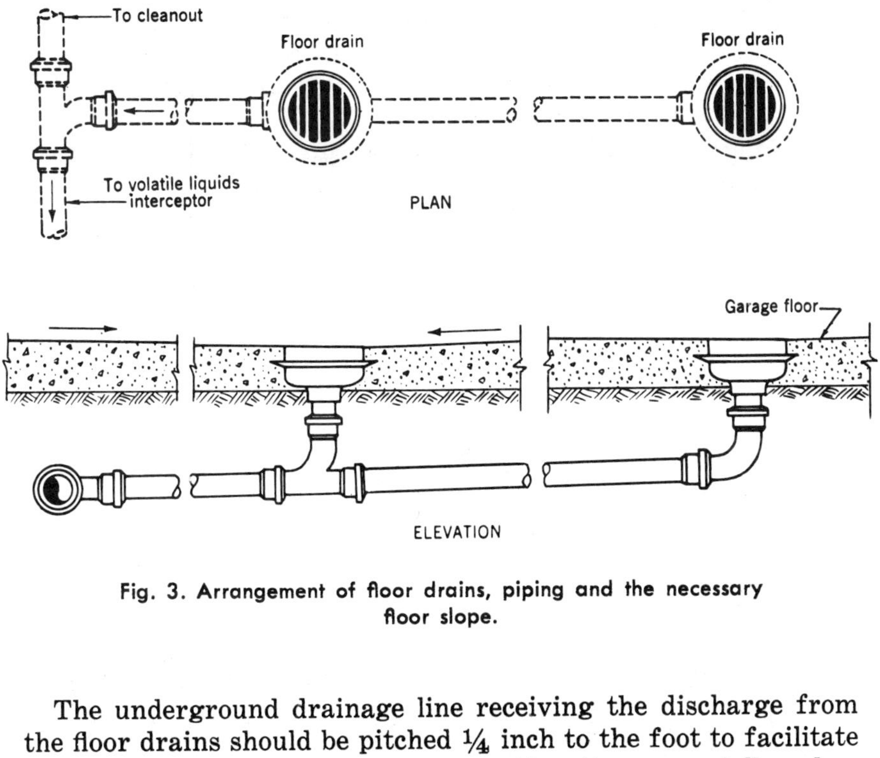

Fig. 3. Arrangement of floor drains, piping and the necessary floor slope.

The underground drainage line receiving the discharge from the floor drains should be pitched ¼ inch to the foot to facilitate drainage. This is an important consideration, especially when car washing facilities are used. The contractor, in running his sanitary line to the street sewer must maintain a slope that will permit his line to connect with the street facilities at a proper elevation. Thus, the slope of floor and underground drainage line are important considerations.

Volatile Liquid Interceptors

When a volatile liquid interceptor is employed, it can be installed similar to that shown in Fig. 4. The interceptor in this instance is equipped with an extension to allow the top of the unit to be flush with the floor surface. If an extension is not desired, then the interceptor must be installed in a pit with a frame and cover at the top of the pit to offer access to the interceptor cover. The usual reason for the extension is the difficulty in attaining the proper elevation between interceptor inlet and underground drainage piping.

The interceptor is equipped with two vent connections at its side. These vent connections are extended up through the roof of the building and terminate at different elevations. The difference in elevation is usually 6 to 12 inches. The vent lines will relieve any explosive gases that accumulate in the interceptor, thereby avoiding a potentially hazardous condition.

A natural circulation of air is accomplished by the difference in elevation of the vent terminals. Wind velocity over the roof of the building will vary at each elevation of vent terminal thus creating an unbalance of atmospheric pressure in the vent lines. Therefore, the lighter-than-air explosive gases will relieve

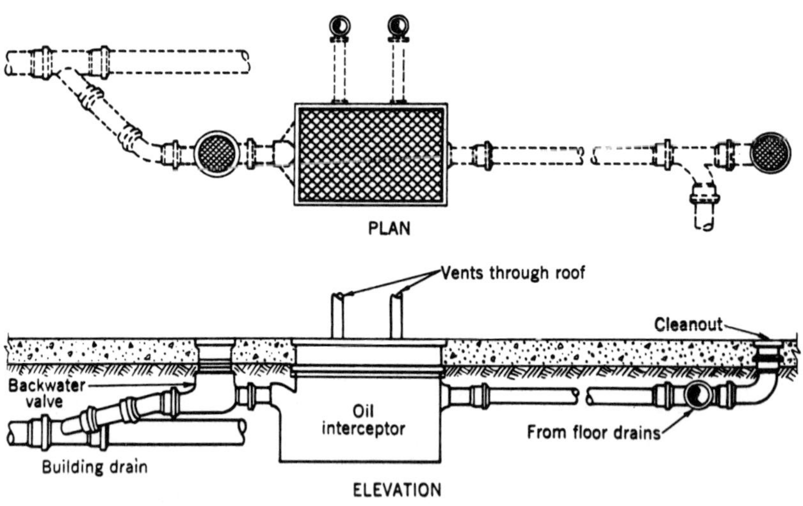

Fig. 4. Oil interceptor installed to prevent volatile liquids from being discharged into the sanitary system.

through one vent and air will be drawn through the other vent.

The waste water discharge from the interceptor passes through a backwater valve. The purpose is to prevent any possibility of sewerage backflow from entering the interceptor. The local plumbing code should always be consulted with regard to the arrangement of the interceptor and required venting before the design proceeds too far.

If possible, there should be a difference in elevation between the outlet from the interceptor and the elevation of the sanitary line to which it connects.

Figure 5 shows a section through the interceptor.

The extension on top of the interceptor that had been included in Fig. 4 has been omitted for the sake of clarity.

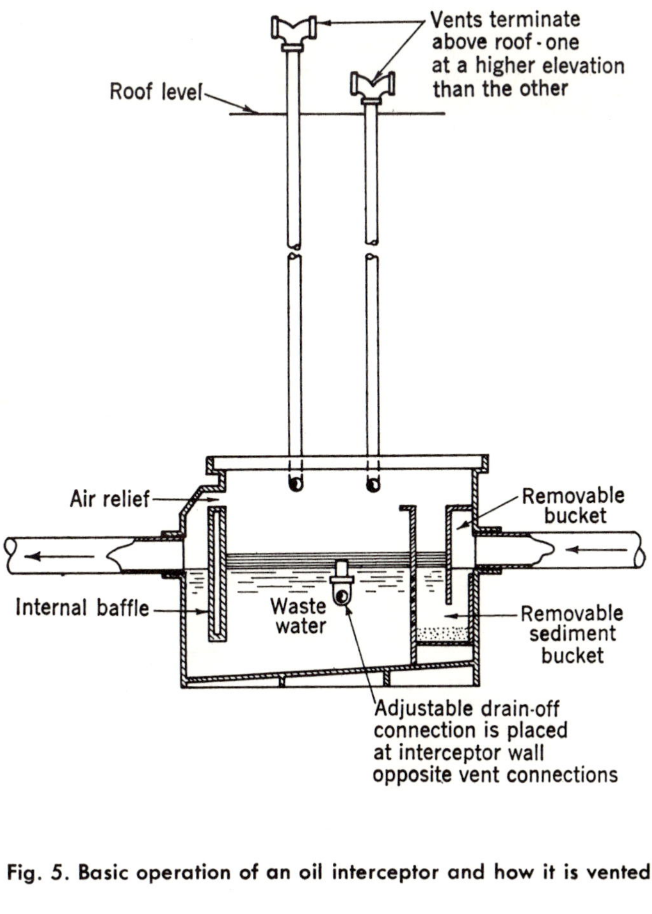

Fig. 5. Basic operation of an oil interceptor and how it is vented.

Here is how the unit functions: Waste water from the floor drains enters the interceptor and a baffle reduces its turbulence. A sediment bucket traps any minute quantities of sediment or debris that is carried by the water. The sediment bucket is perforated to retain the foreign matter and allow the waste water to pass through.

In the settling chamber, the volatile liquids separate from the water and float to the surface while the clear waste water passes under the last baffle and discharges into the sanitary system. The air relief shown at the outlet baffle assures the balance of air needed for the proper discharge of waste water from the unit. The two vent lines extend up through the roof, terminating in a manner shown in Fig. 4.

The draw-off connection shown is used when it is desired to drain the volatile liquids to a volatile liquids receiver tank. An adjustable tube in the fitting is set at an elevation slightly higher than the flood level of the waste water passing through the interceptor. (The contractor can determine the exact level by trial, once the unit is installed.)

The interceptor shown in Fig. 6 is of a generalized shape and is used here to describe the basic operation of the unit. Several

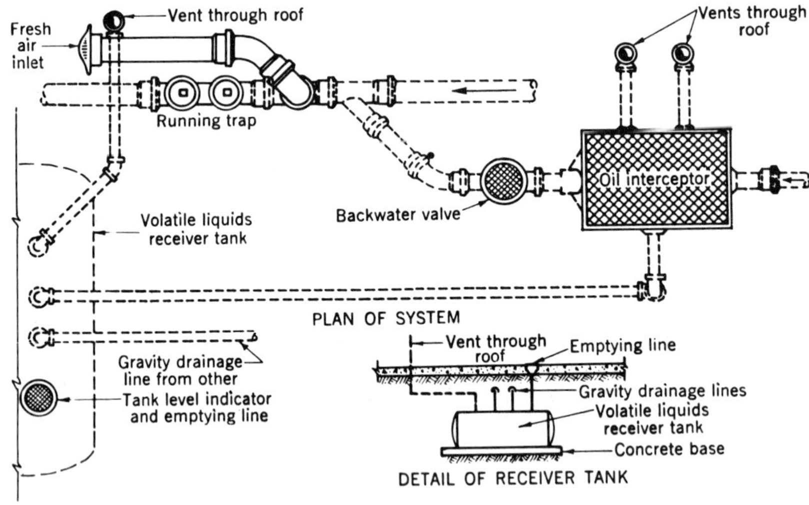

Fig. 6. Arrangement of oil interceptor, piping and volatile liquids receiver tank.

manufactured types, each with specific refinements, are available for the purpose. The covers too, are secured in various ways.

Receiver Tanks

When it is desired to provide a draw-off connection at the interceptor, the piping arrangement might be similar to that shown in Fig. 6. Note that a draw-off line is extended from the interceptor to a receiver tank, which is located just outside of the building structure. This tank, very often, is also used for the waste oil drained from automobiles.

The tank is supported on and secured to a concrete pad. At least one vent line must extend from the tank to the atmosphere. Two vents will serve to the same advantage as that described for the interceptor.

A test line is extended to grade level. When a rod is inserted through it, the liquid level is indicated. The line is also used for sucking out the contents of the tank.

It is very important that the proposed installation be discussed with the local plumbing authorities to ascertain their exact requirements. The capacities of the interceptor and the receiver tank are determined on the basis of the amount of water that will be discharged into the floor drains.

CHAPTER 28

Waste Systems for Laundry Machines

Drainage systems normally used in the laundrette or commercial type laundry establishment are discussed in this chapter. A typical laundrette is illustrated in Fig. 1. Waste piping is shown, including a lint interceptor and a running trap in a concrete pit. The specific roughing arrangement depends upon the layout of equipment and local code requirements.

Drainage Piping Requirements

Three roughing methods, Fig. 2, can be followed for the installation of the drainage lines from the laundry machines. There are other methods.

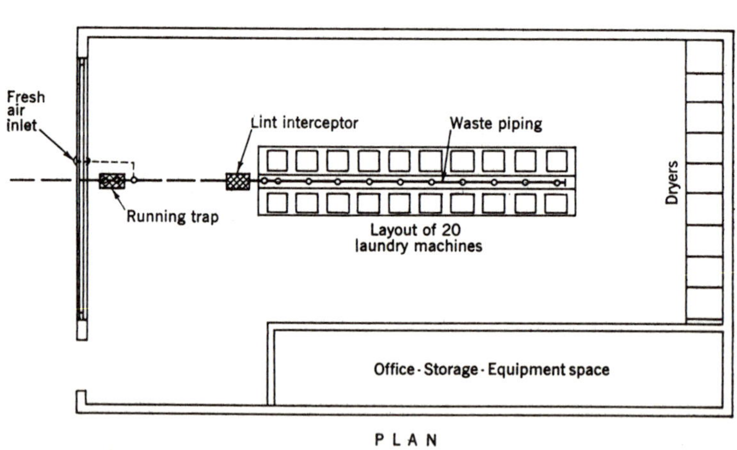

Fig. 1. Typical arrangement of fixtures, including twenty laundry machines, dryers, and basic waste piping, for a laundrette.

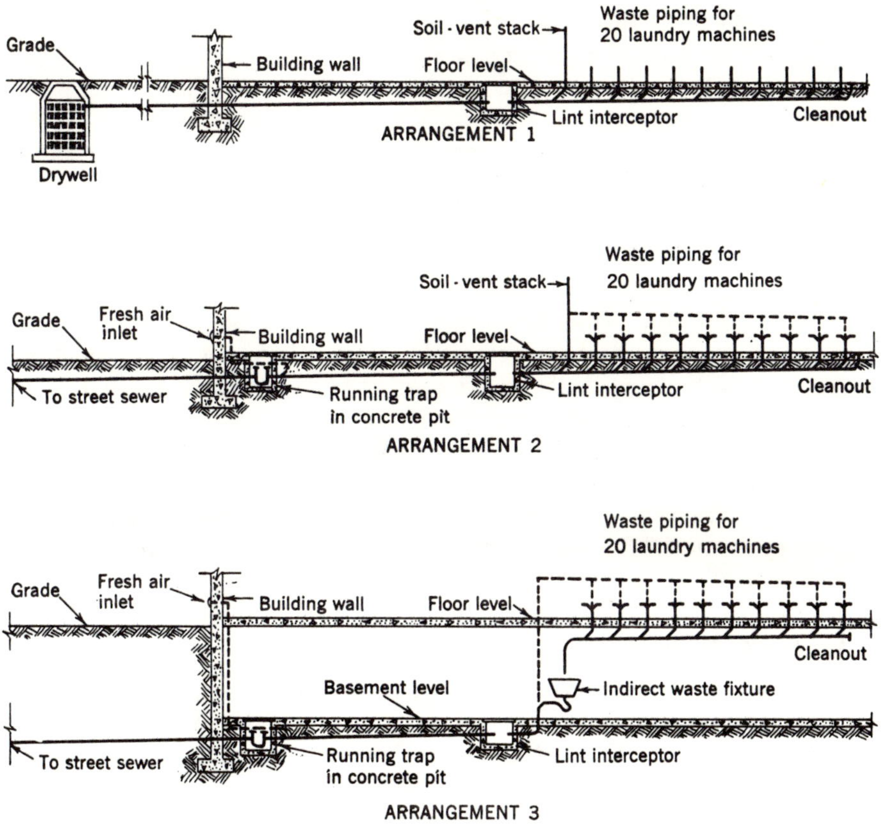

Fig. 2. Three arrangements of waste piping for the laundry machines shown in Fig. 1: discharge into a dry well, to exterior street facilities, and to basement drainage system.

Arrangement 1 indicates how the drainage line can be installed when the waste is discharged into a dry well. Note that the waste first passes through a lint interceptor before it passes on to the dry well. Connections for the laundry machines indicate a standpipe waste connection, but it can be any type favored by the engineer. Furthermore, this arrangement, illustrated in connection with a ground floor installation, is also typical of upper or lower floor installations. The dry well or dry wells should be of a size and capacity that can handle the flow with regard to the specific percolation rate of the soil in which the wells are constructed. When properly designed and installed, this arrangement will function with the most satisfactory results.

In Arrangement 2, waste is conducted directly to the exterior street facilities. In this instance, indirect waste connections must be employed at the sanitary TY connections shown. The TY's have been individually vented. The waste then passes through the lint interceptor, the running trap and then on to the exterior street sewer facilities. This system, too, when properly designed and installed, will work very well.

Arrangement 3 shows how the drainage piping can be handled when the laundry machines are installed at ground floor level where a basement exists below this level. Here you will note that sanitary TY's are again used and are individually vented. The rubber discharge hose from the laundry machines may be connected directly to the TY's, and the waste piping is indirectly run to the fixture or receptacle located in the basement, as shown. This fixture must be properly vented. The waste water then passes through the lint interceptor and through the running trap on its way to the outside street sewer facilities.

Laundry Machine Connections

Waste from laundry machines should not be directly connected to a sanitary system. This precaution must be taken to prevent contamination of the clothes in the laundry machine by a backflow in the sanitary system. Some of the methods which may be employed to accomplish a safe sanitary connection are shown in Fig. 3.

Detail *A* of Fig. 3 illustrates a direct connection to the waste piping. The rubber hose is connected directly to the Y fitting and this fitting is vented. When this type connection is used, the waste piping must discharge into a fixture that is properly vented and trapped. The waste piping in this instance may be regarded as an indirect waste system.

Detail *B* shows a direct connection to the waste line; here a sanitary TY is used. The waste piping must discharge into a fixture that is properly vented and trapped. The waste piping in this instance may also be regarded as an indirect waste system.

Detail *C* shows an indirect waste connection to the waste line. Here, the rubber waste hose discharges into a funnel which is properly vented and trapped. A sufficient air gap should exist

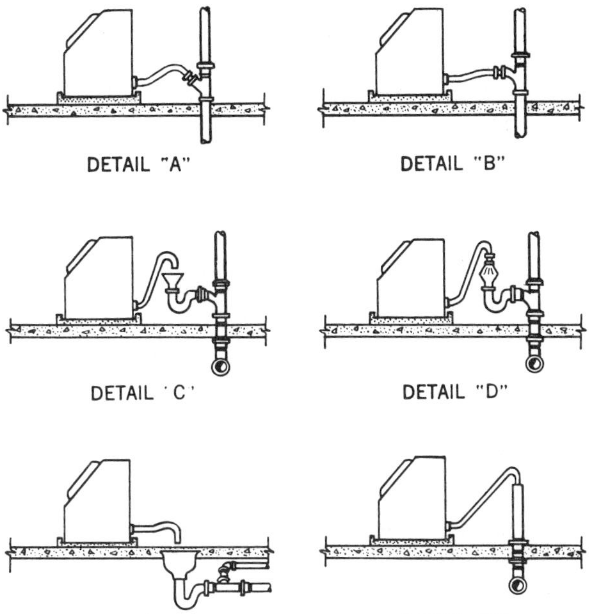

Fig. 3. Six ways to connect a washing machine to the waste system.

between the hose termination and the funnel to safeguard against back siphonage. There are many ways of supporting the rubber hose above the funnel.

Detail *D* also shows an indirect waste connection. Here, the rubber hose is connected to a manufactured indirect waste fitting, which is properly vented and trapped. Open slots in the upper portion of the fitting provide the air gap needed to safeguard against back siphonage. The system in this case is the sanitary system of waste piping while the waste connection is indirect.

Detail *E* shows an indirect waste connection. The rubber waste hose discharges into an open drain which is properly vented and trapped. Several different types of drains are available for this purpose. In details *A* through *E* the drain is part of the sanitary system.

Detail *F* shows a connection to a standpipe waste system. The rubber waste hose in this instance is inserted into the standpipe and must be regarded as a direct waste connection. The drainage thereby must serve as an indirect waste system in that it shall discharge into a fixture that is properly vented and trapped. If the waste piping is intended to serve as the sanitary system, then the rubber waste hose must terminate above the standpipe. Funnels are often employed in this case to avoid splashing when the waste is discharged from the laundry machine. Therefore, this method may serve two purposes, depending upon the manner in which the connections are made.

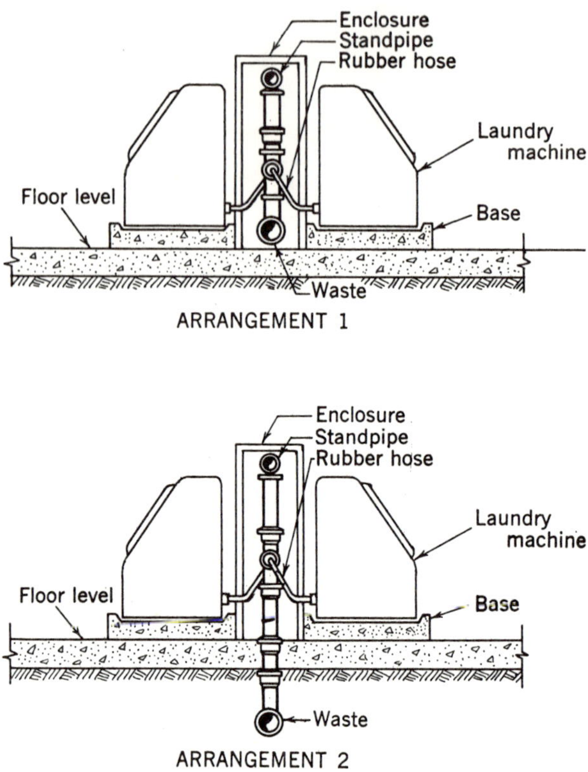

Fig. 4. Waste connections for back-to-back units.

Back-to-Back Installation

Figure 4, a section through the battery of laundry machines shown in Fig. 1, illustrates a typical roughing which may be utilized for two laundry machines that are installed back-to-back.

In Arrangement 1 the roughing is installed above the floor slab.

A double sanitary TY is used and the rubber waste hoses are shown as being directly connected. This means that the waste line must discharge into a fixture that is properly vented and trapped. If the houses are to be indirectly connected, one of the indirect means of connection as shown in Fig. 3 must be used. The roughing is concealed in an enclosure as illustrated.

Arrangement 2 is the same as Arrangement 1 except that the horizontal waste line is installed below the floor. The hook-up regarding indirect waste connections is the same as that described in Arrangement 1.

In both cases, ground floor installations have been shown. However, the arrangements can be applied to upper or lower floor slabs, as well. When the waste piping for the ground floor slab application is to be an indirect system of piping, the waste must be discharged into a drain or pit that is properly vented or trapped. A sump pump installation can also be used.

Installation of Roughing

Figure 5 shows the elevation of roughing illustrated in Fig. 4.

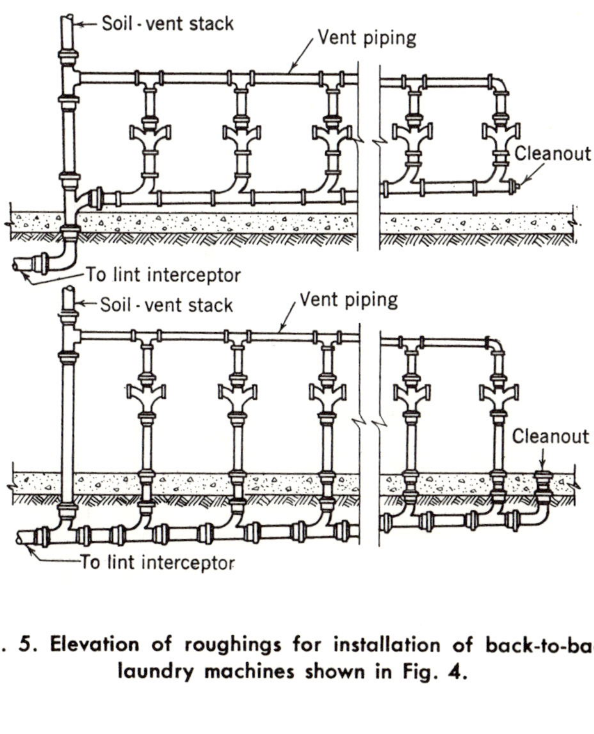

Fig. 5. Elevation of roughings for installation of back-to-back laundry machines shown in Fig. 4.

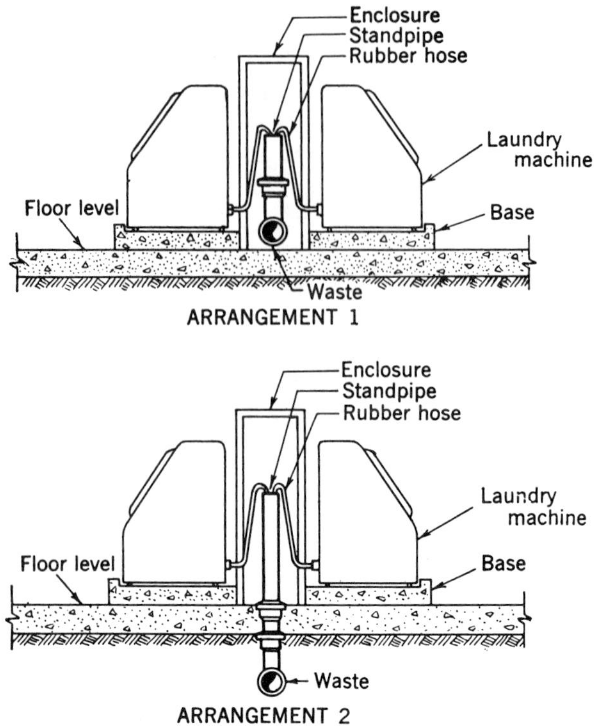

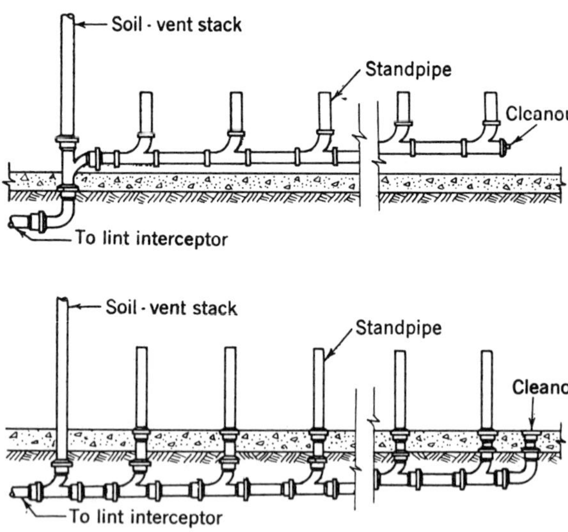

Fig. 6. (top). Roughing and (Fig. 7) elevation of standpipe systems.

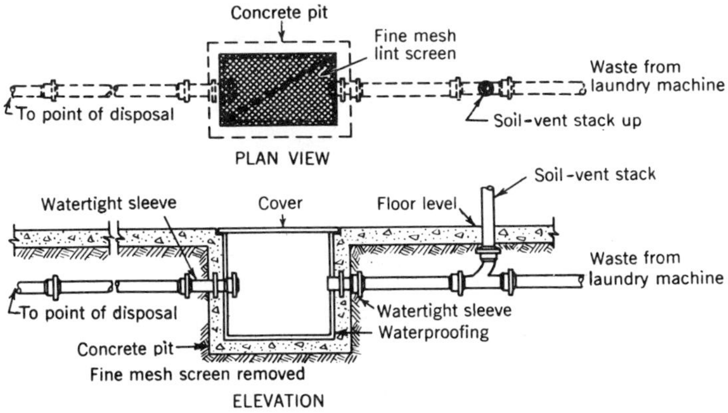

Fig. 8. Lint interceptor with removable screen.

Figures 6 and 7 are analagous to Fig. 4 and 5, except that the standpipe method of roughing is employed.

Lint Interceptors

The lint that is discharged into the laundry drainage system can be troublesome when the waste is discharged into dry wells, for it will impede waste flow from drywell to soil. Many local codes require that lint be intercepted to prevent its discharge into exterior street sewer facilities. Different types of units are available for this in a wide range of sizes, constructed of cast iron, steel or concrete.

Figure 8 shows a typical lint trap. It consists of a watertight concrete pit in which a fine mesh removable screen baffle is placed at an angle in the pit to offer a broader surface for intercepting the lint. The size of the pit is determined on the basis of the number of laundry machines involved. The cover is generally a steel plate secured in place with a gasket between the cover and frame. The sizing, placement and construction of the interceptor for lint must conform to local codes.

Part 3
Building Water Systems

CHAPTER 29

Water Supply for Construction Projects

An adequate water supply is an important need for every construction project for it is required by each of the contractors and serves many purposes. In addition to serving various phases of the work, it is also required for drinking. Therefore, the water supply should be of the highest quality to meet health standards and it should be in abundant supply to serve the demands of the project. The temporary water piping used should be clean and carefully maintained to assure proper supply and avoidance of contamination dangers.

Responsibility for Installing Water Supply

Unless the responsibility for furnishing the water supply is included in the specifications, to be a part of the plumbing contractors' work, it must be handled by others. Often, this work is included in the general contractor's specification. However, he will usually sub-contract it to the plumbing contractor. In any event, the water supply piping and necessary waste system must be properly installed and maintained during the course of the construction project to assure clean and adequate water.

Although the water supply is but a relatively small consideration in any construction project, it nevertheless should be fully covered when the specifications are prepared. This means that the engineer's specifications should contain a paragraph that informs the plumbing contractor that the plumber is to furnish and maintain the proper system.

If this work is to be contained in the architect's specifications, then the engineer should check with the architect to be sure that this phase of the work is sufficiently covered in his specifications. In this way, the engineer is sure that the water supply system is installed in a manner befitting the requirements of a large-scale project.

Misuse of Other Piping Systems

Fortunately, most plumbing contractors will provide an appropriate water supply system even though the scope of work may not be fully covered in the specifications. Unfortunately, however, every now and then when the work is not adequately covered in the specifications, an attempt is made to economize. Then instead of furnishing a separate water system, other piping systems are utilized to accomplish the purpose.

A fire standpipe line is always required in every construction project, with hose stations on each floor of the building. While it would seem perfectly feasible to utilize this line for the water supply in addition to its function for fire protection purposes, this is a dangerous practice for two important reasons:

1. A fire standpipe is intended for fire protection only and should not be used for any other purpose. Once this line is tapped for water supply, its usefulness is destroyed in the event of a fire. Should there be a fire outbreak, all the available water pressure is needed to fight the fire. If the line is being used to furnish water for the construction needs at the various building floors, sufficient pressure will not be available to fight the fire. Besides, in the case of fire, there is not enough time to shut off the supply at each floor to ensure adequate water.

2. The piping specified for this system is usually of black steel or similar material. While this type material is suitable for this purpose, it is not recommended for supplying running water during the course of construction. Running water in this line will cause rapid corrosion of the piping interior and in addition, will cause tuberculation. Tubercles reduce the effective open area of the piping and the increase of friction on the inner lining of the piping due to the corrosion.

A fire standpipe system is an important asset in a completed building and should not have its capacity reduced or its interior encrusted with friction-producing tubercles, just because this line was selected to fulfill the construction needs of the project. If a proper water supply line is installed, this misuse of the fire line will not occur.

Offhand, another piping system that would appear likely for the water supply needs is the gas line. This is a convenient pipe line because it is installed as a vertical riser with a gas connection at each floor. By placing a valve at each connection, it could also deliver water.

However, here again black steel pipe is usually specified as it is not intended for water supply use. When this line is used for running water, tubercles and rust will quickly form.

The result of this misuse is that when the gas line is finally placed in service, its effective open area will be greatly reduced. Friction caused by the rust and tubercles will reduce the gas flow rate. This, too, could have an effect on the amount of gas that reaches the upper story fixtures at a later date when the gas line is placed into service. The rusting of joints, caused by using the line as a water line lowers the piping strength and presents a potential hazard in the form of gas leakage. Moreover, the rust and other solids will settle in the line and will fill the gas drip pockets. Some particles may enter the delicate controls and equipment where they will cause considerable difficulty. Gas lines are intended only to convey gas.

Installation of Water Supply Systems

Even though the water supply and its component waste piping are installed as a temporary system that will be dismantled after the construction work has been completed, such piping should be properly installed and maintained. A typical arrangement of such a system is shown in Fig. 1.

This figure shows that the water and waste piping extend from the lowest level of the building to the highest with water facilities and waste connections at each floor level. Often, two or more systems may be required, depending upon the size of the project. Proper connections must be made to the underground water and sewage systems. The waste system can be connected to the sanitary or storm sewers, or disposal may be made by any other means approved by pertinent local codes.

Pipe for the water line is usually copper or other material recommended to meet the particular characteristics of the water. If the pipe has been used previously, extreme care should be taken to ensure that the pipe was cleaned for the new application. Since the water is to be used for drinking, it may be necessary to sterilize the pipe line before it is placed in service. This can be accomplished by adding a prescribed chemical solution containing chlorine which is permitted to remain in the line long enough for complete sterilization of the piping. When pipe is used to convey water for drinking purposes as well as for the construction needs, every care should be followed to assure a clean supply.

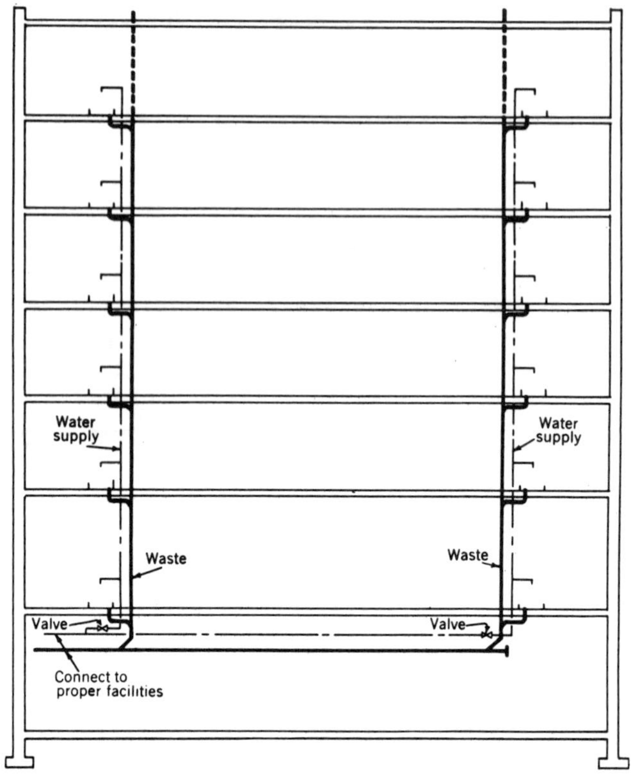

Fig. 1. Water supply and waste for a typical project.

Water Supply and Waste Connections

There are any number of ways in which the water supply and waste connections can be arranged at each floor level. Each plumbing contractor has his own particular method which he uses. Figure 2 shows an arrangement that has been used on many construction projects.

This arrangement consists of a water supply, waste piping, a valve, vacuum breaker, hose bibb, a pail and a shallow pan with its waste connected to the waste line. Both the water and the waste line should be sized in accordance with the number of outlets that may be utilized in the building at one time. A safe method to use in sizing this line is to figure at least two-thirds of the outlets being used at one time. It is a recommended practice to install vacuum breakers at each of the water connections. This will prevent the damage that

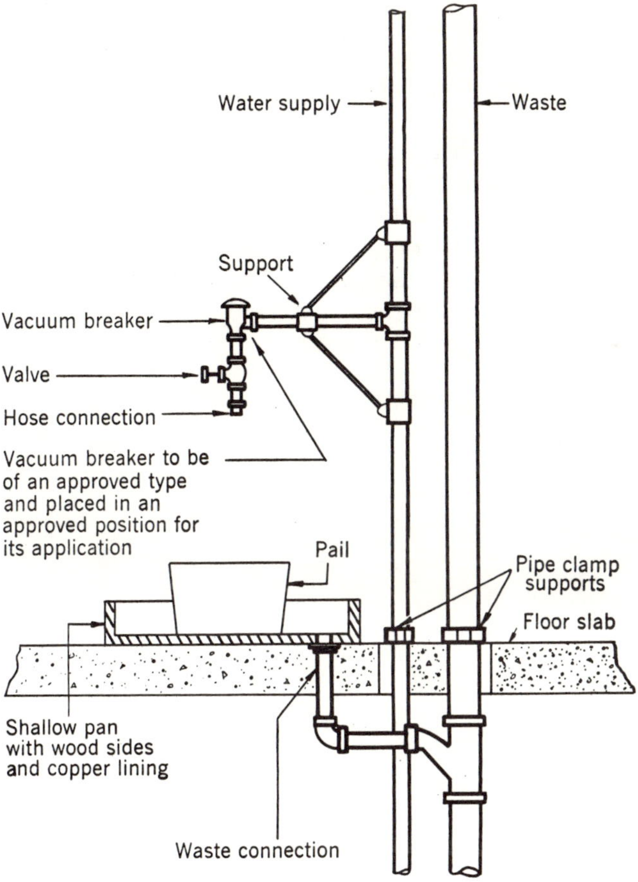

Fig. 2. Detail of water supply and waste connections.

back siphonage can cause in the event of a drop in water pressure. Should the water pressure be insufficient to furnish the proper amount of water at the highest building floors, then an approved method with mechanical means should be employed to boost the water pressure to the desired limits.

If the project is being erected in climates subjected to freezing weather, it is necessary to prevent damage from freezing by wrapping the pipe with an insulating covering. Here again, each contractor has his own particular method he favors to safeguard the piping.

The object of the pail is to intercept the water and avoid water splashing into the pan. Water splashing on the concrete slab develops a slippery condition that could cause accidents to construction workers.

Proper supports are a very necessary piping provision. Such supports should be placed on the vertical risers, as shown, to rigidly secure the pipe in position.

Support may also be required for the horizontal branch to the hose bibb, as shown, to keep this line from being bent out of place. The support also serves to hold this line in place when the construction hoses are pulled to the various floor areas.

CHAPTER 30

"Wet Columns" in Multi-Story Buildings

Multi-story buildings, such as large office buildings, are unlike most similar structures in that they are designed for varying types of occupancy. Other buildings, in general, are designed for a specific type of use which will more or less remain the same for the life of the building. And although alterations, after a period of years, might be made to this structure, the occupancy will remain the same.

Multi-story buildings are especially designed to permit different types of occupancy or a change in the initial occupancy. For example, when the building is first planned, an insurance firm may sign a long lease for the entire fourth floor space. Therefore, this particular floor is laid out and arranged for the convenience of the insurance company. With the passage of years, the insurance company may decide to terminate its lease and relocate elsewhere, leaving the owner of the building with an entire floor to rent.

If the owner finds it too difficult to rent this entire floor he may subdivide this space to suit two or more tenants. To accomplish this, it is often necessary to erect partition walls adequately to divide the space for the tenants' needs. This is the reason why office buildings are different from all other building types. Each floor area is so designed to include the special arrangement or space subdivision. The speedy erection and removal of partitions is an asset to the owner and is a procedure that must be considered to suit the changing needs. Flexibility in arrangement of the spaces is one of the key factors to be considered when a multi-story office building design is contemplated. It is this flexibility that insures the owner an economical and modern building.

Water and Waste Facilities

Toilet rooms originally provided on each floor of the building will not alone suffice for the individual needs of certain tenants. Therefore, consideration must be given to water and waste connections to be installed in specific locations on each floor, preferably at the concrete or building columns. Such facilities will permit the easy hook-up of water and waste to fixtures desired by the tenant. Figure 1 shows a typical layout of the floor area including the toilet rooms. In addition, the various columns have been spotted to show their proximity to the rentable space. Certain columns were selected to include water and waste facilities and these are known as "wet columns."

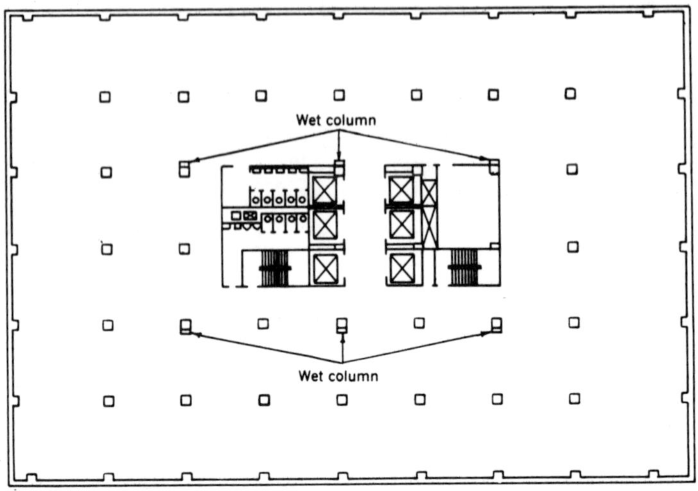

Fig. 1. Typical arrangement of floor area in a multi-story building, showing location of "wet columns."

These particular columns were chosen because they are the ones most favorably situated for the overall space conditions. Moreover, it is at these columns where partitions are most generally erected. Water and waste facilities at such locations offer facilities that can be readily adapted to satisfy any desired arrangement of floor space or partitions. If necessary, this piping can be arranged to suit as many as four adjoining office spaces.

This is the type flexibility needed in multi-story buildings to serve the needs of the tenants and the expected functions required by the owner. It is not too difficult to determine just which columns would be ideally suited for this purpose and the engineer on cooperation with the architect can usually work out the most practical locations. It is possible that the owner may have definite thoughts on this subject.

Type of Facilities

To provide adequately for the ultimate number of fixtures installed, the engineer should first make sure that the facilities are properly arranged.

Figure 2 illustrates a typical layout of the required facilities. Five lines are shown — a vent line, waste line, cold water line, hot water line and a hot water circulating line. All of these lines are plugged as shown. When these connections are to be tied in, the plugs are removed and the waste, vent and water lines extended to the fixture to be served. It is good practice to consider the use of plugs that can be removed easily. Water lines are equipped with valves that are plugged to assure positively against leakage. The valves can be either exposed or concealed in the pipe space. Since it is generally considered better practice to conceal the valves, it is also good design to include an access door to gain entry to the valves. When valves are concealed behind a locked access door, there is little chance for tenants to tamper with the valves.

The arrangement of facilities in Fig. 2 shows an upfeed cold and hot water system of piping. This arrangement is for the purpose of example only. If the building were any taller, a downfeed system of cold and hot water piping would be in order. In general, these lines are arranged in accordance with the particular water system design that the engineer contemplates using for the special type and size of multi-story building.

If a downfeed system of cold and hot water is planned, then the hot water circulation line is not needed. In any event, sufficient room must be provided in the pipe space at certain floor levels for the installation of sectional valves to control the water in the vertical risers as well as for pressure reducing valves that are required for the vertical risers in tall buildings.

All of these conditions must be carefully studied during the design stages of the project.

Spacing for Facilities

Proper spacing for the facilities is a factor that also requires study. This is quite an important design phase because it determines the amount of room required in the pipe space for the proper installation of the vertical lines as well as for the connections provided on the risers. Since the columns are limited as to size, a little study is needed for the proper placement of the lines in the space allowed. Both the architect and the structural engineer must know the amount of room

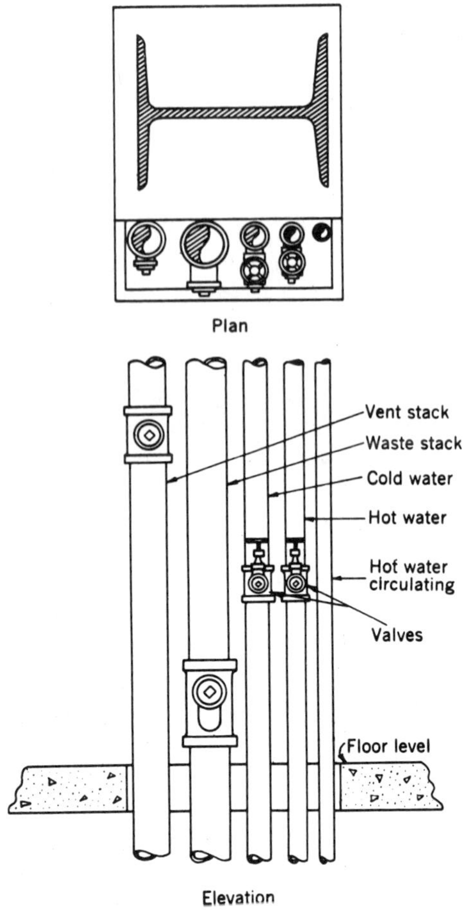

Fig. 2. Plan and elevation of typical arrangement of facilities at a wet column.

to allow at the columns for the installation of the necessary facilities.

Another important consideration is provision for the passage of piping through the slab. This is accomplished by either a slot or individual holes cut in the slab. The structural engineer should be consulted on this detail to avoid weakening the slab. The particular type of slab construction used determines whether proper reinforcing must be provided. Reinforced concrete slabs, beam and arch slabs and preformed steel slabs all require special attention to insure adequate slab strength.

Size of Facilities

The size of the vertical risers and the individual connections provided at each floor depend upon the greatest anticipated fixture load expected on each floor and the height of the building. If only drinking fountains are to be installed at each floor, it is necessary to know the number that will be connected at each floor level. If several fixtures, including lavatories or small sinks, are to be installed, the piping must be sized for this enlarged capacity.

Water closets are not usually installed at wet columns; however, if the owner feels that he would like to include water closets at some later date, then it is necessary for the engineer to size his system accordingly. The wet columns must be designed to meet the present and future ultimate needs of the specific installation. Each project is an individual one and demands individual treatment. The size of the vertical waste, vent and water risers, therefore must be sized to conform with the particular needs of the project as well as with the requirements of the governing local code. Only in this manner will the proper system of facilities be included.

Vertical Arrangement of Piping

Facilities that comprise wet columns must be handled in the same manner as the sanitary system that serves the general building fixtures.

Figure 3 is a schematic diagram of a typical arrangement of wet columns. On this illustration are the water, waste and vent lines. They run from the highest story of the building down to the level where they connect with the horizontal water and drainage lines in the building. For the purpose of example, a 12-story building was used. If the building were any higher, the system would be of the downfeed type and the hot water circulating line deleted. The water system design selected depends upon the engineer.

For the system shown, note that valves are placed at the base of the cold, hot and circulating risers. These control the risers themselves. If it is desired, because of the height of water pressure in the building, to sectionalize the water supply, then valves could be placed at, say, the fifth story to control the upper stories. If such control valves are needed, the space within the wet column should be large enough to contain them.

The same attention must be given to the piping contained in the wet columns as is given to the piping systems. Therefore, supports should be placed at least at every other floor as shown in Fig. 4.

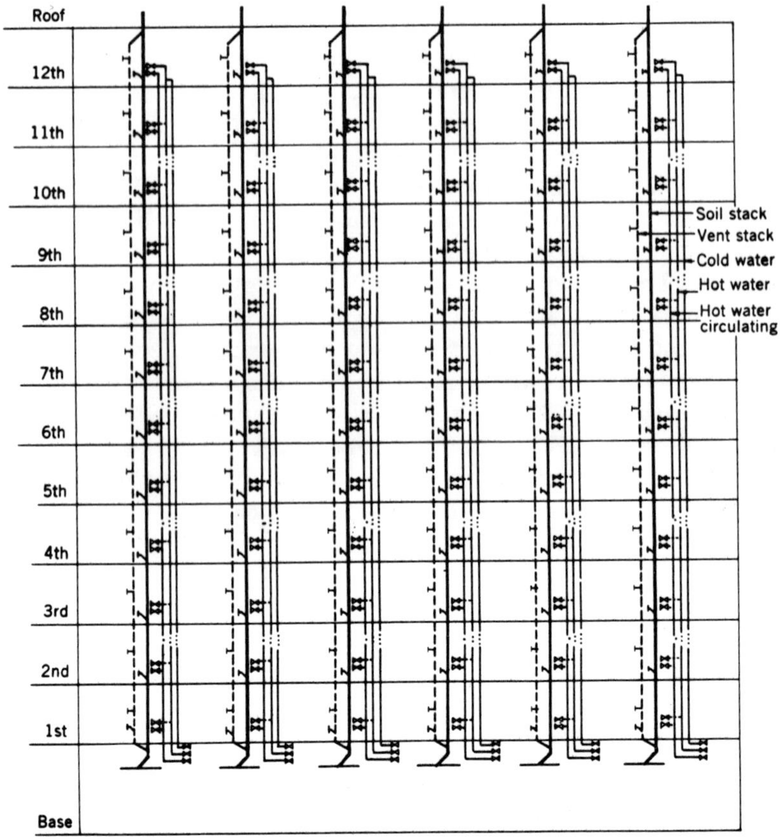

Fig. 3. Riser diagram showing arrangement of facilities in a wet column.

Facilities are shown in Fig. 4 passing through holes cut in the slab. Therefore pipe clamps are used to support the individual lines. The pipe clamps generally are placed at every second or third floor level to carry the weight of the piping, dependent of course on the diameter and weight of the piping itself. If slots in the slab are provided for the passage of the piping, then pipe clamps would again be used, except that the clamps would be so placed to straddle the slot. In any event, it is very important that some means of support be used in order to prevent the stress due to the weight of the main piping from being transmitted to the branch piping.

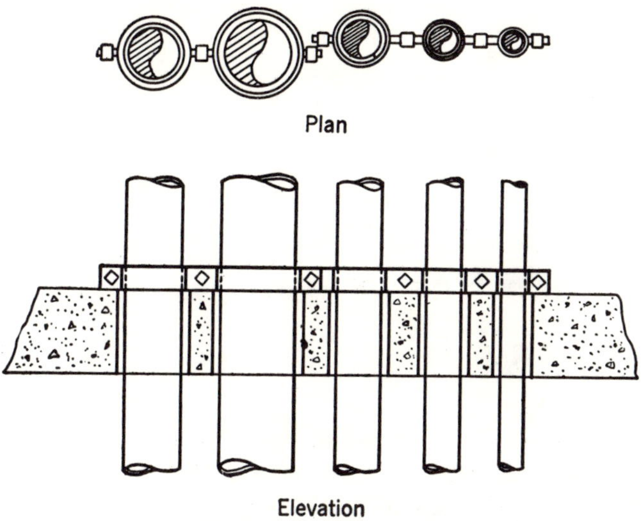

Fig. 4. Supports for pipes passing through slab.

Piping Material

A selection of materials is available for the waste, vent and water piping. Choice of type and material for water piping depends upon the particular characteristics of the water. The waste and vent stacks can be cast iron, galvanized wrought iron, or copper tubing. The relative economy of the project, as well as the expected life of the building, are important factors to guide the engineer in determining the material he will use. In every case, the matter should be carefully studied so that the system will enjoy long life and trouble-free operation.

CHAPTER 31

Service Hot Water for Schools

An abundant supply of service hot water is an absolute necessity in every school. This need has become of still greater importance with the more general usage of the school.

Over a period of years, the hot water requirements for the average school have been steadily increasing. In the past, the hot water was furnished only to the lavatories in the toilet rooms and to the slop sinks on each floor. As the design and function of the school changed so has the hot water demand. At present considerable amounts of hot water are needed for showers, kitchens, laboratories, special classrooms and training rooms as well as for many other varied uses. This demand exists from September through June, which is the normal school year in most municipalities. Moreover, the modern school has departed from its conventional two-term use. It now serves as a community center, it is used for evening and summer events, basketball, football, baseball games as well as other sporting events. In addition, many schools sponsor evening adult education classes. This use continues throughout the year so that the school has become the accepted place for community dances, sporting events, plays, lectures and a local meeting place for civil discussions and affairs. Therefore a good deal of thought should be given to the service hot water to afford a system that is flexible enough to satisfy the varying requirements.

Factors That Determine Service Hot Water Needs

Since a number of different methods may be employed to generate the service hot water it is well to first determine the overall needs of the school. These needs are usually determined in the following manner at the time the plans of the school are first being formulated.

1. Initial discussion between the engineer and architect establish the school's intended function and the amount and type of fixtures and equipment to be served with hot water.

2. Early discussions between the engineer, architect and school board help to establish the contemplated use of the school's facilities including evenings and summer application. Any known increased activity in the near future that will affect the design and installation of the service hot water system should be considered in advance. It is better judgment to make plans for possible additional needs than to make costly alterations after a few years to fulfill the increased capacity needs.

3. After the average, peak, and future demands have been determined, it becomes necessary to provide a system that will satisfy these needs in the most economical manner.

Hot Water Generation Equipment

Economy in usage of one fuel over another because it is abundantly available is usually a decisive factor in selection of the particular type of heating means to be used. Fuel is important in the warmer areas of the country where it is used mainly for the generation of hot water. It is even more important in colder areas where it is used both for space heating and for the generation of service hot water. In this case, the selection of fuel to be utilized is generally the responsibility of the heating engineer and contractor.

Where the space heating plant is to be used for the generation of hot water, submerged coils, indirect heaters, tankless heaters, and hot water storage generators may be used to advantage.

For example, assume that a hot water storage generator, similar to that shown in Fig. 1, has been selected for application in the school.

This type of system is practical and is rather widely used. Steam of 5 or 10 pounds from the heating plant passes through internal coils in the storage tank to transmit heat to the water.

A mixing valve is installed at the tank to ensure controlled temperature of hot water circulating through the system to the fixtures and equipment. The purpose of the small diameter by-pass line is to permit return hot water to reenter the distribution system when the hot water port in the mixing valve is closed due to the temperature of the water from the storage generator.

The various thermometers shown offer a quick reading of water

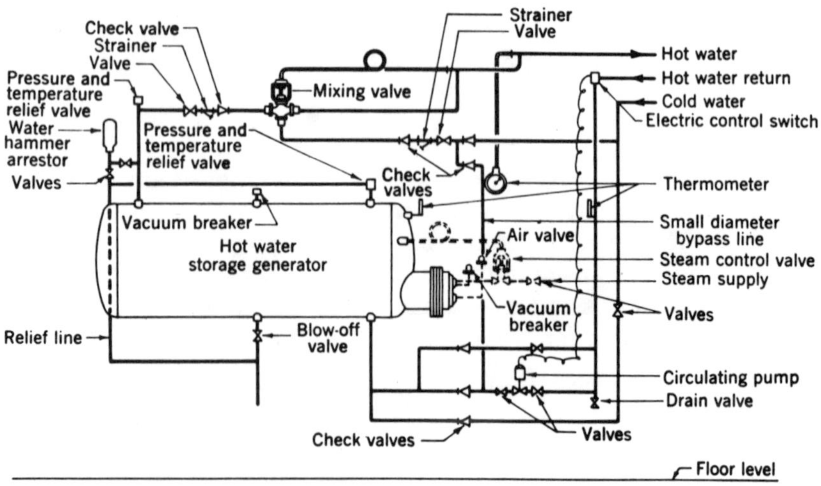

Fig. 1. Arrangement of hot water storage generator and piping to furnish a circulated system of hot water at constant temperature.

temperatures in various parts of the system so that the maintenance men can adjust for the desired water temperature in the system.

The pressure and temperature relief devices safeguard the storage generator and component piping in the event of excessive water pressures and temperatures.

Dual Hot Water Storage Generators

Very often it is advisable to offer a more flexible means of hot water generation in schools. This is accomplished by providing two hot water storage generators instead of one. A dual system permits one unit to be operated when water demand is reduced as during the summer months. A dual system also permits one unit to be operated while the other is being maintained. The dual system should be given serious consideration since it offers many advantages in a school. Although two tanks are required they do not occupy much more overall space than just one tank.

Figure 1 illustrates a hot water storage generator that depends upon steam from the space heating plant for its operation. In most areas, however, the heating season lasts from October to April, after which the heating plant is shut down for the duration of the warmer weather. The heating plant can of course be utilized to furnish steam to the hot water storage generator during this period if it is sc desired. Installation of special devices and control valves will permit the heating plant to function for the service hot water without heating the building.

Supplementary Heaters

If the school desires to shut down the heating plant during the summer months, then a separate boiler or supplementary heater must be provided as illustrated in Fig. 2.

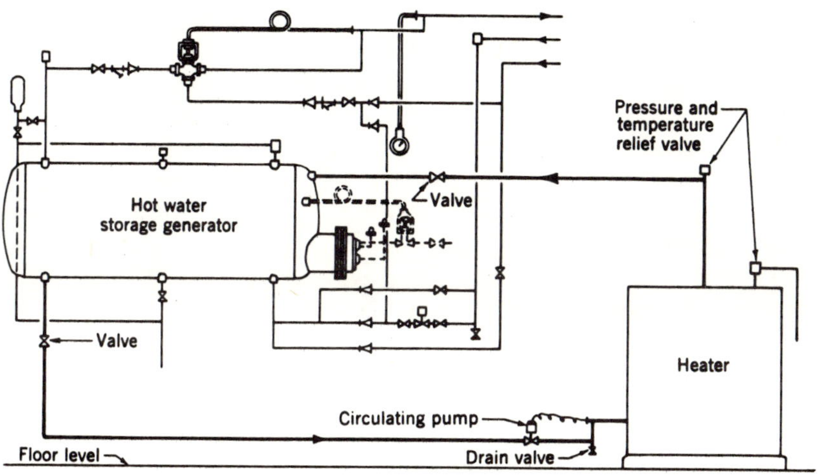

Fig. 2. Diagrammatic arrangement of hot water storage generator, heater, and component piping.

In Fig. 2, the normal circuit of piping as required for the hot water storage generator is shown in light lines while the circuit of piping for the supplementary heater is shown with heavy lines. Supplementary heaters may be gas, oil, coal or electric type and if more than one heater is required the heater supply and return mains can be combined for satisfactory operation of the system. This means that the supply headers of the heaters are joined together to function as one and the return headers are joined together for the same purpose.

Using one heater for example, by closing the two valves shown on the supplementary heater piping the system can be operated during the heating season with the heater shut down. Then during the summer the valves are opened and the heater is utilized. Additional economy in operation can be accomplished during the summer months by closing the hot and circulating valves that govern the hot water supply to those areas of the building that will not be in use.

When dual hot water storage generators are used, the supplementary heater piping is so connected that either one or both of the generators can be operated at one time.

A circulation pump is generally installed on the heater circuit of piping as indicated to ensure a positive circulation of water between the storage generator and the heater.

Two Temperatures of Hot Water

Where large kitchens are required for school cafeterias, an abundant supply of hot water in excess of 180-deg. is needed. The water can, of course, be heated to the desired temperature by means of supplementary heaters in the kitchen itself. Nevertheless, it is a generally accepted practice to produce the water in the storage generator as suggested by Fig. 3.

In Fig. 3, note that the 180-deg. water circulates through its own system of piping to the kitchen fixtures while the 140-deg. water

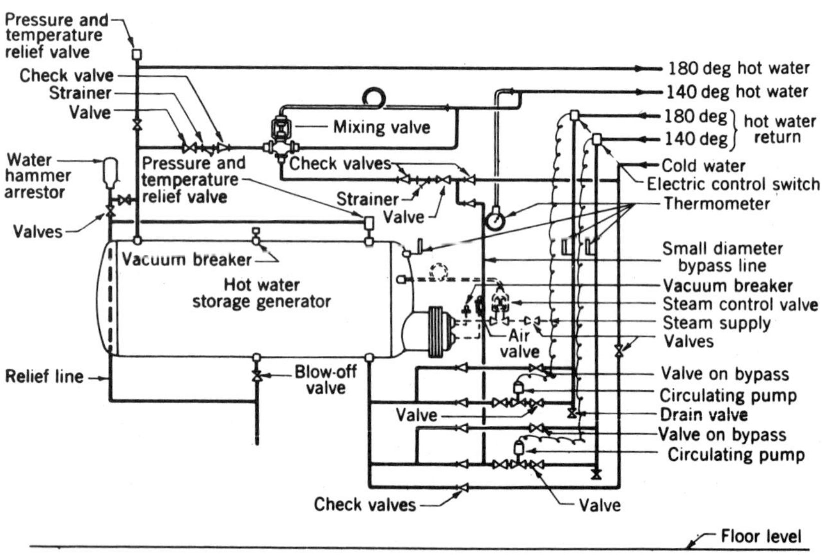

Fig. 3. Diagrammatic arrangement of hot water storage generator and piping which shows how two temperatures of hot water can be furnished.

circulates through its system of piping. If the heating plant is operated year round then this system is adequate as shown. If the heating plant is shut down during the summer then a supplementary heater of adequate capacity is required to generate the hot water. It is timely to mention here that the storage generator serves as a storage tank when a supplementary heater is used. Here again it is good practice to shut off the hot and circulating valves for those sections of the building that will not be used.

Also the 180-deg. water circuit may be shut down since the kitchens will be inoperative during the summer months. If dual hot water storage generators are used the hook-up of piping may be similar to that described for Fig. 2.

Summary

A hot water storage generator is a very flexible means of generating service hot water and may be depended upon to serve in a most efficient and economical manner. With just a little planning the system can be modified to satisfy any specific need in a modern school.

CHAPTER 32

Two-Temperature Service Hot Water

All new buildings under construction require a service hot water system of sufficient capacity and recovery to fulfill the hot water demands by the building occupants; an adequate service hot water system is a very important asset in all types of buildings.

Often, however, more than one temperature of service hot water may be required to suit the particular requirements of a building. Besides furnishing hot water for the general fixtures of a building, it is frequently necessary to supply 180-deg. hot water for certain kitchen fixtures. In hospitals it is not unusual to furnish an additional hot water system which supplies water at a temperature different from the main system for special equipment. Schools, for example, require 180-deg. water for kitchen fixtures plus the regular service hot water system. In general, where a building may require service hot water at more than one temperature, consideration must be given to the various methods that can be employed to furnish water at the needed temperatures before deciding on the system.

Methods of Furnishing Required Temperatures

There are many ways to generate service hot water at varying temperatures and different combinations of hot water generation equipment may be used. However, this chapter discusses only the hot water storage generator and the manner in which it may be used to supply the necessary hot water temperatures. Regardless of the particular method of hot water generation chosen, a separate system of hot water piping must be installed for each water temperature.

Hot Water Storage Generator

The hot water storage generator is one of the more efficient methods that can be used to generate hot water. With the proper arrangement of piping, valves and controls, a dual system of service

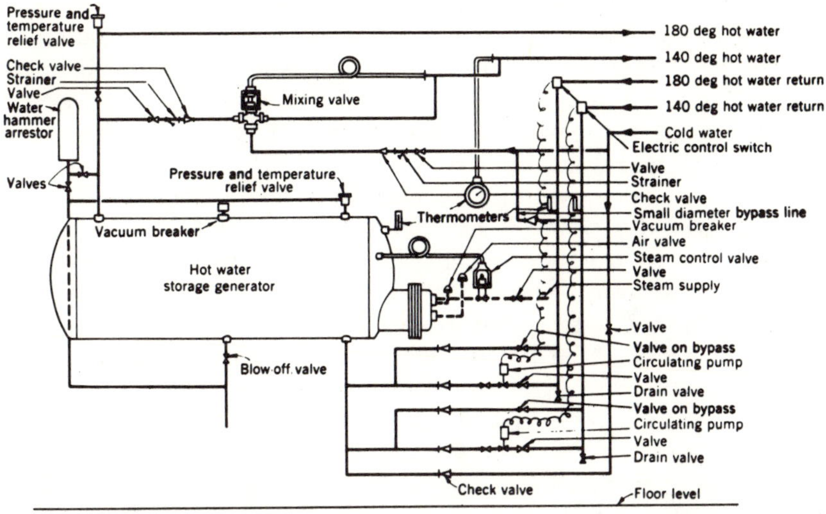

Fig. 1. Detail of hot water storage generator with piping arrangement for two-temperature hot water.

hot water can be provided and a typical arrangement of this system is illustrated in Fig. 1.

Steam is used as the heat source for hot water in the storage generator of Fig. 1. Steam passes through the coils within the tank and transmits its heat to the water outside the coil. The steam control, actuated by a thermal element set in the storage tank, controls the range of temperature desired in the storage generator.

In this case we shall assume that a large volume of 180-deg. water is demanded by the numerous kitchen fixtures. Therefore, all the water in the tank is heated in excess of 180-deg. and this temperature is accurately maintained by the steam control valve.

Pressure and temperature relief valves guard against scalding or equipment damage should the steam control valve fail and excessive pressures and temperature develop in the storage tank. Other safety devices of equal value and importance protect the piping system as shown in the detail.

Water in the storage tank is heated in excess of 180-deg. to compensate for heat lost in traveling through the pipe to kitchen fixtures. Thermometers set in the water line in the kitchen will help find the temperature at which the water should be regulated in the storage generator to compensate for this pipe line heat loss. After the 180-deg. water completes its circuit, it returns to the storage generator via the hot water return line. A circulating pump, installed in the return line and actuated by the electric control switch shown, speeds the movement of the water in this return line when the temperature in the hot water return pipe drops below a desired degree.

For the 140-deg. water, a connection is taken from the 180-deg. line. The 180-deg. water is reduced to 140-deg. or any other desired temperature by the mixing valve. A thermometer set in the line as shown, and located to be easily read, facilitates the proper temperature settings. The temperature setting for this supply is also set in excess to provide the desired water temperature at the fixtures; account is taken of a temperature drop of the water in its travel through the pipe. After the 140-deg. water has completed its circuit, it, too, returns to the storage generator by way of the hot water return line. A circulating pump on this line performs a similar function as on the 180-deg. line.

Bypasses around the circulating pumps permit water circulation in the event of maintenance or repair to the circulating pump. The cold water supply is introduced to the system as shown to temper the cold water with the hot return lines before it reaches the storage generator. This avoids mixing cold water directly with hot water in the tank.

Water hammer eliminators are placed in the piping system at the locations suggested to absorb shock waves set in motion by quick-closing valves and faucets. These devices avoid the noise and damage to the piping system caused by water hammer as this reaction is known. There are several excellent patented shock arrestors that may be used. Water hammer eliminators, of course, must be placed at other points in the piping system to afford complete protection to the piping system.

When this type of system is properly installed, it can be counted upon for efficient operation. Furthermore, it will offer both, a flexible as well as a safe method of furnishing two separate water systems with each service water a different temperature to meet specific building requirements.

Two Hot Water Storage Generators

Often, because of the particular hot water demands of the building, one may desire to design a system that includes two hot water storage generators. The larger storage generator will contain the

140-deg. water for the general fixtures of the building while the smaller storage generator will contain the 180-deg. water to be used expressly for kitchen fixtures and equipment or perhaps other types of equipment.

Such a system would be appropriate in a school or any other building where 140-deg. hot water is needed all year round while the 180-deg. water may be required only part of the year or at certain seasons. A typical arrangement of this system is shown in Fig. 2.

Although, the arrangement of the storages generators in Fig. 2, shows the smaller unit underneath the larger unit, it must be remembered that the illustration is only diagrammatic. The smaller unit can be placed in any position with regard to the larger unit.

For the larger unit, steam is the medium used to generate the hot water. The water is heated in excess of 140-deg., leaves the unit and by means of the mixing valve is maintained at a temperature that will deliver 140-deg. or any other temperature that may be desired for the general fixtures of the building. The mixing valve, in addition to maintaining the proper water temperatures, also prevents the passage of superheated water to the fixtures in the event that the steam control valve should fail to function.

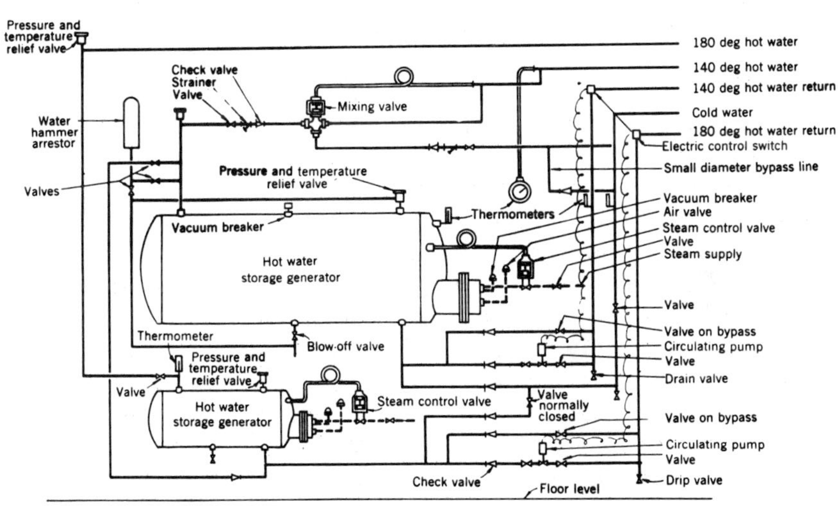

Fig. 2. Piping arrangement for two hot water storage generators and piping arrangement for two-temperature hot water.

After the 140-deg. water completes its circuit through the piping system it returns to the storage generator via the hot water return line. The circulating pump assures a positive circulation when the water in the return line drops below a prescribed temperature.

The smaller storage generator operates in a manner similar to that of the larger unit with the exception that the 140-deg. water from the larger unit is introduced into the system. The 140-deg. water is raised in excess of 180-deg. to compensate for the heat loss of the water in its travel through the piping to the kitchen fixtures and equipment.

After the 180-deg. water completes its circuit through the piping system, it returns to the smaller storage generator, via the hot water return line. The hot water return line is also equipped with a circulating pump to provide positive circulation when the water in this line drops below a prescribed temperature.

The valve on the auxiliary cold water supply to this storage generator is normally closed. It is opened only when the 140-deg. supply from the larger unit is shut off due to maintenance or repair of the larger unit.

This system is quite efficient and affords the flexibility that is needed in certain types of buildings. Since the smaller unit can be shut off during periods of the year when it is not needed and since the larger unit is required only to furnish 140-deg. water, a considerable saving is realized in the generation of hot water.

Hot Water Storage Generator and Instantaneous Heater

Another method of furnishing hot water at two different temperatures is by the arrangement of a hot water storage generator and an instantaneous heater as shown in Fig. 3.

Water at 140-deg. is generated in the storage generator while 180-deg. water is generated in the instantaneous heater. Here again is a flexible system that is ideally suited for buildings where the 140-deg. water is required the year round while the 180-deg. water might be required only at certain times of the year for special use.

The 140-deg. water leaves the storage generator, and passes through the mixing valve where its temperature is regulated for the general fixtures of the building. After the water has completed its circuit, it returns to the hot water storage generator. The circulating pump installed on the hot water return line, operates the same as that explained for the previous system.

For the 180-deg. water, a branch is taken from the 140-deg. water line, as shown, and is introduced into the instantaneous heater where

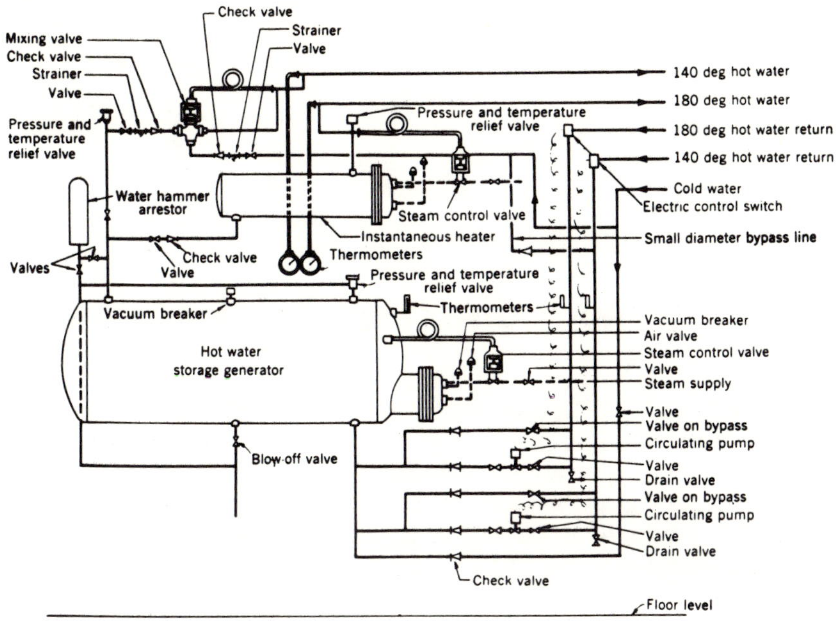

Fig. 3. Piping arrangement for hot water storage generator and instantaneous heater.

the temperature is increased to 180-deg. The temperature of this water is controlled by the steam control valve, placed on the steam supply line. After this water completes its circuit to the kitchen fixtures it returns to the storage generator. A circulating pump is placed on the hot water return line to ensure positive circulation, when required. Thermometers for the 140- and 180-deg. water, assist in maintaining the proper water temperatures.

With a system such as shown in Fig. 3, a great deal of flexibility is assured for the particular building requirements.

The systems described are economical in operation in that they offer full utilization of fuel and service. Manufacturers of the proposed equipment should be consulted with regard to the capacity, recommended installation and arrangement of their equipment.

CHAPTER 33

External Indirect Water Heaters

There are numerous methods and varying types of equipment that may be utilized for the generation of service hot water in a private residence or a moderately sized building. One means that has been employed with a good deal of success is the external indirect heater. An example of the equipment and piping arrangement for this type of system is shown in Fig. 1 and Fig. 2.

In these illustrations, it will be noted that one system uses a steam boiler while the other uses a hot water heating boiler. Both systems are similar in that they use boiler water for the heating medium, even the piping arrangements are similar with the exception of the indirect heaters which are connected to the heating plants in a manner that best utilizes the output of the heating plants.

The prime advantage of the systems shown is that since they depend upon boiler water for heat generation, a year-round supply of hot water in abundant quantities and at relatively low cost is possible. Thus, supplementary heaters for use during the summer months are not required. The heating boilers shown may be fired by gas, oil, coal, or other means.

The purpose here is not to recommend this system over other systems in use but rather to discuss its merits, its limitations, the function of its component parts and to offer design suggestions that will provide efficient and economical operation. Some of these suggestions may also be helpful to correct faults in existing systems.

As in all types of service hot water systems, the major factors that influence the selection and installation of the means of hot water generation are the hot water demand requirements of the building, peak periods of demand, the type of space heating apparatus used and the particular judgment of the engineer designing the system or the contractor installing the system.

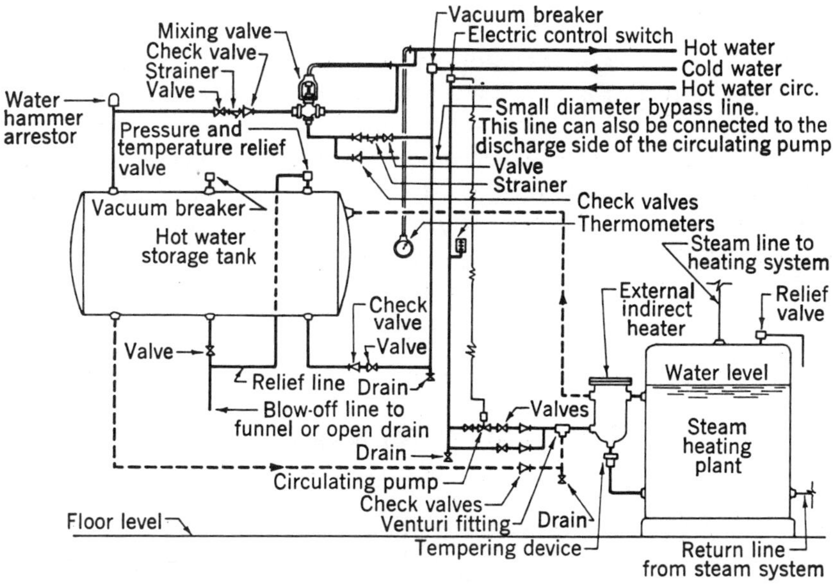

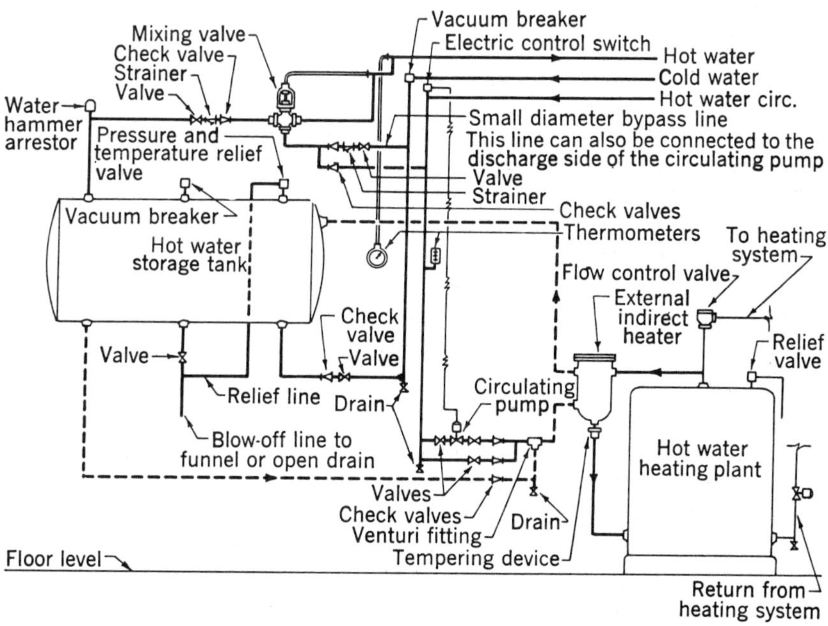

Fig. 1. (top) System utilizing a steam heating boiler with an external indirect heater and a storage tank to generate hot water.

Fig. 2. Same system, but utilizing a hot water heating boiler.

Capacity of Indirect Heater & Storage Tank

Once the hot water demand requirements of the building have been determined, it is necessary to select the proper size heater and storage tank. The table below is a typical one. It lists two types of external indirect heaters commonly used and their rated capacities.

Although the figures offered in the table are approximate, they have been presented for the purpose of example. In actual work, the manufacturers of the proposed units should be consulted in order to obtain accurate rated capacities.

Rated Hot Water Capacities For Indirect Heaters

Indirect Heater	Gallons of Water Heated from 40 to 140 F in 3-Hour Period			Recommended Size of Storage Tank, Gallons, for Heater Using Boiler Water at 180 F
	Steam at 0 PSI Gage Pressure	Boiler Water at 212 F	Boiler Water at 180 F	
Single Coil	100- 300	50- 150	30-120	40-100
Multi Coil	475-1,620	250-1,000	150-600	120-400

The capacities shown in the table represent the number of gallons that may be generated in a three-hour period. Values listed under 180 F. temperature ratings are generally used for year-round systems since this is the actual output capacity of the boilers during the summer months.

The storage tank is intended as a reserve for the hot water generated by the indirect heater. It offers the means to meet the hot water demand during the peak hours of the day. For example, if a building has a hot water demand requirement of 400 gph during one of the peak hours of the day, an indirect heater with a rated capacity of 600 gallons per three-hour period would be used in conjunction with a storage tank of a minimum of 400-gallon capacity. This arrangement will satisfy the peak demand requirement and easily handle the average demand, provided that the average hourly demand is less than 200 gph, (600/3). If the peak demand exceeds 400 gph, a larger storage tank may be used.

Limitation of Supply

There are however, certain limitations to the amount of hot water these systems can furnish and yet operate efficiently and economically. With reference to the capacities listed under Table 1, it is recommended that the limit be set at a peak demand requirement of 500 gph. This demand would require an indirect heater with the highest rated capacity of 600 gallons per three-hour period and a storage tank with a maximum storage capacity of 500 gallons. This requires that the average hourly demand be less than 200 gph. If this limit is exceeded, it is not practical to use larger indirect heaters or storage tanks. Therefore, other means of hot water generation should be considered, such as a separate heating boiler used expressly for the generation of service hot water without the use of an indirect heater.

With the steam boiler, additional recovery is available as shown in Table 1, by the capacities under boiler water at 212-deg. and steam at 0 psi gage pressure. These higher boiler water temperatures may be used to supply 180-deg. water for kitchen use. This benefit is only available when the boiler is used during the heating season.

Operation of System

Assuming that the indirect heater and the storage tank have been properly sized for the hot water requirements of the building, the first items to be considered is the location of the storage tank with respect to the indirect heater. If the storage tank is so elevated that the bottom of the storage tank is higher than the top of the indirect heater, a good natural circulation of hot water between the indirect heater and the storage tank will be assured. If space conditions do not permit the elevation of the storage tank, it is then necessary to install a circulating pump on the return line from the storage tank to the indirect heater to provide a positive circulation.

Indirect Heater

The next item to be considered is the piping connections between the indirect heater, storage tank and heating boiler. Although, there are several different piping arrangements that may be used, the arrangements used in Fig. 1 and Fig. 2, are shown because they illustrate a method that fully utilizes the component parts of the system.

The boiler water circulates through the indirect heater and transmits its heat to the water contained in the internal coils of the heater.

The heated water will circulate to the storage tank and back to the indirect heater until all the water in the storage tank is heated to the prescribed temperature.

In Fig. 1, the boiler water supply to the indirect heater is taken from below the boiler water level where the temperature is between 170- and 180-deg. The tempering device installed on the return pipe to the heating boiler controls the circulation of boiler water through the indirect heater. If it is desired to maintain a water temperature of say 140-deg. in the storage tank, then the tempering device is adjusted to a setting that will achieve this temperature. When the desired service water temperature is reached, the tempering device will partially close and retard the circulation of boiler water through the indirect heater until the service water temperature drops below a prescribed value. Since boiler water aquastats control the burner operation, the system may be used for year-round performance.

In Fig. 2, the boiler water to the indirect heater is taken from the hot water header at the boiler as shown. The tempering device functions in the same manner as that described for the steam boiler. The weighted flow control valve or other approved equipment device allows the burner to operate and maintain the boiler water temperature for circulation through the indirect heater and yet prevents the movement of hot water through the heating system. When hot water for the heating system is needed, the circulation pump located on the heating return line to the boiler provides the pressure flow needed to open the weighted flow control valve. This permits the boiler water to circulate through the heating system. This arrangement also affords year-round performance.

Hot Water Mixing Valve

A hot water mixing valve has been added on the service hot water line as it leaves the storage tank, since it may be desired to maintain a certain temperature of hot water to the fixtures. The mixing valve provides an accurate means of controlling the water temperature. Both hot and cold water supplies to the mixing valve are equipped with a valve, check valve, and a strainer. The strainer is installed to prevent foreign objects in the water from damaging the delicate mechanism in the mixing valve. It might be mentioned here that there are a number of different types of mixing valves available on the market for this function.

The mixing valve action is controlled by a temperature bulb placed in the hot water line at the location shown. The force that actuates the mixing valve is transmitted by the fluid in the temperature bulb and tubing. By enlarging or reducing the hot and cold water port in the mixing valve, the temperature of the water to the fixtures is regulated. The temperature bulb should be placed at least 6 to 8 feet or more away from the mixing valve to assure accurate temperature regulation. The small diameter by-pass line is included to permit the circulation of service hot water in the event that the port in the mixing valve is closed to the hot water leaving the storage tank. This possibility could occur during prolonged periods when the service hot water is not drawn or used.

Circulating Pumps

To speed up the return of water in the service hot water piping system, a circulating pump is installed. The operation of the circulating pump is governed by an electric control switch set in the return line as indicated. The temperature of the hot water returning to the circulating pump will be a few degrees cooler than when it left the storage tank, depending upon the overall length of the piping runs. Therefore the control switch is set at a temperature range that takes into consideration the heat loss of the water passing through the piping.

For example, if one desires to furnish 135-deg. water to the fixtures, then the mixing valve at the storage tank is so adjusted to furnish water at this temperature. Hot water from the tank is automatically mixed with cold water to provide water at 135-deg. Water in making its circuit through the piping system, will give up some heat so that by the time it returns to the circulating pump, it may be 135-deg. or less. It is possible to set the electric control switch for the proper range of circulating pump operation. For example, using a 5-deg. differential for the pumping cycle, the circulating pump will operate when the water in the return line drops to 125-deg. and will shut off when the temperature again reaches 130-deg. If desired, a larger temperature difference may be used for the upper and lower limits.

The thermometers shown aid in obtaining the proper setting for the system.

Return water from the storage tank is connected to the hot water circulating line as shown by means of the special tee connection. The purpose of the tee connection is to speed up the movement of the water in the storage tank by venturi action while the circulating pump is in operation. When the circulating pump is not operating, the water in the storage tank will still move by natural circulation which is benefited by the elevation of the storage tank.

Cold Water Supply

The cold water is connected to the storage tank in the position shown because this is the farthest point from the hot water outlet and the return outlet to the indirect heater. In this location, the cold water will spread out more evenly along the bottom of the tank without mixing with the hot water in the tank. A vacuum breaker is placed on the cold water line to prevent siphonic action, in the event of a drop in water pressure.

The piping arrangements shown in this chapter are only diagrammatic and intended for the purpose of example. It is recommended that the manufacturers of the proposed equipment be consulted for their suggestions in order to provide a system that will furnish abundant quantities of hot water in a most efficient and economical manner. Arrangements other than those shown can be used with equal success depending upon the specific equipment or devices used.

CHAPTER 34

Economy in Service Hot Water Generation

When a new building project is proposed, one of the systems which requires careful design, is the service hot water system, for there are numerous methods that may be used for the generation of service hot water. The important consideration, however, is to select the method and available source of heat that will best satisfy the particular hot water demand requirements of the building. Not only should the system be efficient, but it must be so designed that its operation is economical.

A certain amount of energy is required to heat a prescribed amount of water to a desired temperature. When the price of fuel and the Btu content of the fuel are known, it is possible to estimate the cost for generating the hot water. Occasionally, however, there will exist at the site, certain factors or benefits that may be utilized to affect a savings in the cost of operation.

Such a factor of economy is available when steam is used for space heating and its condensate is either wasted or is reduced in temperature for return to the space heating plant. This condensate by means of a condensate cooler, may be used efficiently to bolster the generation of hot water.

Condensate Cooler

A condensate cooler is, in theory, a heat exchanger. The condensate passes through the interior of the shell and cold water passes through coils placed inside the shell and submerged in the condensate. By means of this unit, the temperature of the condensate is reduced to a degree suitable for return to the space heating plant.

To accomplish this, the condensate cooler is installed with typical piping connections as shown in Fig. 1.

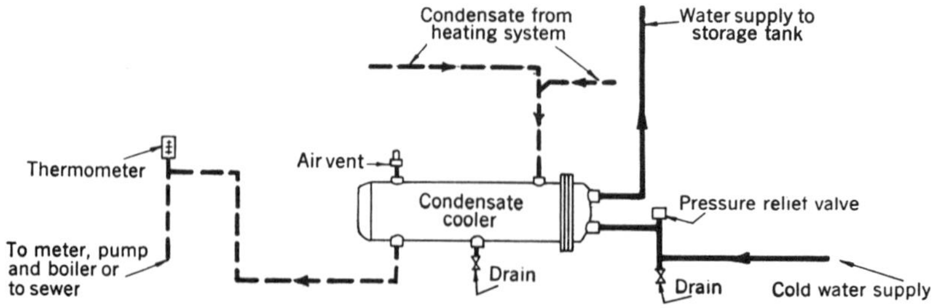

Fig. 1. Hookup showing basic piping arrangement for condensate cooler.

As indicated in Fig. 1, the steam condensate piping is installed to a connection at the top of the shell. At the bottom of the shell, diagonally opposite to the point where it entered, the condensate leaves the shell to be wasted or returned to the space heating plant. A bundle of copper coils are inserted in this shell so that there are two flow circuits within the shell. Cold water enters the coils, passes through the coils, and then leaves from the top connection. By this process, the temperature of the condensate is reduced and the temperature of the cold water is raised. Depending upon the amount of condensate, size of condensate cooler, and the volume of cold water, the final temperature of the condensate can be maintained at any desired degree.

Hot Water Storage Generator

A condensate cooler has many varied applications. At present, consideration is only given to its use with a service hot water system which depends on a hot water storage generator for the heating and storage of the water. A detail of such an arrangement, illustrated in Fig. 2, indicates the various fittings and controls.

It will be noted that the condensate cooler is located near the hot water storage generator. The cold water supply to the storage generator is first passed through the cooler, with the result that the condensate is cooled and the water supply is heated. Because the water supply is heated, the amount of steam required for generating the hot water is reduced, thereby affecting a savings in fuel consumption.

It is important to mention at this time, that the application of a condensate cooler will only assist in the generation of hot water and should not be considered as a primary source of heat. The hot water storage generator should be designed for the hot water demand requirements of the building, with steam as the primary source of heat. The system is not sized to include the benefit derived from the condensate cooler. The reason for this is that the condensate cooler will only pre-heat the water supply to the storage generator to a variable degree.

As an example of the benefit that may be obtained from a condensate cooler, the following is given: Condensate can be cooled from 200- to 100-deg. by using 2 lb. of cooling water to each pound of condensate. As a result, the cold water supply to the storage generator can be heated as much as 50-deg. above the cold water inlet temperature.

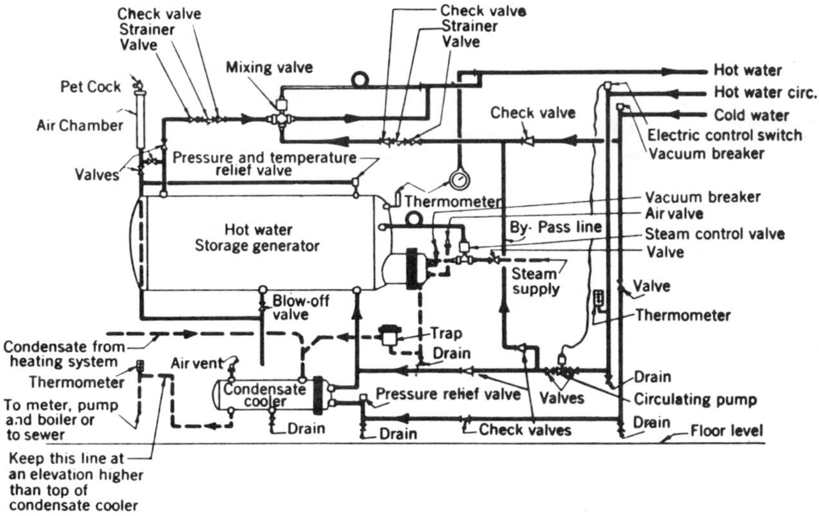

Fig. 2. Detail of piping hookup for hot water storage generator and condensate cooler.

The general piping arrangement, Fig. 2, is a typical one for hot water storage generators. The purpose of the mixing valve shown, is to maintain controlled hot water temperatures for the fixtures. The bypass line, between the hot water circulation line and the cold water supply to the mixing valve, is installed to assure a positive circulation of hot water through the system at times when normal circulation through the hot water storage tank is retarded.

For any system where steam is used, there is always the danger that the steam control valve may fail. This occurrence, especially during the off-periods of the day, would allow the water in the storage tank to exceed safe prescribed temperatures. The mixing valve, reacting to this high temperature, would tightly close its hot water port and since there was little or no water being drawn at the fixtures, cold water would not enter the system. The bypass line, however, affords positive protection by allowing the return water to bypass the storage tank and recirculate through the piping system, thus preventing injury from scalding water at the fixtures and avoiding, to some extent, damage to the mixing valve. The thermometer at the storage tank will indicate, by its unusual reading, that the steam control valve needs repair or replacement.

Controlling Condensate Temperature

When a condensate cooler is used in the manner shown in Fig. 1 or Fig. 2, the final temperature of the condensate can be readily controlled during the average hot water demand periods of the day. However, during the off-periods, when the amount of cold water passing through the cooler is reduced, it will not be possible to maintain the desired condensate temperature. If it is important that the temperature of the condensate be controlled at a uniform degree for return to the space heating plant, or for wasting to the sanitary system, then provisions for this must be made by a piping arrangement, similar to that shown in Fig. 3.

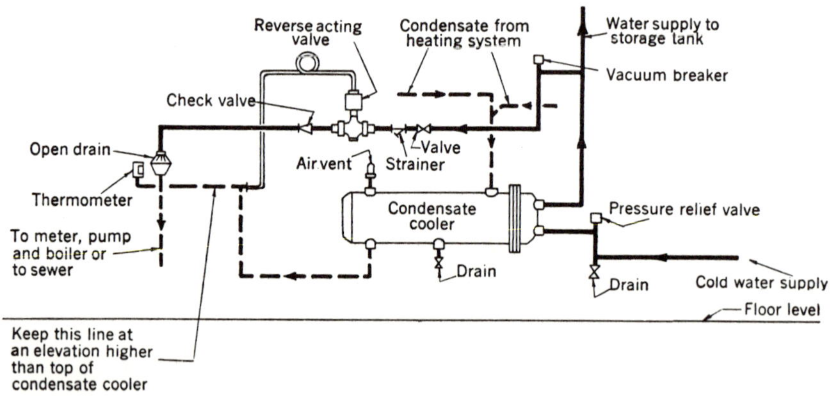

Fig. 3. Condensate cooler and piping arrangement for controlling final temperature of condensate.

As illustrated by Fig. 3, a branch is taken from the preheated water supply line to the storage tank, at a point after it leaves the condensate cooler. A reverse acting valve is placed on this line. This valve will remain closed while the desired temperature of the discharged condensate is maintained. When this temperature is exceeded the valve will open to allow the necessary volume of cold water to pass through the cooler and the branch to cool the condensate. This system will serve its function best during the off-periods of the day, when the flow of cold water to the storage tank is reduced. When a system such as this is used, it is very important that a direct connection between the water and waste lines be avoided. The suggested piping connections shown will prevent such cross connections.

Maintenance of Condensate Cooler

A condensate cooler should be so located and installed that it is accessible for proper maintenance. To assure efficient operation of the unit, it is necessary to keep both the interior of the shell and the inside of the copper tubing clean of sediment. The drains shown, one under the condensate cooler and one under the cold water supply to the cooler are provided for this purpose. By opening these valves periodically, it is possible to flush out any accumulated sediment. This procedure is very important when cement lining has been applied to the interior of the storage tank, as small particles of the cement will settle in the copper coils for a period of time, after the system is first installed. Likewise, scale and foreign matter carried by the condensate lines will settle inside the shell of the condensate cooler.

If the service hot water system, including the piping, hot water storage generator, condensate cooler and all accessories are properly sized and installed, it is certain that the generation of the hot water will be both efficient and economical.

CHAPTER 35

Water Hammer Arrestors

One of the most damaging factors to a water piping system is water hammer action. In addition to its effect on the piping system, water hammer causes banging noises in the system that are very disagreeable to occupants in the building. Water hammer occurs when a column of water flowing through a pipe line and discharging at an open outlet, is suddenly stopped by closing the outlet. Since flowing water has force, tremendous pressures result at the point of closure and pressure surges move along the pipe. The manner in which water hammer occurs is illustrated in Fig. 1.

Steps that produce water hammer, shown in Fig. 1, follow.

Phase 1. The valve on the line is closed and the water contained in the line is at rest. Water pressure is thus exerted in all directions in an equal way.

Phase 2. The valve on the line is open and water flows freely through the open outlet. Now the water pressure is utilized to force the water out of the open end of the pipe. Arrows indicate the direction of force in the column of water.

Phase 3. When the valve is quickly closed, the column of freely flowing water is suddenly stopped; excessively high pressures are generated at the point of stoppage. This reaction is the same as would result if a steel bar moving through the line at the velocity of water were suddenly stopped by the valve.

Phase 4. In an effort to equalize the pressure build-up of the water, a shock wave will travel back along the branch line until a larger diameter pipe is reached. This will allow the shock wave to dissipate itself. Arrows denote the direction of force toward the valve and then its reversal as a shock wave toward the point of relief. Since the shock wave travels at speeds in excess of 4000 fps, it causes a piping clatter all along its route. Often the shock wave will oscillate back and forth between the valve and the point of relief until the pressure is stabilized within the branch line.

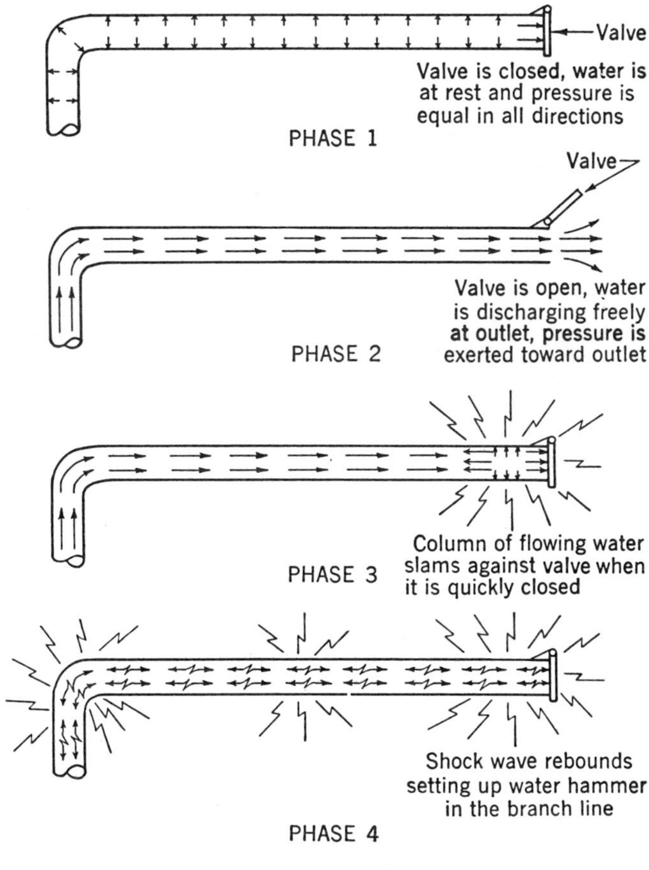

Fig. '1. How water hammer can develop in a branch line.

The pressure generated by the shock wave can expand and often rupture the piping. Although piping clatter is normally associated with water hammer, you cannot assume that when these noises do not occur, that the shock wave is non-existent. Quite often, water hammer takes place without any physical sounds. Therefore, it is very important that piping systems be designed with all due consideration given to the means that compensate for the action of water hammer.

Causes of Water Hammer

Not all the noises heard in the piping system can be attributed to water hammer. Loose faucet or valve washers can cause a pounding or chattering in the piping. Improperly supported and secured piping can create noises as the flowing water causes the piping to vibrate and thus rattle against steel members. This noise is easily transmitted through the piping system. Undersized water piping with excessive pressures will produce shrill sounds. Then, of course, there is certain equipment, as pumps, that will produce noises unless the pumps are isolated and the piping connections are equipped with flexible piping connectors.

As for water hammer, it is generally impossible to predict just where in the piping system that water hammer will occur. There may be a small diameter branch line in the system which, by the nature of its length and the fixture it supplies should produce water hammer, yet it does not. On the other hand, a larger diameter line which should not cause water hammer, often will cause such noise.

One of the basic causes of water hammer is quick closing valves. These include any valve or faucet that is closed rapidly either by hand, spring loading or solenoid. If the same valve were closed slowly, then the flowing water would have the chance to stabilize without producing the shock waves of water hammer. However, since these valves must function in the manner intended, other means must be employed to minimize the violent action of water hammer.

Other factors that contribute to the positive occurrence of water hammer are excessive water pressures, inadequate piping sizes and water piping that is improperly installed.

Methods of Controlling Water Hammer

In order to eliminate the danger and piping clatter that results from water hammer, it is important that certain steps be taken in piping system design to compensate for the excessive pressures that are generated when a column of flowing water is suddenly stopped.

The consideration needed is some means or device that will provide flexibility in the system to absorb the initial shock wave of water hammer thereby confining the action to a given section of piping.

Air is the most effective medium for absorbing the shock wave caused by water hammer for (1) water is non-compressible; (2) air can be compressed to considerable pressure. When the water compresses the air, it also fills the void offered by the displaced air. Because water has this flexible means to expend its force, the shock wave that would otherwise result, is quickly absorbed. The manner in which air serves to eliminate water hammer is shown in Fig. 2.

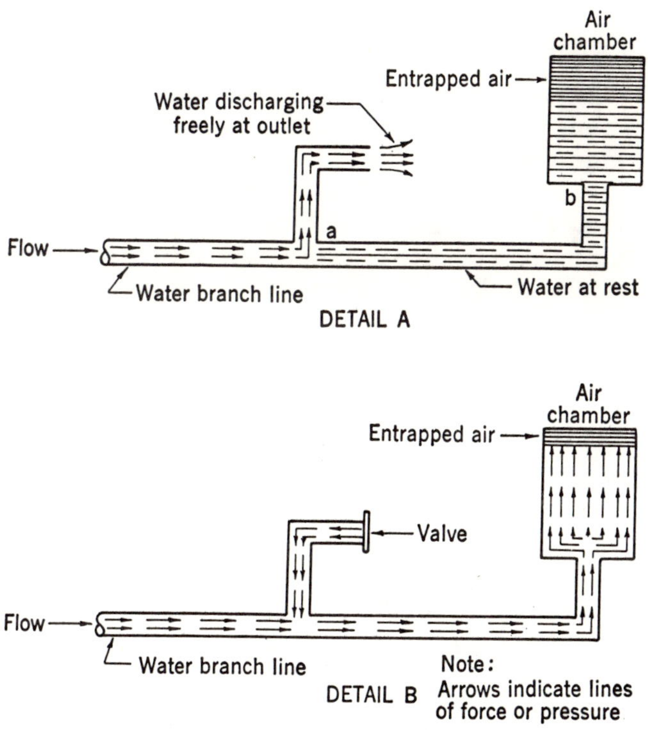

Fig. 2. How air chambers cushion the initial shock wave generated by water hammer.

Detail *A* in Fig. 2 shows water flowing in a branch line and discharging at an open outlet. Water in the line from *a* to *b* is at rest with very little pressure exerted on the air that is contained in the chamber.

Detail *B* illustrates the condition when the flowing water is suddenly stopped by the valve. The shock wave generated rebounds but is absorbed by the air in the chamber. Thus by absorbing the initial shock wave, the water pressure is stabilized and the occasion of water hammer has been removed. Both details are diagrammatic and have been used only for example in showing how the phenomenon of water hammer can be properly controlled.

The extent of water pressure in a given line and the required sizes of air chambers to absorb the resultant shock waves can be computed. Much research work was done and published by F. W. Dawson and A. A. Kalinske of the Iowa Institute of Hydraulic Research, on this subject of water hammer.

Simple Air Chambers

One of the methods that has been used for many years in the average building to protect the water system is the simple air chamber. This device consists of a short length of pipe installed in a vertical position at the top of a water supply to a fixture.

The upper portion of the chamber fills with air during installation. The entrapped air in the line is compressed until it is equal in pressure to the water contained in the line. As excessive pressures occur due to the quick closing of the valve or faucet at the equipment or fixture, the air will compress still further, allowing the water to fill the void left by the displaced air. Thus water hammer is avoided.

Figure 3 represents the manner in which this type of air chamber can be employed.

In Detail *A* of Fig. 3 the air chamber is placed at the top of the cold and hot water supply to the lavatory, which has been used for example here. The length of the air chamber is determined by the engineer. Usually it is between 12 in. and 24 in. long.

Detail *B* illustrates how the air chambers are installed on the branch hot and cold water lines that serve a battery of lavatories. When the piping system is first installed, the chambers contain air at atmospheric pressure which compresses when the water is turned on. The chambers will function until the entrapped air becomes water logged. It usually takes but a short time for the water to absorb and displace the air in the chamber. When this happens it becomes necessary to recharge the chamber with air if it is to serve its purpose in an effective way.

Recharging is accomplished by shutting off the branch valves and opening the fixture faucets. This procedure will allow air to bubble into the supply lines and enter the pipe chamber again. Unfortunately, this method of recharging is not positive. Therefore, these air chambers, which may be regarded as an extra expenditure, will not cushion water hammer unless they are continually main-

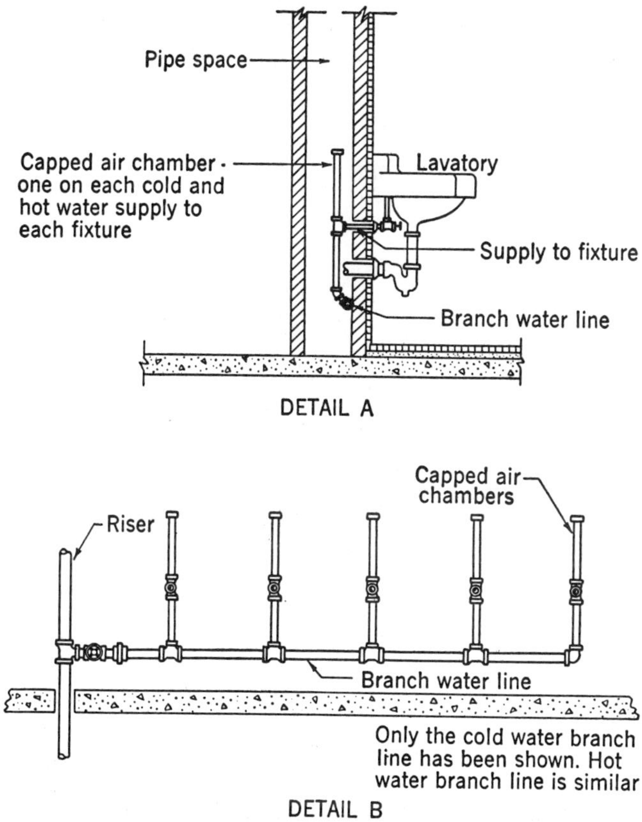

Fig. 3. How the capped air chamber is installed on branch water lines.

tained. When you estimate the cost required to place such air chambers at each cold and hot supply to every fixture, you might find it more economical to consider other types of chambers or devices that are more positive in operation.

Rechargeable Air Chambers

A more effective means of absorbing water hammer is the device shown in Fig. 4, which comprises an air chamber equipped with a valve, hose bibb and pet cock. This device can be placed in any location on the piping system where water hammer is likely to develop. It is recharged by closing the valve, opening the pet cock and allowing the water to drain out of the hose bibb. Once filled with air at atmospheric pressure the chamber is ready to function again.

Detail *A* shows the air chamber installed in an access corridor to serve a branch water line which supplies a group of fixtures. Both the cold and hot water branches would each require one of these chambers.

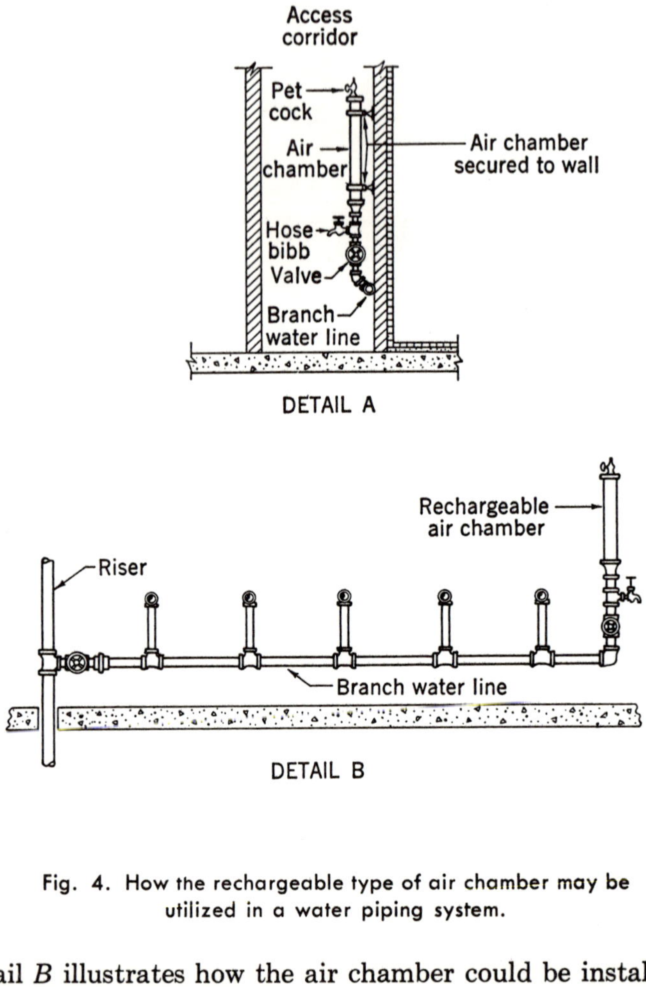

Fig. 4. How the rechargeable type of air chamber may be utilized in a water piping system.

Detail *B* illustrates how the air chamber could be installed on a branch water line that supplies several water closets. When this type of air chamber is installed in a space other than an access corridor, it is important that an access door be provided to permit recharging of the chamber. Since such air chambers are rather heavy devices, they should be supported so that its weight will not be imposed on the piping.

Manufactured Devices

There are on the market a number of different devices that have been developed to control water hammer. Figure 5 represents several of those devices used in large quantities.

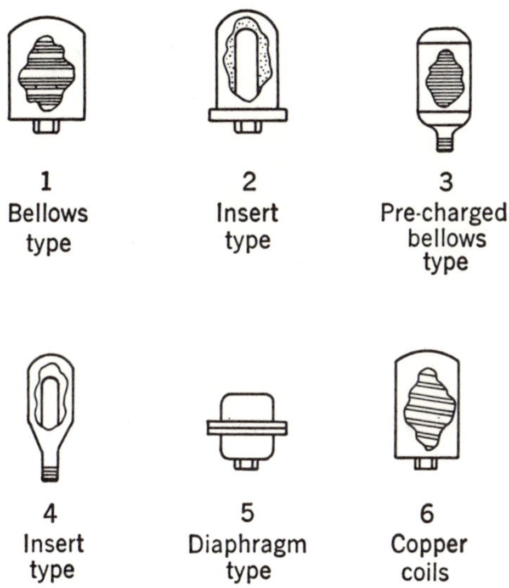

Fig. 5. Several different types of manufactured water hammer arrestors now on the market.

These devices may be classified as follows:

1. bellows type; 2. insert type; 3. precharged bellows type; 4. insert type; 5. diaphragm type; 6. copper coils.

In addition to the devices shown in Fig. 5, there are of course, other types available. Manufactured water hammer arrestors reflect the tremendous strides made in this field. Such devices require no maintenance or recharging and therefore can be placed in any type of a concealed pipe space. Moreover, since these units offer a neat appearance, they can also be placed in exposed locations. It is both practical and economical for the engineer to consider such devices for his water hammer applications.

Advice on type, specific capacity and the recommended placement of the device can be obtained from the manufacturer of the particular unit selected.

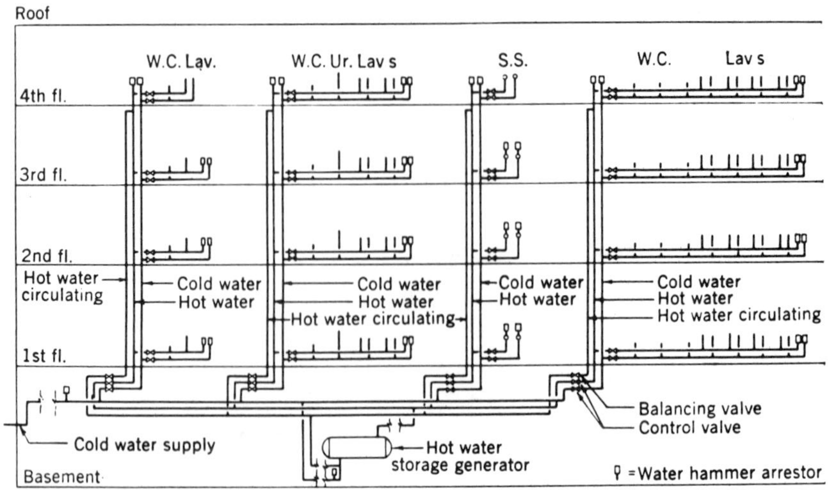

Fig. 6. Typical layout of water distribution system in an average building. This is one of several arrangements that may be followed for the location of water hammer arrestors.

Typical Water Piping System

Water hammer arrestors may be placed on a piping system in a number of different locations depending upon the particular design the engineer favors. One manner of placing the water hammer arrestors is shown in Fig. 6. The piping system shown is an upfeed system and consists of cold, hot and hot water circulation piping. The locations given for the placement of the water hammer arrestors is for the purpose of example only. Other locations can be used with equal success. You will note, however, that these locations will protect the overall piping system from the damage that can result from water hammer.

Water hammer arrestors are placed on the water service into the building at the building side of the valves, meters and check valves. They are also placed at connections to various types of equipment. Water hammer arrestors are especially important at the discharge side of different types of pumping equipment to prevent damage. These devices are assurance that the water system is properly protected.

CHAPTER 36

Balancing Service Hot Water Systems

Of all the systems installed in a building, the service hot water system, often requires the greatest amount of adjustment before it finally delivers water at a uniform degree of temperature to all the fixtures in the building. Unless the service hot water system is completely balanced, it is not unusual to have an ample supply of hot water in one portion of the building and an inadequate amount of hot water in another.

There are a number of factors that affect the operation and efficiency of a service hot water system, and the factor that can ensure a balanced system is proper circulation. Service hot water systems of varying sizes will now be discussed and the means will be pointed out that may be employed to furnish the even distribution of hot water throughout the building.

Principle of Hot Water Circulation

Heated water rises. When the water in a service hot water system is heated, it also rises as illustrated in Fig. 1 which depicts a hot water storage generator and the distribution piping. The storage generator is the means for heating the water. The heated water then passes into the hot water main and up into the hot water risers to supply the various plumbing fixtures. The hot water return permits the water to circulate throughout the entire system. Moreover, the return line conducts the water back to the storage generator where it is reheated. When properly installed, a system of this type will operate with a good deal of efficiency.

For the sake of example, only three risers are shown, labeled 1, 2, and 3. Riser No. 1, offers the path of least resistance to the flow of water, therefore, the water will circulate most rapidly through this

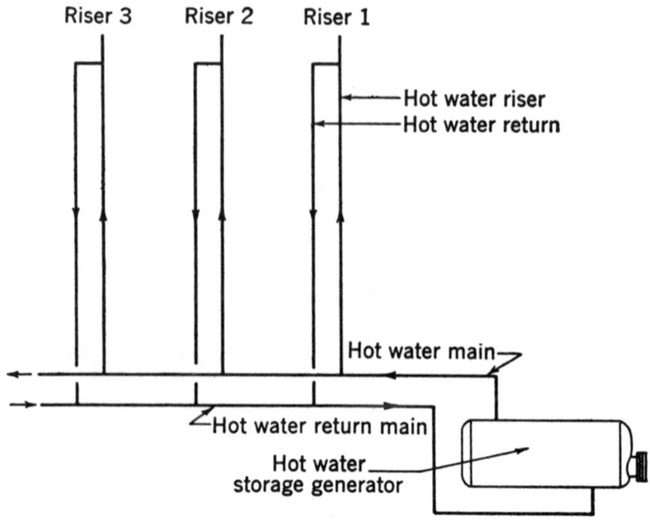

Fig. 1. Piping hookup shows how hot water circulates through a piping distribution system.

riser. Riser No. 2, offers the second path of least resistance and riser No. 3, the next. If more risers were shown, the rate of flow would decrease proportionately. Thus, in a system containing many risers, unless adequate controls are provided, the first riser could easily short circuit the system and deprive the succeeding risers of adequate circulation.

Typical Upfeed System

Figure 2 shows a system capable of serving three stories of fixtures. Water heated in the storage generator will circulate up through the hot water risers and return back again to the storage generator via the hot water return piping.

In this system, too, the first riser represents the path of least resistance. Even though this system consists of but a few risers, the riser farthest from the storage generator will not receive the same benefit of hot water circulation as do those risers that are nearer to the storage generator. Since there are but a few risers involved, it is doubtful that there will be much of a temperature differential between the first and last risers. However, to make certain that the system operates as a balanced one, globe or balancing valves are placed at the base of each of the hot water return risers. This valve,

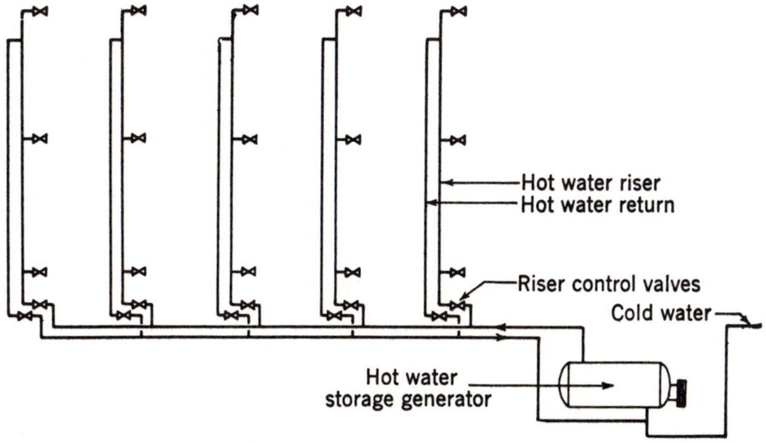

Fig. 2. Typical upfeed service hot water system.

in addition to serving as the shut-off valve for the riser, also is the means of balancing the entire system.

With the proper throttling of each valve, the rate of flow in the return risers can be adjusted to favor those risers through which adequate circulation is more difficult. Actually, the valves on the risers nearest the storage tank are partially closed to retard the circulation. Each valve is progressively adjusted to favor those risers through which adequate circulation is more difficult. This flexibility permits an even distribution of hot water for all the risers.

For the sake of clarity, the circulating pump, valves and other controls, normally included as component parts of the hot water storage generator have not been indicated in Fig. 2, or any of the other illustrations.

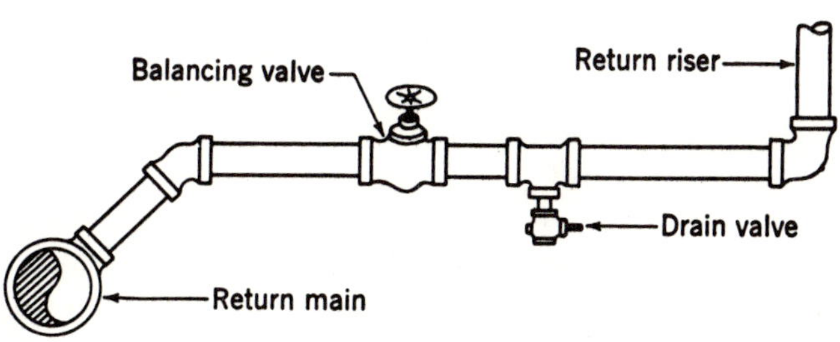

Fig. 3. Installation of a balancing valve.

Valve Arrangement

The arrangement of the balancing valve on the return riser would be similar to that shown in Fig. 3. The valve is placed on the horizontal branch from the hot water return main. The drain valve shown between this valve and the riser is used in the event that it becomes necessary to drain the water out of the riser after it is shut off.

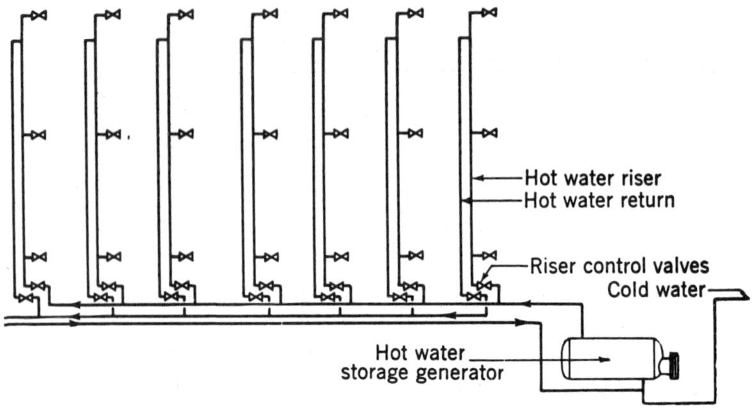

Fig. 4. Typical upfeed service hot water system with a reverse hot water return main.

Enlarged Upfeed Systems

Figure 4 represents an enlarged version of an upfeed service hot water system. For such systems, where numerous risers are involved, it becomes even more important to consider the means that will ensure a balanced system. Balancing valves placed at the base of each hot water return riser are a necessity. Often, when many risers are contemplated, the valves alone may not be sufficient to balance the distribution system. In this event, a reverse hot water return main can be quite beneficial. Here the direction of the hot water return main is reversed as shown to assist those risers farthest away from the hot water storage generator. For each riser the total length of the supply line and return is the same.

The risers nearest the storage generator will of course be first to receive the hot water from the main. The returns for these risers however, will be the farthest from the storage generator. This means that the return for the hot water riser at the end of the run will be nearest to the storage generator. As the water circulates through the return main, its passage stimulates the flow of water in each return riser thereby benefiting the entire system. Patented venturi fittings are often employed on the return main to serve as connections for the return risers. These fittings are designed to improve the circulation in the return risers.

More Extensive Systems

For systems that are even more extensive with regard to the number of risers involved, a method similar to that shown in Fig. 5, may be used.

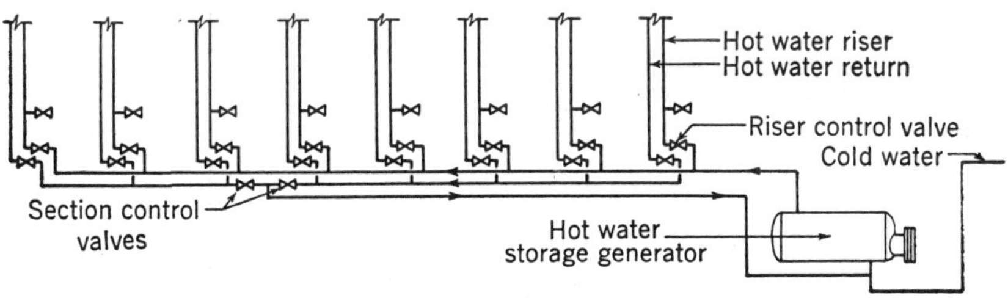

Fig. 5. Upfeed service hot water system with a divided hot water return main.

As illustrated by Fig. 5, the control valves are included at the base of each hot water return riser. To facilitate a more accurate means of balancing the water distribution system, the hot water main has been divided into two sections. By dividing the return main in this manner, a more precise adjustment can be made with the balancing valves contained in each of the sections. Moreover, the direction of water flow in both return mains is to the center of the main as shown. From this point, the return main is extended back to mix with the cold water entering the storage generator.

To provide an even greater assurance in balancing the system, section control valves are placed on the return main as noted. With the proper adjustment of these valves it is possible to equalize the circulation in the two sections containing return risers.

Taller Buildings

The same sort of reasoning is given to taller buildings in which an upfeed system of piping is utilized. The system shown in Fig. 6, will supply water to seven stories of fixtures. This number of stories has been used for example only. A taller building could be served with an upfeed system depending upon the pressure in the street main. Also for example and clarity, the distribution system has been divided into two sections with an even number of risers in each section.

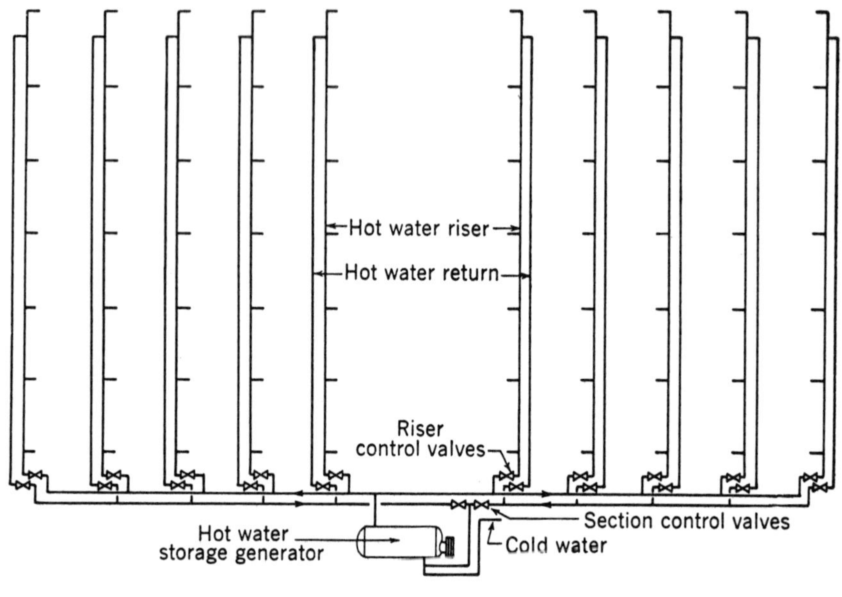

Fig. 6. Sectionalized upfeed service hot water system.

Controls for Taller Buildings

Here again, the path of least resistance for the water will be through the first riser and return on either side of the hot water storage tank. It follows then, that the resistance will increase progressively in the risers toward the end of each section. It is therefore recommended that balancing valves be placed at the base of hot water return riser. With this provision, it is possible to balance the circulation of hot water in each section. In addition, section control valves are placed on the return mains as indicated. These valves enable one to balance both sections of the distribution system. Thus, regardless of the size or scope of the upfeed system of piping, when sufficient attention is given to the matter of controls, a perfectly balanced system of hot water circulation will result.

Very Tall Buildings

Very tall buildings will be regarded here as buildings that cannot be supplied with an upfeed system of piping but instead, depend upon other methods for furnishing water to the upper floors. Figure 7 is typical of the type system frequently used for such buildings.

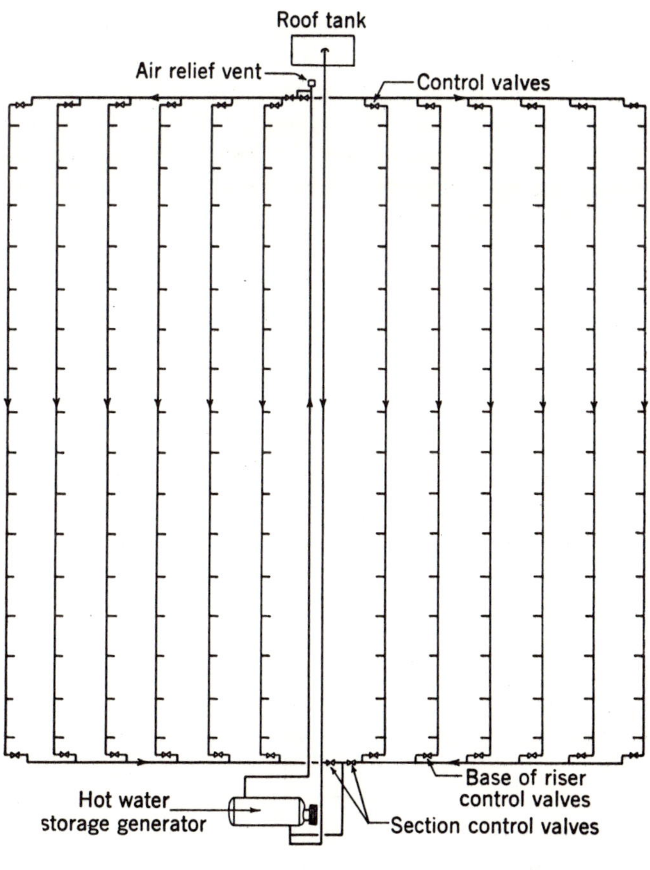

Fig. 7. Typical downfeed service hot water system.

Cold water, by means of house pumps is lifted to the roof tank. The piping for the pumps has been omitted from the drawing. The roof tank, by virtue of its height, provides the pressure needed for the hot water to adequately serve all the fixtures in the building. As shown in Fig. 7, a cold water line extends directly downward from the roof tank to the hot water storage generator. The hot water from the storage generator is elevated to the ceiling of the highest floor in the building, where it is distributed to supply each downfeed hot water riser. After the lowest fixture branches are taken off, the risers then continue on to connect with the return mains in the basement. The return mains conduct the water back to the storage generator.

Even in tall buildings, the first downfeed risers on either side of the storage generator offer the path of least resistance for the circulation of hot water. Balancing valves are placed at the base of the return risers to offer the flexibility needed to equalize the hot water circulation in each section of risers. Moreover, section control valves are placed on the return mains to have the sections operate equally.

An air relief vent has been included on the high point of the hot water line as shown. Air pockets present considerable resistance to the circulation of water, therefore, devices that will release such air pockets are generally placed at those locations in the system where the air is most likely to accumulate.

Large Scale Projects

For buildings that cover several acres of ground, the need for balancing the service hot water system becomes a problem of major consideration. In such buildings, the piping runs are very long and unless adequate attention is given the initial design, it is rather easy to end up with a section of the building that does not receive its supply of hot water. For example, we have selected a typical system, Fig. 8, which shows the general shape of the building as well as the arrangement of piping that might be installed to service all areas of the building with hot water. The building shown could be a school, hospital, research laboratory or similar types of structure.

For buildings with extensive runs of piping, there are a number of factors that would cause sections of the system to be bypassed and not receive sufficient quantities of hot water. Water circulation can be retarded by air pockets, improper slope, traps created by offsets in piping, as well as other minor reasons.

The major problem is to design the piping in such a manner that complete control is possible in all sections of the system. Thus the

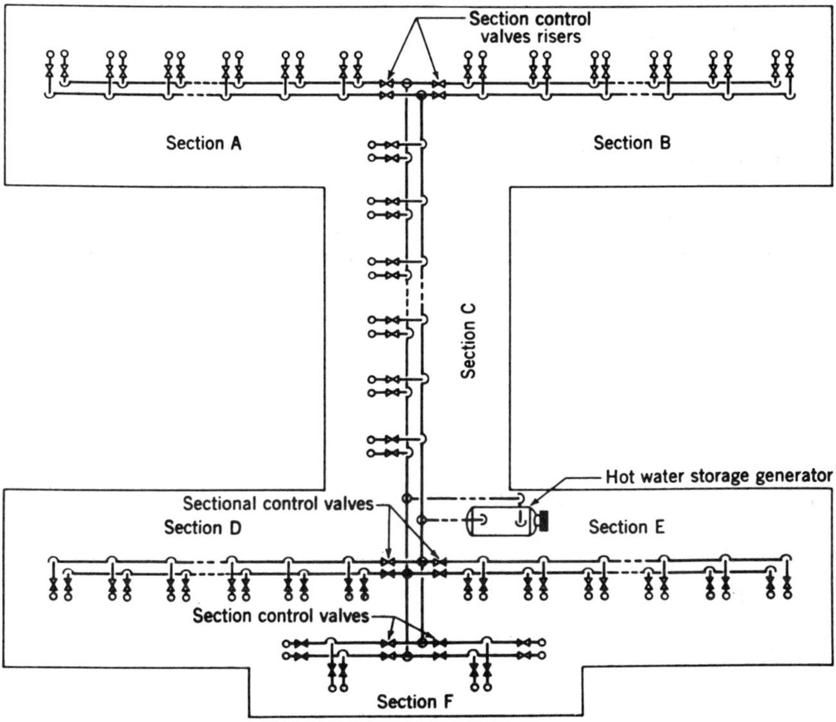

Fig. 8. Plan of lower level in a building, showing the general arrangement of hot water and hot water circulation distribution system.

entire system will be a balanced one. Referring to Fig. 8, the various sections of piping are designated by Section *A*, *B* and up to Section *F*. Balancing valves are placed at the base of each return riser. Section control valves are placed on each return main. By these means, each section of piping can be adjusted to favor one or retard the other, that hot water in abundant quantities may reach all the fixtures in the building.

Check Valves

Check valves have an important role in water distribution systems because they permit the water to flow in one direction but prevent its backflow. Care should be exercised to avoid misplacement of the check valves, thereby removing any possibility of impeding circulation. Check valves with sticky or misaligned flappers can also be troublesome and retard circulation. It is a good policy to specify only quality materials for the controls so vital to a properly balanced hot water distribution system.

Pumping

Pumps with a greater capacity than the average circulation pump are sometimes used to move the water through the system a little faster. It must be remembered however, that such pumps, although they are beneficial to the system will not assure positive circulation through all the risers. Good circulation can be accomplished only by the careful placement of balancing valves as suggested.

CHAPTER 37

Expansion in Hot Water Piping Systems

An important consideration in the design of piping distribution systems, especially hot water, is that of expansion. Unless this factor is given adequate attention, the flexibility needed to compensate for the linear movement in the piping will not be available. Since piping material will expand to some degree when heated, serious water leaks can result from cracked fittings or piping connections.

Fortunately, in most installations there is a certain amount of give to the piping systems which offsets the effects of linear expansion. For example, when expansion takes place, the piping may sag or arch to compensate for the linear movement. Many times, because of the short runs of piping and the number of offsets in the piping, the linear expansion will have no serious effect. However, in long piping runs, although no serious leaks develop initially, a tremendous strain is imposed upon certain fittings and pipe connections. When this condition is prevalent, it is merely a matter of time before leaks or cracked fittings will result from the strain. One way in which expansion can impose such strain on fittings and connections is shown in Fig. 1.

This figure shows a length of horizontal piping with vertical branches, and what takes place when pipe, for example, is installed in an ambient temperature of 70-deg., and then conveys water at 140-deg.

The total linear expansion of this line has placed a tremendous strain at the elbow on the end of the run. Moreover, the branch lines are no longer perpendicular to the main but are now at an angle to the main in proportion to the length of piping run. Swing connections to the top of the main have been shown to portray the amount of movement. If fittings such as these are used, the swinging of the fittings could easily cause leaks. If the branches are connected to

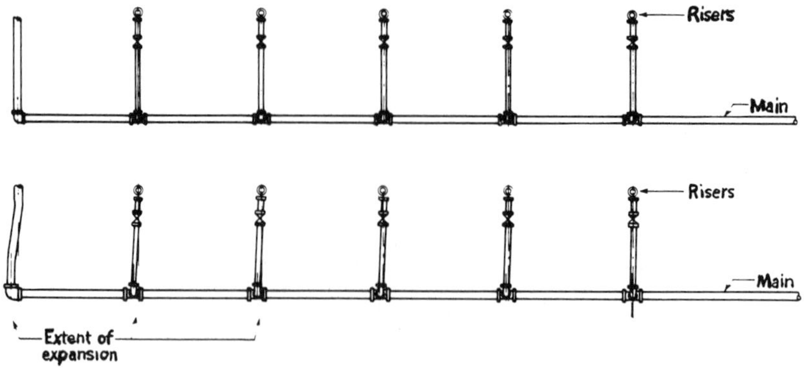

Fig. 1. Action of linear expansion in a hot water line and its effect on branch connections.

the main in a rigid manner, severe strain will result at the connections. Cracked fittings can result.

The more extensive the piping distribution system, the greater the danger from the effects of expansion. Rather than leave the matter of expansion to chance, it is much safer to include devices that will compensate for this natural function in heated pipe.

Expansion in Different Piping Materials

Galvanized steel, galvanized wrought iron, brass and copper are materials most generally employed for interior water piping systems. Each of these materials has a different linear expansion when heated to specific temperatures. The table on the following page lists the amount of expansion above zero degrees datum.

The table of linear expansion is only a partial one but shows the factor of expansion for the different piping materials listed. It shows that steel expands the least while brass expands the most. Knowing the range of minimum and maximum temperatures the line will be subjected to, the limits of expansion can be determined.

Expansion in Horizontal Piping Runs

The more extensive the piping distribution system becomes, the more important the need for proper expansion control. Figure 2 represents a horizontal piping distribution system as may be found in the basement or ground floor of a research laboratory building. This type building has been used for example because of its size and

Linear Expansion For Various Metals When Heated From Zero Degrees F

Temperature, Degrees, F	Linear Expansion, Inches per 100 Feet			
	Steel	Wrought Iron	Brass	Copper
0	0	0	0	0
20	0.15	0.155	0.25	0.26
40	0.30	0.31	0.45	0.45
60	0.455	0.475	0.67	0.65
80	0.61	0.63	0.92	0.87
100	0.77	0.80	1.17	1.10
120	0.915	0.96	1.41	1.35
140	1.075	1.13	1.69	1.57
160	1.235	1.29	1.91	1.77
180	1.40	1.46	2.15	2.00

spread-out nature. Furthermore, because the piping runs are so long in this type of building, it offers a good example for the complicated problems created by expansion.

Referring to Fig. 2, you will note the specific locations where expansion devices have been placed. These devices will confine the linear expansion of the piping to the sections of pipe installed between the expansion devices. Since several hundred lineal feet of piping run is involved in Fig. 2, the total expansion, unless restricted, could be rather extensive, thereby imposing considerable strain on the piping connections.

Expansion Devices

There are a number of different devices that can be utilized to handle expansion in piping systems. Such devices may be classified as the pipe loop type, piston type or the bellows type. Figure 3 shows the first two.

Detail *A* of Fig. 3 illustrates a typical expansion loop which can be constructed in a number of different ways and shapes. The required dimensions of the loop for a given installation is easily calculated or can be found in the engineering data tables contained in piping handbooks. This type of device is very effective in solving expansion problems; its only disadvantage, in some instances, is the amount of space needed for the installation of this device.

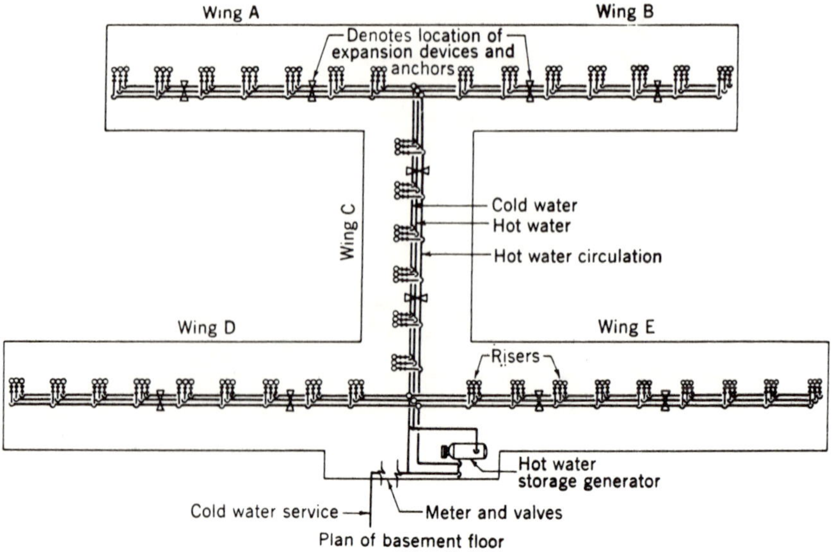

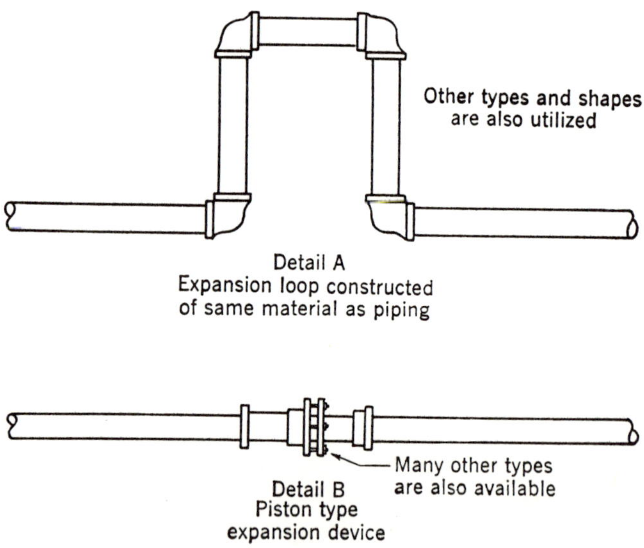

Fig. 2. (top) Typical water piping distribution system in a research laboratory showing location of expansion devices and anchors.

Fig. 3. Typical expansion devices.

The piston type expansion device shown in Detail *B* is a manufactured device available in a number of different types. When expansion takes place, one section of the unit telescopes into the other. The flanged unit serves to tighten the device and avoids leakage. This device is very practical but must be carefully installed to ensure that the device is perfectly aligned with respect to the piping.

The bellows type of device, not shown, is also available in a number of different types and shapes. This device consists of some form of bellows that will contract when expansion takes place.

Trapeze Hangers

The piping distribution system at the basement or ground floor must be supported in a substantial manner. Devices most frequently used for this purpose are called trapeze hangers, and one type unit is shown in Fig. 4.

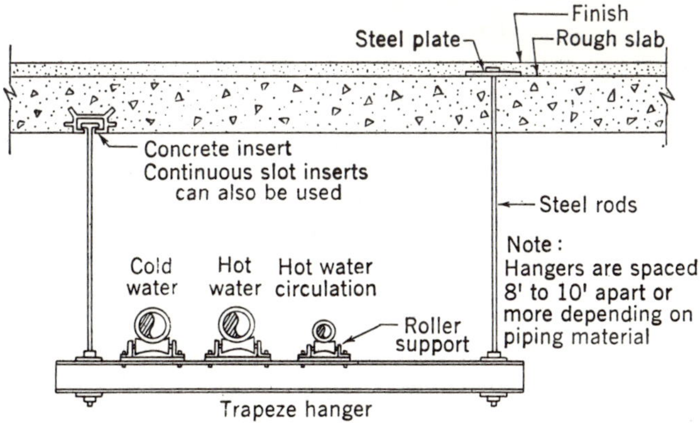

Fig. 4. Detail of a trapeze hanger for supporting water piping.

This unit may be constructed in a number of different ways. It consists of two steel channels mounted back to back and supported by steel rods anchored to the slab construction. The steel rods may be supported either by a concrete insert or a steel plate resting on the rough slab as shown. The steel rods pass between the channels and are rigidly secured in place. Manufactured roller supports on which the piping will rest are placed on the trapeze hanger. Since the water lines are generally insulated, covering protective devices are set into the underside of the covering where it comes in contact with the roller. These rollers provide adequate support but yet permit linear expansion of the lines. Very often, U-bolts are fitted around the piping to remove any danger of the pipe slipping off the rollers, in cases where severe expansion takes place.

The branch water lines are taken from their respective mains at a 45-deg. angle. The cold and hot water branches are equipped with riser control valves while the hot water circulation line is equipped with a balancing valve. The balancing valves ensure an even distribution of hot water throughout the entire hot water piping system.

Anchoring

In order to limit the expansion of the hot water lines to certain confined areas, suitable expansion devices must be employed. To make certain that the expansion devices will function as they should, anchoring devices are utilized. Figure 5 suggests a method of anchoring that can be used with considerable success where anchoring is accomplished at the trapeze hanger following the expansion device. The piping is secured to this anchored trapeze in any of a number of different ways. In Fig. 5, although it is not shown, the piping is anchored to the trapeze by means of a U-bolt placed around the bare piping and secured to the trapeze. Some expansion devices, Fig. 5, are so made that they can be fastened directly to the trapeze.

The function of the expansion device and anchored trapeze is to prevent the entire run of piping from expanding as a unit, but

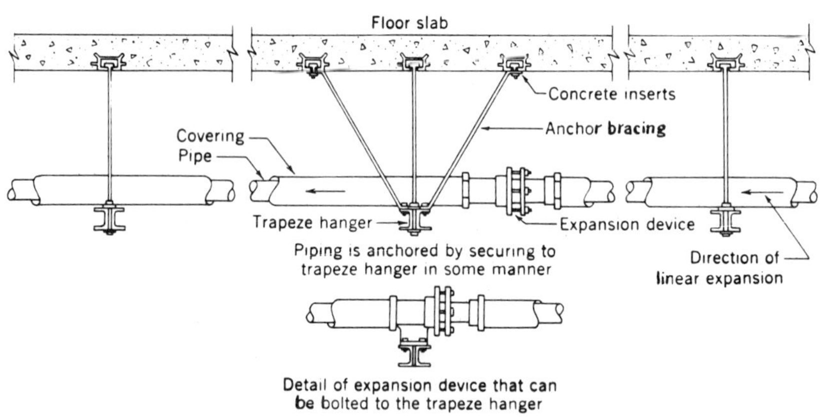

Fig. 5. Detail of expansion device and anchoring methods.

instead to confine it to prescribed sections of piping. For example, let us suppose that a hot water line of a given material is about 200 ft. long, and when heated will expand approximately 2 in. This increase in length of piping can be sufficient to impose a strain on all the branches connected to the run. The branches near the end of the run must endure the greatest strain.

If one expansion device were placed at the center point of the 200 ft. run, the total expansion in each section of this run would only be 1 in. If expansion devices were placed every 50 ft., the total expansion in each section of pipe would be reduced to ½ in. Depending upon the specific requirements of a design, the amount of expansion can be kept within any desired limit.

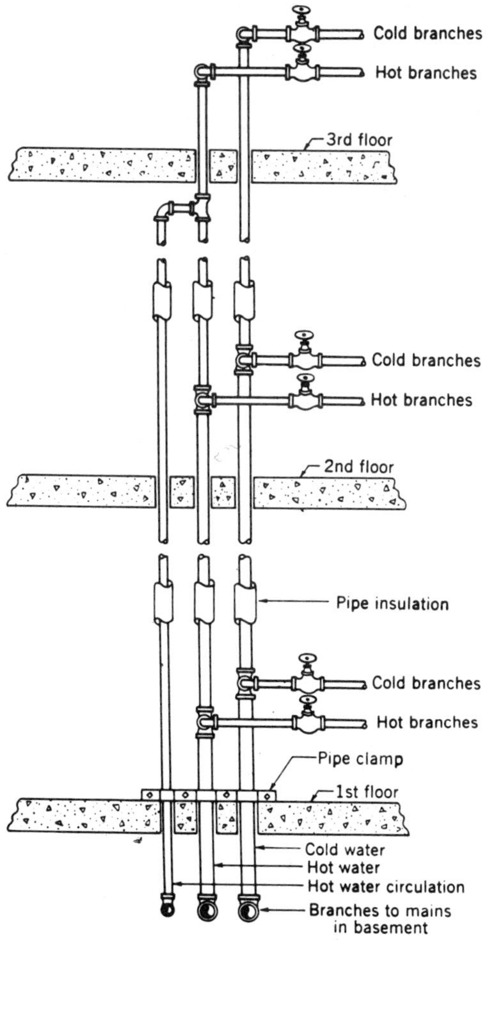

Fig. 6. Arrangement of risers for laboratory building that is shown in Fig. 2.

Risers

Figure 6 illustrates typical water risers that might be used with the system of piping shown in Fig. 2. These risers would serve the upper floors in the research laboratory. Figure 6 shows the cold, hot and hot water circulation lines along with the branch takeoffs to the fixtures. Since the height of risers does not exceed 25 to 30 ft., there is no need for expansion devices, but it is necessary in this installation to avoid rigid branch connections. Branch piping should be installed with enough flexibility to allow for the slight movement in the vertical piping.

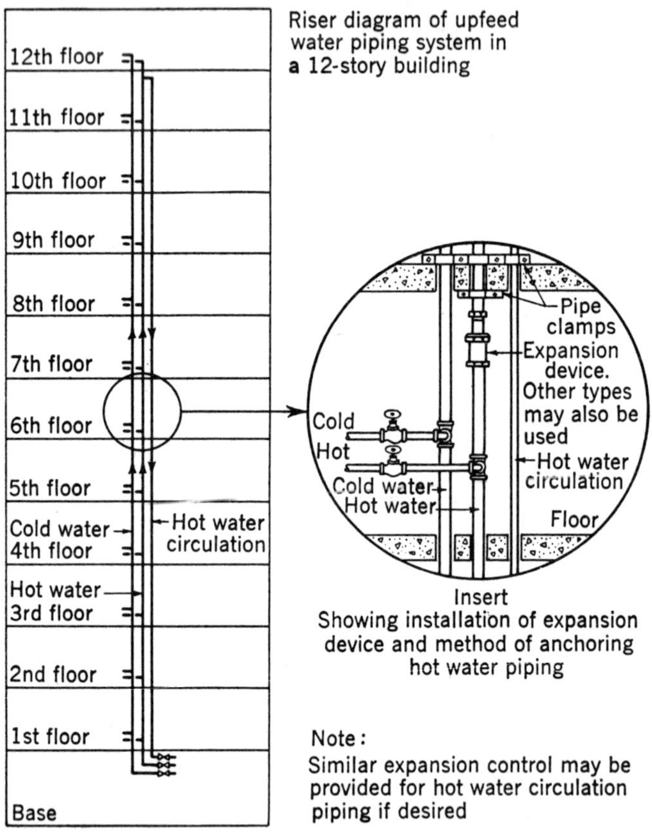

Fig. 7. Arrangement for expansion and anchoring devices on hot water piping in a 12-story building.

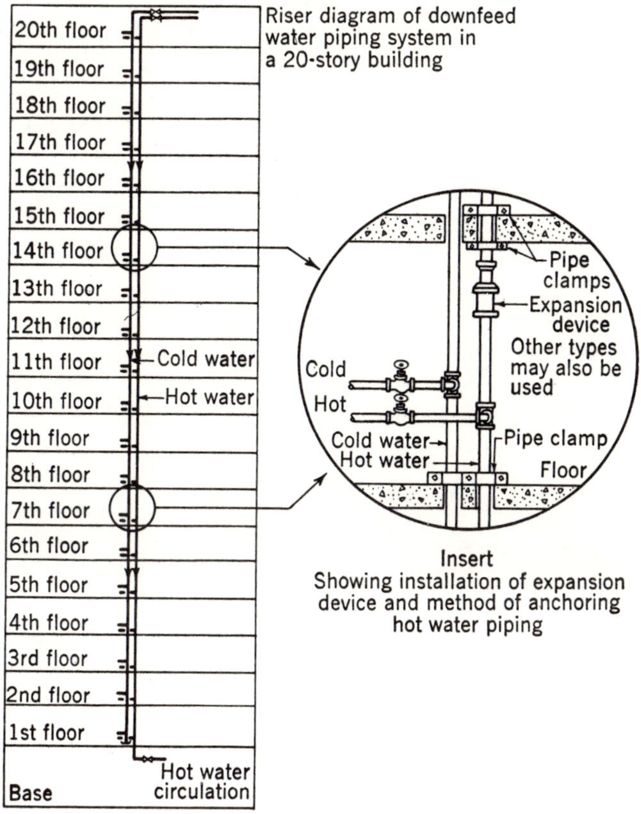

Fig. 8. Arrangement for expansion and anchoring devices on hot water piping in a 20-story building.

Tall Buildings

Now let us study the condition of expansion in a building, Fig. 7, that is 12 stories high. A typical set of water risers, as may be designed for this building, is illustrated. This type of supply is known as an upfeed water distribution system. Assume that the total height of the piping is 100 ft. and it is desired to place an expansion device at the middle point of the hot water piping.

The insert shows the placement of the expansion device. The lower 50 ft. of piping will expand and compress the device while the upper 50 ft. of piping has freedom to expand in an upward direction. It is always advisable to provide an access door or other means for inspection and care of the expansion device.

Two sets of pipe clamps are included for the support of the piping at the expansion device. These combined pipe clamps are the anchor. The lower pipe clamp prevents the piping from expanding any further in an upward direction while the other pipe clamp supports the upper 50 ft. of piping so that its weight will not be carried by the expansion device.

Expansion device protection is afforded the hot water circulation line in a similar manner.

Very Tall Buildings

Figure 8 shows typical water piping risers as they may be installed in a 20-story building. This piping system is the downfeed type, meaning that both cold and hot water are distributed at the highest floor in the building and extend downward to supply each floor of fixtures. The expansion device is installed in a like manner to that shown in Fig. 7. Because of the building height, two expansion devices are utilized with similar anchoring procedure.

Since pipe clamps are placed at every second or third floor to support the piping, the piping itself can only expand in an upward direction. Therefore, the piping above these two expansion devices has some freedom to move in an upward direction. Sections of piping below these devices will expand upward and compress the expansion device.

Although the pipe clamps at the various floors support the piping, the uppermost ones in each section of piping will be lifted clear of the floor with the expansion of the piping contained in that particular section. Similar consideration must be given to the installation of expansion loops or any other type of expansion devices that may be used.

CHAPTER 38

Increasing Water Supply Pressure in Buildings

Every new building under construction, regardless of its size, demands a water supply of good quality and ample pressure. Good quality means water that is bacteriologically pure and of acceptable appearance and taste. Ample pressure means water supplied to the various fixtures and equipment in the building at a pressure that will provide a steady flow to fulfill the needs of the fixtures and equipment despite the number of fixtures that may be used at the same time.

Since water of sufficient pressure is not always available at a given site, consideration must be given to means that can be employed to boost pressure within the building. Several methods that are more commonly used to serve this purpose will be described.

Street Mains With Inadequate Pressure

An engineer frequently finds that a project which he is helping to design will be served by a street main with pressure so low that the main cannot supply the entire needs of the building. When this occurs, the underlying reason for the low pressure in the street main can usually be attributed to one of the following factors:

1. Existing low pressure water distribution system.
2. Street main of variable pressure, undersized piping.
3. Mains overtaxed by the increased number of new buildings.
4. Mains that are old and low pressure is purposely maintained to prevent rupture of piping.
5. Partial stoppages by foreign objects or tuberculation in the water distribution system that tend to reduce the pressure.
6. Summertime load — large volumes of water used by the public and by equipment such as air conditioning units.

Importance of Adequate Water Pressure

In modern construction, just the mere flow of water, no longer suffices the building needs. A dependable flow and an adequate pressure is a necessity (1) because it lessens the possibility of back-siphonage in a part of the interior water supply system; (2) the greater number of flush valves that are installed require prescribed pressures for their proper function; (3) the public uses more water and the various fixtures and equipment utilizing water are dependent on a water supply that has steady flow and constant pressure to produce the desired results.

For these and many more reasons, careful consideration must be given to the water supply for a proposed project, especially when an inadequate water pressure prevails in the street main.

When low water pressure is experienced, the engineer has several methods from which he may choose to increase the pressure. The method chosen of course is dependent upon the specific demand requirements of the building or nature of his proposed design.

Although there are numerous ways in which this may be accomplished, we shall confine ourselves to just three of the more generally used methods. These follow:

1. Tankless water supply system.
2. Pneumatic tank water supply system.
3. Roof tank using house pumps.

A tankless water supply system serves its function best in a building where small quantities of water are used but the water pressure in the street main is insufficent to fulfill the demand requirements.

By means of a pump tied into the line, the required pressure is built up and maintained. Figure 1 shows a typical layout of the pump and piping hook-up necessary for such a system.

It will be noted in this figure that the water supply line leads directly to the fixtures, with a branch taken from the line and connected to the pump. The system operates as follows:

Suppose a building requires a water pressure of 40 psi and the available water supply varies between 30 and 40 psi, depending upon other factors that draw on the water in the street main. While a pressure of 40 psi is flowing through the interior piping system, the

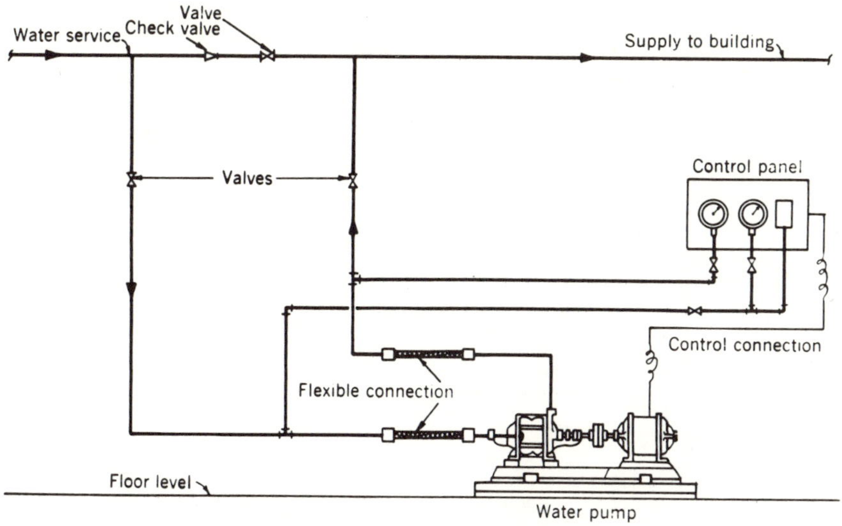

Fig. 1. Typical arrangement of tankless water supply system.

pump is inoperative. However, once the pressure drops to say 35 pounds, a pressure switch is activated which starts the pump and this builds up the desired water pressure. A check valve is placed on the bypass line to prevent the pumped water from backing up into the supply line. Flexible connections are installed between the pump and the piping to prevent the pump vibration from being transmitted through the piping system. In addition, the pump is supported on a cushioned base and this is in turn anchored to the floor slab.

The pump capacity is selected on the basis of the building needs. If a more dependable pumping arrangement is sought, duplex pumps are used. When this system is carefully designed and adequately sized for the needs of the particular building, it can be counted upon for satisfactory service. A disadvantage is that the water will be furnished to the fixtures as a throbbing supply due to the action of the pump.

Pneumatic Tank Water Supply System

For larger buildings with greater water demands, where a constant pressure is desired, a pneumatic tank system is employed. Here again, the water is supplied directly to the fixtures with a branch taken from the water main and connected to the pneumatic tank

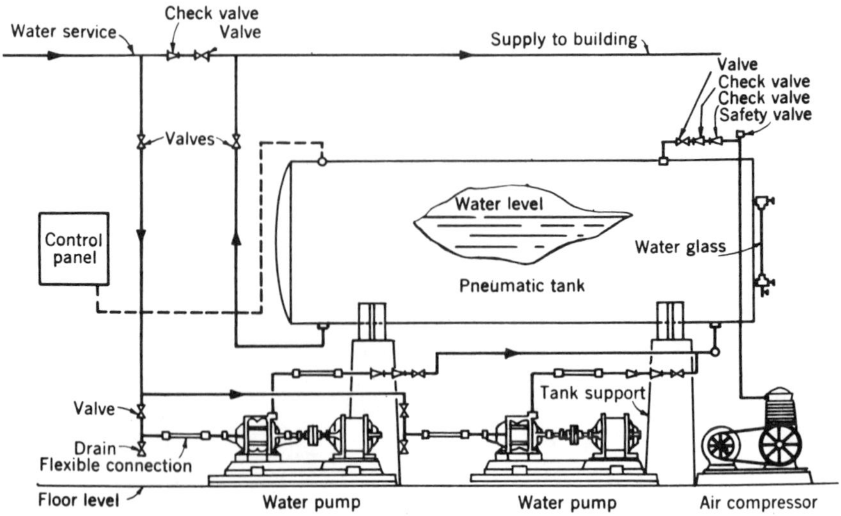

Fig. 2. An arrangement of a typical pneumatic tank water system.

system. The water pressure is built up by the pumps and tank, thus assuring a steady flow of water throughout the building. Figure 2 shows a typical arrangement of tank, piping, air compressor and component apparatus.

As illustrated, the water supply bypasses the pneumatic tank system but a branch from the water supply is connected to the pump suction. The discharge from the tank is reconnected back to the supply for the fixtures.

The pump or pumps force the water directly into the storage tank which is partially filled with a volume of air at a prescribed pressure maintained by the air compressor. The water in the tank occupies approximately two-thirds the capacity of the tank. The water level in the tank is shown on the water glass. Since the water is pumped into the tank under pressure, it will further "squeeze" the compressed air which is the force that acts to furnish the water in the amount and pressure needed for the height and water demand of the building where it is installed.

The size of the pneumatic tank, pumps and related equipment is determined by the size or height of the building and potential water demand.

A pneumatic system has the advantage that it can be installed in the building at basement level and is used for an upfeed water distribution system for the cold water as well as the hot water. This system when properly designed and installed can be counted upon to service efficiently the needs of the building. In very tall buildings an intermediate pneumatic tank system is installed at an upper floor level to sectionalize the water distribution systems. A disadvantage of the system is that a considerable amount of space is required for the equipment to fulfill its function. Moreover, if a fire standpipe system is contemplated, a fire pump must be installed.

Roof Tank System

A roof tank system, like the pneumatic tank system, is used in buildings where large volumes of water are needed for the demand requirements. The tank is located above the roof level of the building and is filled by means of house pumps located on the ground floor or in the cellar. A typical system is shown in Fig. 3.

The roof tank is located at a level above the roof and at a height calculated to maintain the needed pressure at the top floor level of fixtures and equipment. The size of the roof tank is determined on the basis of the volume of water in gallons per minute that will be used in a given time.

House pumps are usually of the duplex type and are sized for the capacity that will furnish the required amount of water to satisfy the desired pumping cycle.

If the roof tank is to be used for both the fire and domestic water needs, the tank must be sized to hold at least 3500 gallons of water for fire reserve in addition to the domestic water reserve. The domestic water reserve is estimated by either the pumping cycle desired of the pumps or by the gallons of water to be held in reserve, should the water supply fail or drop. Some engineers in their design figure reserves that will last anywhere from 15 to 30 minutes of normal consumption, depending upon the type of building served.

The water level is maintained in the tank by means of a ball float valve. Many varied types of valves are available for this purpose. An actuating switch, an integral part of the valve, starts and stops the pumps. Here again, various types of switches are available for this purpose.

The roof tank can be constructed of steel or of wood. Also, it consists of just one compartment or several compartments, depending upon the particular design favored by the engineer. A properly designed roof tank system is an efficient system and will serve, in a troublefree manner, the needs in a building of several stories or tall buildings of many stories.

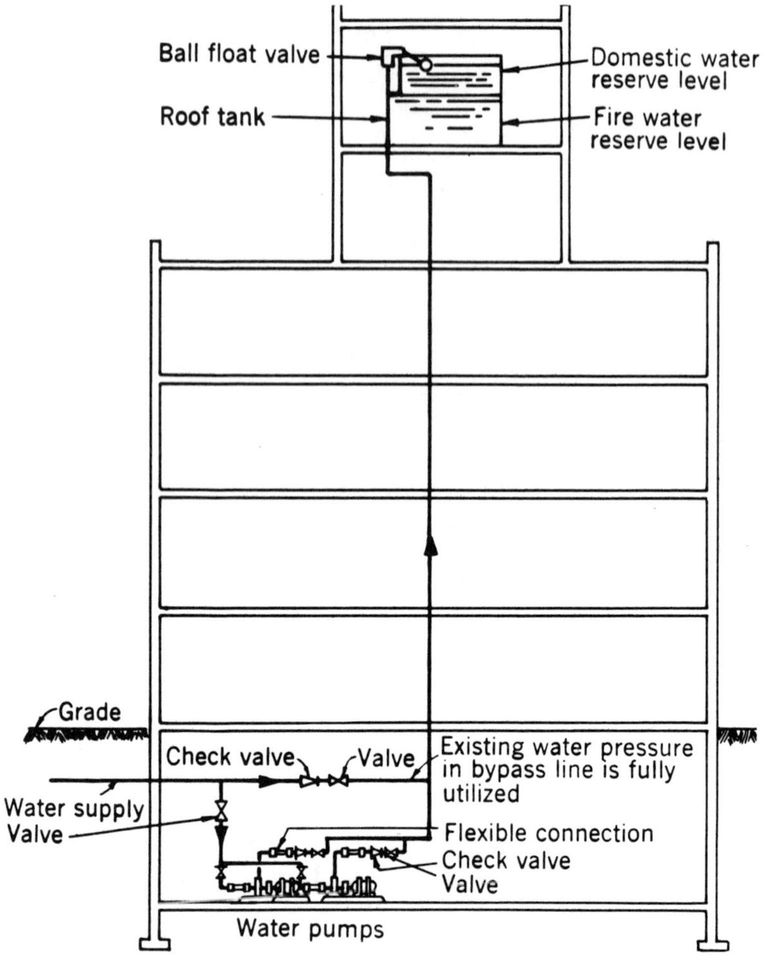

Fig. 3. Arrangement of roof tank water supply system.

The disadvantage of a roof tank system is the periodic cleaning required to maintain a taste-free water. To this end the water in the tank will remain cleaner for longer periods if the entire water reserve in the storage tank is circulated. This circulation may be accomplished by the proper arrangement of piping comprising the down feed supply in the tank.

When a system is selected for use in a particular application, the equipment manufacturers should be consulted with regard to the proper type to use and the hook-up recommended.

CHAPTER 39

Drinking Water Systems

Adequate drinking water facilities have an important place in every new building being erected. The proper placement of fixtures for dispensing water is given considerable attention by the architect and the engineer. In most cases, drinking fountains or drinking water stations, as they are sometimes called, are located within easy reach of the major groups of persons working at each floor level.

In general, the drinking fountains are placed in corridors, kitchens, work spaces or other prescribed areas where they will be centrally located for the groups of persons using them.

Classification of Drinking Fountains

Drinking fountains are available in many different types and forms to fit the most diversified needs. In addition, drinking fountains are manufactured from many materials, including vitreous china, enameled cast iron, stainless steel and aluminum. To limit the scope of this article, attention will be directed only to drinking fountains that are normally installed in the corridors of say a school building. Figure 1 illustrates the three basic types of fountains that might be utilized in such areas.

Exposed Type. Detail *A* of Fig. 1, represents the exposed type of drinking fountain. This unit is fully exposed and projects into the corridor. It is supported by means of bolts fastened to the wall construction.

Semi-recessed Type. Detail *B*, shows a drinking fountain that is partially recessed in the wall construction. Only the bowl portion of the fountain projects into the corridor.

Fully-recessed Type. When it is desired to avoid any projection in the corridor, the fully recessed type of drinking fountain as shown in *C* is used. This particular unit is completely recessed in the wall

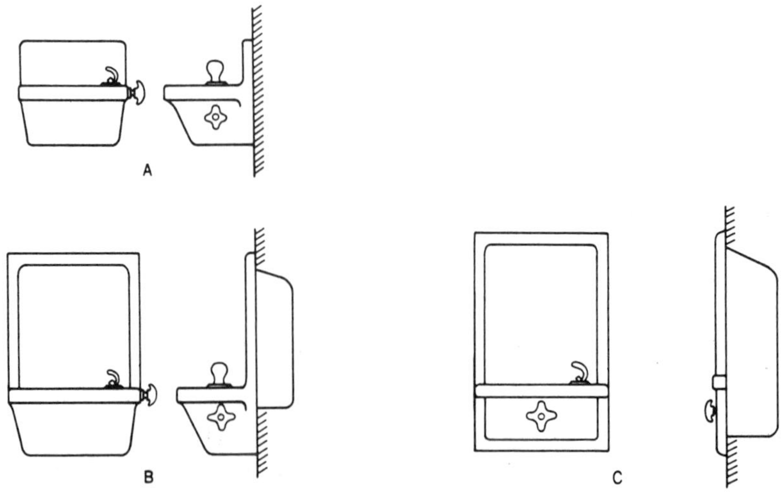

Fig. 1. Three styles of drinking fountains — exposed, semi-recessed and fully-recessed.

construction with nothing more than the flange or edge of the unit projecting into the corridor.

The units illustrated may be used with the regular cold water system or a chilled water system, as will be described later.

Electric Water Coolers

Water is most palatable when it is chilled. In climates where it is warm year-round, the temperature of the water supply is also high. In temperate climates, the water becomes palatable during the winter months but loses this desired feature during the summer. Therefore, in order to maintain the drinking water at its most suitable temperature, water coolers are used. A basic type of water cooler is featured in Fig. 2.

This unit consists of a 5 gal. bottle of water and a self-contained refrigeration unit. A removable bucket is attached to the front of the unit to catch water spillage. No connections to water or drainage systems are required. The unit is operated, merely by plugging the electric cord into a power outlet. The big advantage of this drinking fountain is that it can be set in any desired location.

The next step in electric water cooler development is the type that connects to the water and drainage systems of piping, Fig. 3. This unit is somewhat similar in shape to the one shown in Fig. 2, with the exception, that the 5 gal. bottle has been omitted and the means of dispensing water has been replaced with a bubbler. This unit must be placed where water and drainage facilities are available. Different styles are offered in such a unit. The unit can be supplied with either hand or foot controls to activate the bubbler. In addition, several modifications are available in the manner of dispensing the water. Glass fillers are a typical example. Further modifications include a storage space in the unit for food or bottled liquids. More recently, a new innovation has been added that will dispense hot water for coffee and tea.

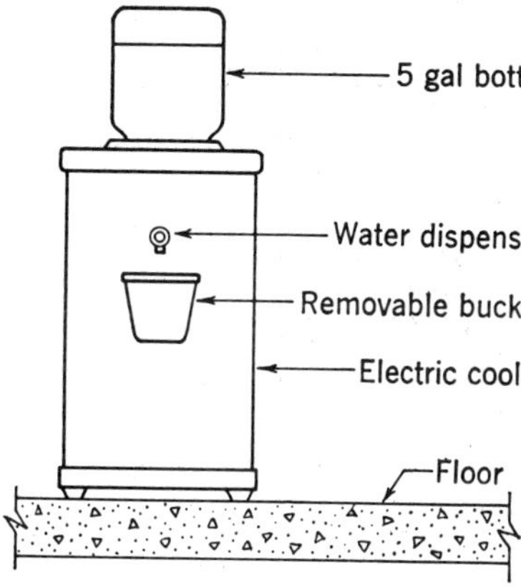

Fig. 2. Detail of self-contained water cooler.

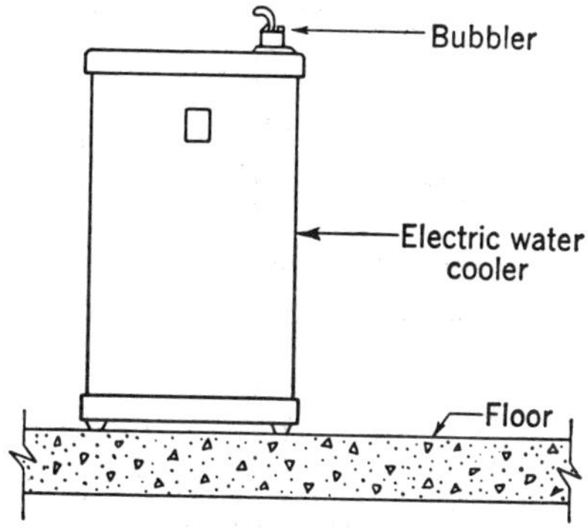

Fig. 3. Detail of electric water cooler.

This type of drinking fountain can be placed in an exposed position in the corridor or can be placed in a recess when it is provided. When required, the unit can be so arranged to supply another station with chilled water. To do this, a unit like Fig. 3 is placed on one floor and one of the units shown in Fig. 1, is placed on the floor above or below. The electric water cooler will supply the other station with sufficient chilled water.

Roughing

As detailed in Fig. 4, the drinking fountains can be connected to either an indirect system of waste piping or to a system of piping that also serves other plumbing fixtures.

Detail *A* of Fig. 4 shows the components of an indirect system of waste piping while *B* shows how the fountains can be connected to a vented drainage system. The piping, of course, may consist of cast iron, galvanized steel, galvanized wrought iron or copper mate-

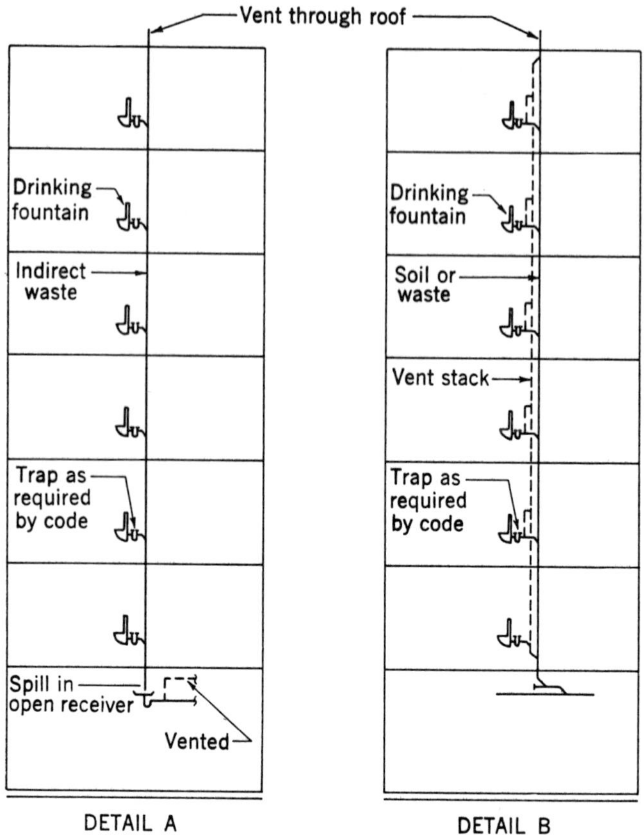

Fig. 4. Methods of installing drinking fountain waste systems.

rials. For the indirect waste system, an open sink, funnel drain is installed in the basement or cellar to receive the waste from the drinking fountains. The receiver must be vented to comply with code requirements.

Concealed Cooler Units

To fulfill a pressing need in many new structures, manufacturers have devised a cooling unit that can be concealed in a wall space. Figure 5 illustrates this new design trend.

Any one of the three designs shown in Fig. 1 can be used. The cooling unit is concealed in the wall space. The necessary cooling air needed for the refrigeration apparatus is drawn through the screen or louvers placed under the drinking fountain.

It might be timely to mention, that other types of refrigeration equipment is also available for remote location. However, since we are only interested in corridors for this chapter, the unit shown shall be used for example.

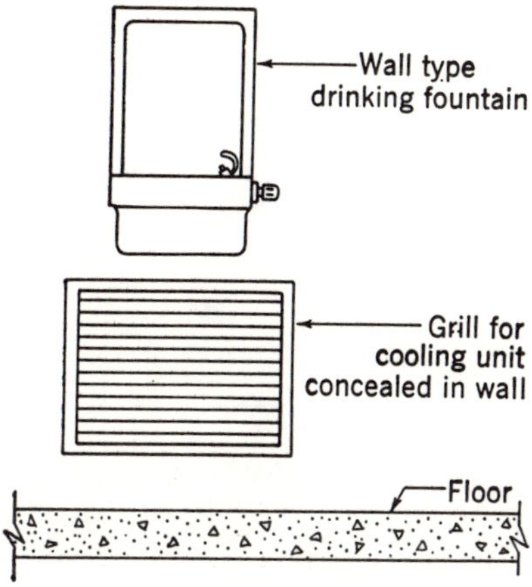

Fig. 5. Typical arrangement of drinking fountain and concealed water cooler.

This particular unit has a very distinct advantage for specific locations where it is desired to supply chilled water to drinking fountains on the floor above and below. Figure 6 is a typical example of this application.

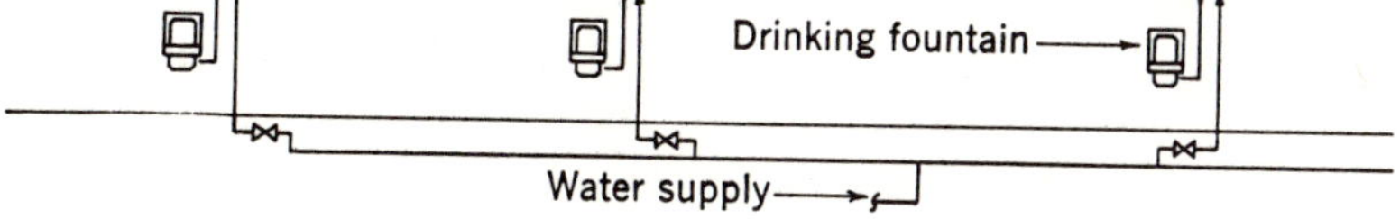

Fig. 6. Arrangement of cooling unit used to serve drinking fountains on floors above and below the installation.

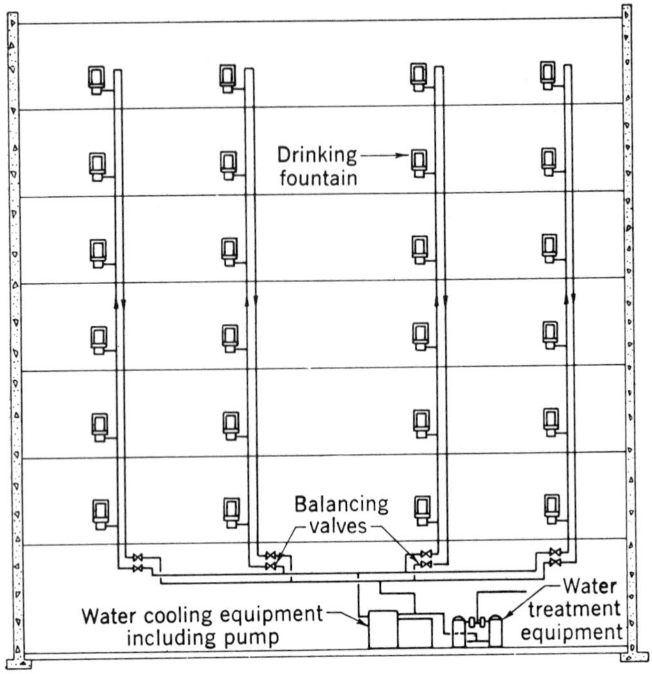

Fig. 7. Typical arrangement of a central system.

Suppose a 3-story school or similar building is contemplated. The cooling water units can then be placed at each of the drinking water stations on the second floor as shown. Water piping can then be extended from these units to the drinking fountains above and below. The drinking fountains are one of the three types illustrated in Fig. 1.

The economy of such an installation offers a good deal of merit especially when the building is expansive in size. Should a taller building be contemplated, a similar installation can be used. For example, if a 6-story building were involved, the cooling units could be placed on the second and fifth floors. The drinking fountains served would then be placed on the floors above and below the cooling units. For given conditions, this system may be regarded as a practical means of dispensing chilled water.

Central System

In many buildings it is both practical and economical to utilize a central system for the circulation of chilled water to the various

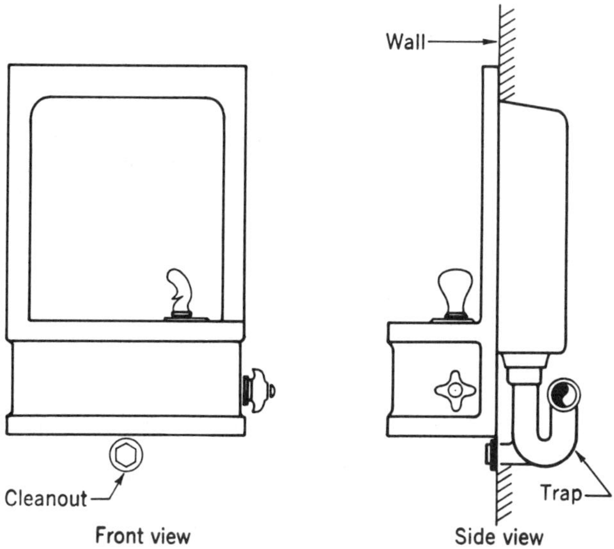

Fig. 8. Detail of drinking fountains used in Fig. 7.

drinking fountains in the building. For example in this case, we shall consider an office building. A typical arrangement of piping and equipment for a building of this type is detailed in Fig. 7, which shows the location of the drinking water stations on each floor of the building. The water chilling equipment is located in the cellar. When required, this equipment can be located at the top floor or an intermediate floor. The chilled water distribution system extends from the equipment up to the drinking fountains on each floor. At a point at the top of the riser and above the highest fixture, a return line is installed to allow the water to return back to the refrigeration equipment.

In a central system, the chilled water is pumped to circulate the water to each of the drinking fountains. To assure proper balance in the distribution system, balancing valves are placed at the base of the return risers. By the proper adjustment of the balancing valves, it is possible to prevent short circuiting of the chilled water through the risers nearest the refrigeration equipment. Therefore, each valve is throttled until the chilled water circulates evenly through the entire distribution system. This is an important factor to remember because it affords the flexibility in control needed for the equal distribution of water.

There are several ways in which the piping and equipment may be installed for a system of this kind. It is always good practice to consult with the manufacturers of the chilled water equipment in order to make the best use of the various types of equipment they have available for this purpose. When properly installed, this system can be depended upon for many years of satisfactory service.

Figure 8 illustrates the type of drinking fountain that has been used in Fig. 7. It shows the particular arrangement of the trap for this semi-recessed drinking fountain. A cleanout for the trap is placed flush with the wall. Although a water connection is not shown, it is connected to the back of the fountain.

Insulation

In order to retain the desired temperature of drinking water in a central system, the entire piping distribution system should be insulated. Several types of insulation are available.

CHAPTER 40

Noise Transmission in Piping

When designing a water distribution system for a new building, it is important to design for minimum noise transmission. Piping noise generated in one portion of the building is easily conveyed to other parts of the building. Sometimes these noises even seem to be magnified by the piping system. It is the responsibility of the engineer to determine just where piping noises will most likely be generated and to prevent them. It is equally important for the contractor to install the component parts of the piping system with due regard for those sections in the piping distribution system where noises are prone to develop.

1. Fixtures and Equipment.

When a water closet is flushed it creates a noise that is quite audible. Some water closets produce more noise than others. Often the noise is generated by the particular arrangement of the jets in the closet bowl. Again, it may be caused by the type of ball cock used in the tank closet combination. When flush valves are employed, a noisy flush generally indicates that some adjustment must be made on the flush valve or that the diaphragm strainer must be cleaned. One way of assuring a quiet operating water closet is to specify a siphon jet unit instead of the blow-out type. Where a tank-closet combination is used, it is important that water supply to the tank be adjusted by the control valve to the point that is most quiet in filling the tank. The float in a tank combination must not be allowed to rub against the side of the tank. Flush valves for water closets and urinals should be adjusted to operate quietly.

A vibrating loose faucet washer in sink, lavatory or bath tub can create a chatter in the piping system. The noise generated by the vibration is a loud one and is generally transmitted through a large portion of the piping system. The fault is easily corrected by using a screw driver to secure the washer in place.

If a faucet spindle is bent, a squeal will develop when the faucet is opened to a certain degree. It occurs when the flow of water reaches sufficient velocity to cause the spindle to vibrate against the side of the faucet shank. This condition is especially noticeable in hot water faucets where the heat of the water aggravates the occurrence. The solution here is to replace the faucet stem or spindle and if necessary to replace the faucet or faucet combination.

As the flapper in a check valve slams closed, it can be heard throughout the piping system. It is rather difficult to pinpoint a solution because each situation has a different cause. A water hammer arrestor is often an effective remedy. Otherwise, a check valve with a soft seat, or a spring loaded check valve might be considered as a way to cure the difficulty.

When a gate valve is open or partially open, a surge of water will cause a loose fitting wedge to vibrate and thus transmit noises. A solution is to specify good quality valves. In an existing system, it is rather difficult to locate the noisy valve. It often becomes necessary to shut off parts of the piping system until the section of piping where it originates is located. Once the faulty valve is discovered, it must be either adjusted to prevent vibration or be replaced.

A globe valve can create noise when a loose washer vibrates with the surge of the water passing through the piping. Here again, the faulty valve must be located and its washer tightened.

Any type of equipment that vibrates can transmit its sounds to the water distribution system unless the equipment is properly isolated to avoid direct connections with the piping. An example is house pumps. Each time the pump starts up, a considerable vibration takes place. Flexible connections placed at the inlet and outlet connections of the pump will prevent the pump vibration from being transmitted to the piping.

The same type of consideration must be extended to all other equipment that can project its noises directly to the piping system.

2. Water Hammer

One of the most common causes of noise in piping is water hammer. The basic cause is quick closing valves and faucets. When water hammer occurs, it is signified by a loud banging, pounding or rattling of the piping in which a shock wave is traveling. Because the piping system is a good conductor of sound, the noise can be heard in many parts of the building.

In addition to being very annoying, water hammer may rupture the piping or cause pipe connections to leak. Water hammer arrestors are the best means to rectify or reduce the effects of water hammer. The water hammer arrestors should be placed where the water shock waves are most likely to originate.

3. Excessive Pressures

Excessive water pressure can be the cause of noise in the piping system. It is especially apparent in small diameter piping when the result is noise generating, high-velocity flow when a faucet is turned on. Also, the causes of water hammer, when the faucet is closed, are exaggerated.

Throttling of main and riser valves sometimes is sufficient to alleviate the problem. Often, it is necessary to install a pressure-reducing valve on the water service line as it enters the building. If the building is rather large or spread out, it may be necessary to partially control the pressure at the water service line and then install individual pressure regulators at prescribed locations in the building to control the pressure in sections of the piping.

4. Undersized Piping

A system that is undersized, or contains sections of undersized piping, will react in a manner that is similar to a system containing excessive pressures. The only positive solution in this type of a situation is to design a piping system that is adequate for the building in the first place.

If the building is an existing one and the problem is that of undersized piping, it becomes necessary to determine which sections of the piping seems to be the noisiest and to replace these sections with larger diameter piping. In many cases, corrosion will reduce the inside diameter of the piping to an extent that it may be regarded as undersized. When this situation prevails, it might be timely to check the condition of the piping in various parts of the system to determine the extent of corrosion that has taken place. If the piping is sufficiently corroded it may be advisable to replace these sections of piping and at the same time, utilize a larger diameter piping that is more suitable for the needs of the building.

5. Piping Supports

Improper piping supports are often the factor that is responsible for a noisy piping system. This includes the supports used for the main piping runs as well as the supports used with the vertical piping risers or the horizontal branch piping.

For main piping runs in the basement of a building, it is necessary that the piping not only be adequately supported but also that provision be made for the expansion and movement of the piping. To reduce the noise factor, it is imperative that expansion devices be installed in the lines at proper locations to control the movement of the piping and confine its action to predetermined limits.

Where long runs of piping are involved, it is good practice to use roller supports on which the piping will rest. The roller supports will prevent the pipe from rubbing against or moving the trapeze hangers.

The piping mains should be so designed and installed that they do not come in contact with another section of the piping. Moreover, different systems of piping should not be installed so that one line might rest on another. In addition, the piping should not be installed too close to any wall or structural member. If the piping is installed too close in any of these instances, any movement of the piping will cause it to rattle or vibrate against the other piping or parts of the building and thus create noises that are transmitted through the distribution system.

Similar consideration must be extended to the piping risers where two lines can rattle or vibrate against each other if they are installed too close. In tall buildings, sufficient attention must be given to the expansion in the piping to avoid the occurrence of noise.

Small diameter branch piping requires special attention to ensure that it will not create a noise condition. This piping will move considerably with the passage of the water through the piping and closure of faucets or valves. Therefore, it should be carefully supported and spaced far enough apart from other piping or building construction against which it can rattle or vibrate.

CHAPTER 41

Suction Tanks

It is not uncommon today to find that the increase in water consumption in cities has outstripped the expansion of municipal water services. The result is often inadequate pressure in the street mains. When an engineer finds that an existing water supply pressure is unreliable, he has several means at his disposal to ensure that a new building will be adequately supplied:

1. If the building is not too high, a booster pump supply system can supplement the water pressure when it drops below a prescribed limit. For example, if 50 psi is needed to amply serve all the fixtures in the building, including those on the highest floor level, it becomes imperative that this pressure be maintained at all times. The booster pump will operate when the water pressure drops in the line serving the building until the street pressure is back to normal.

2. For a larger building, a pneumatic tank system can be used. In this system, water is pumped into a tank to a predetermined level. Compressed air in the upper portion of the tank provides the force required to lift the water at the desired pressure and quantity to all portions of the building. Often, several pneumatic tanks are used in the building and may be placed at different building levels.

3. For tall buildings and buildings with heavy demands, a roof tank is generally employed. The tank is usually sized to hold a water reserve of 20 to 30 minutes or more. The roof tank, because of this reserve, compensates for any fluctuations in the street water pressure and thus provides a constant water supply.

These methods can be used where the street water pressure is inadequate to serve the needs of the building. In addition, these methods described are also used where some fluctuation exists in the street pressure so that the engineer may regard the service unreliable and attempt to compensate for the fluctuations.

Inadequate Water Service

In some areas, the water supply and pressure are so uncertain that it is impossible for the engineer to depend upon the supply alone. Moreover, he cannot draw water from street mains with his pumps to any reliable degree because other nearby buildings are also depending upon their pumping equipment to deliver their water needs from the street mains. To safeguard his client and the persons that will occupy the building, the engineer must consider a different means of providing an adequate water supply. Figure 1 illustrates this typical condition, which generally calls for the use of a suction or surge tank. Since this is a diagram, equipment, valves and other necessary devices that are part of this type installation, are not shown.

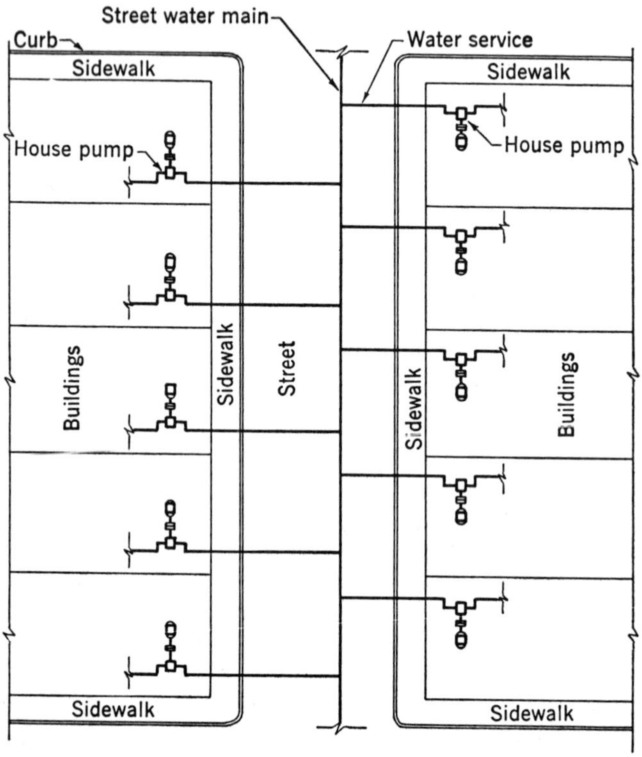

Fig. 1. Manner in which tall buildings facing a street have the house pumps connected to the street water main.

Suction Tanks

There are a number of types of suction tanks that may be used. Figure 2 represents just three types, the open suction tank,

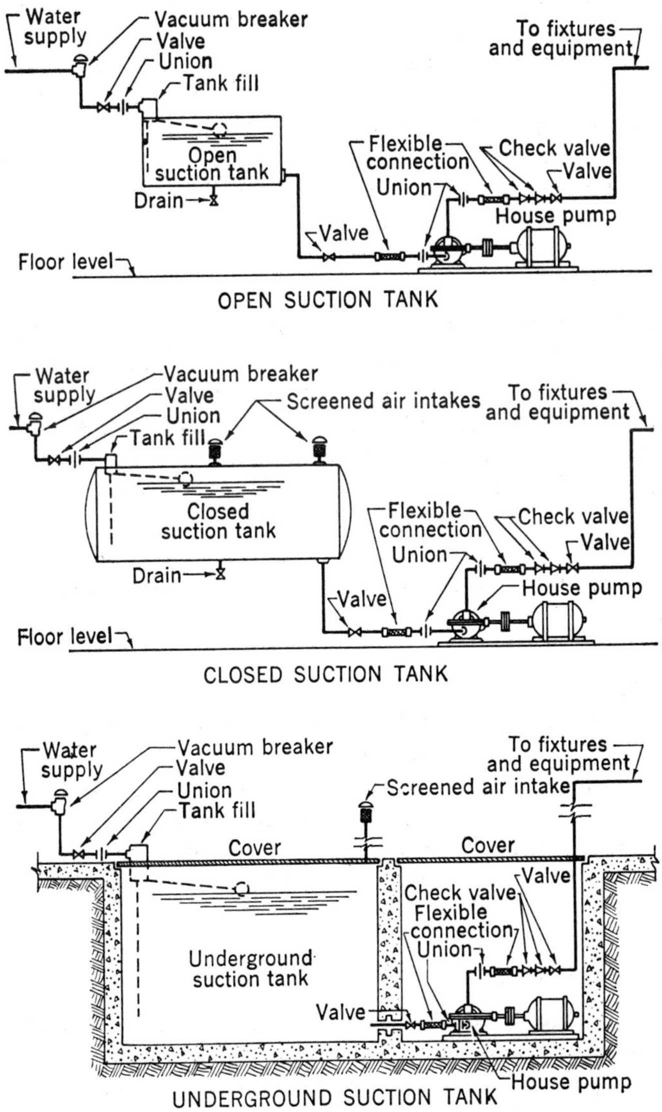

Fig. 2. Three different types of suction tanks that may be installed in a building.

the closed suction tank, and the underground suction tank. Piping arrangements shown are diagrammatic and they are not intended to include all the component equipment or devices that may be utilized with each system.

The open suction tank, as its name implies, may be open on top or may include a cover. If a tight fitting cover is used, some such provision as a screened air intake must be added to assure a balance of air in the tank. Screened air intakes are also used where a closed type suction tank is installed.

The underground suction tank is used when needed to satisfy the requirements of the project or where it is desired to save space in the equipment room.

Purpose of Suction Tanks

Suction tanks hold a reserve of water for the pumping equipment. Instead of the pumps drawing water directly from a street main, they will pull the water from the tank.

The water service fills the tank to a prescribed level. A ballcock or system of floats automatically controls the height of the water. Various control systems are applicable. The suction tank is commonly sized on the basis of information regarding the adequacy of the street water mains that the local water company can supply. Otherwise, a more detailed investigation is required.

Application of Suction Tanks

Three generally accepted methods for providing a constant water supply were covered earlier, using either booster pumps, a pneumatic tank system or a roof tank. Now it will be shown how each one of these systems can be installed in conjunction with a suction tank in areas where the water service is inadequate, and it is not practical to pump water from the street mains.

Booster Pump System

A typical arrangement of a booster pump system is shown in Fig. 3. The water service, installed to bypass the suction tank and the pump, will serve the building needs as long as its pressure is adequate. A connection from the water service replenishes the reserve of water in the suction tank. A supply line from the suction tank is connected to the suction side of the

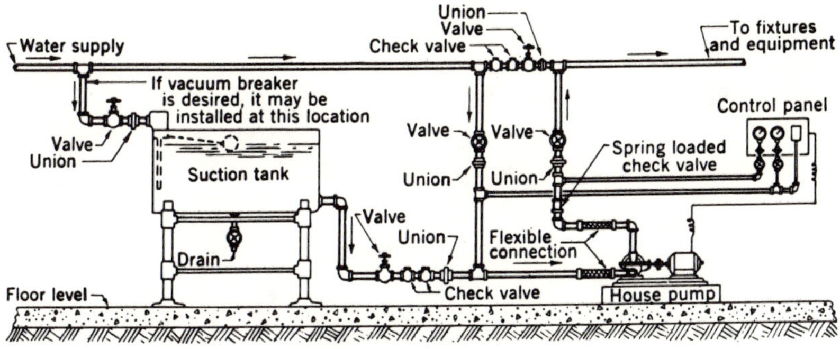

Fig. 3. How a suction tank can be used with a booster tank system.

pump. The check valves placed on this line prevent water from the water service line from backing up and overflowing the suction tank. At the same time, these check valves are held in a closed position by the water service pressure acting upon them and prevent the water in the suction tank from draining into the system until needed.

When the pressure in the water service line drops below a prescribed limit, the pump operates and will draw upon the reserve of water in the suction tank. The size of the suction tank, in this instance, need be only large enough to make up the difference in the quantity of water required during periods of peak demand in the area. For example, this low ebb in water supply may occur during the noon hour when all the buildings in the area are exacting the heaviest toll from the street mains. Another peak period may occur from 4:30 to 5 p.m when employees are preparing to leave.

Pneumatic Tank System

Figure 4 shows how a pneumatic tank system can be installed, using a suction tank in conjunction with this arrangement. In this system, the pumps draw water directly from the suction tank. Therefore, the suction tank must be of sufficient capacity to fulfill the demand needs of the pumps during their pumping cycle. For example, if the pumps supply water at a rate of 100 gpm to the pneumatic tank and the pump operates for say 10 minutes, a reserve of at least 1000 gallons is needed in the suction tank. However, since some factor of safety is desirable, it

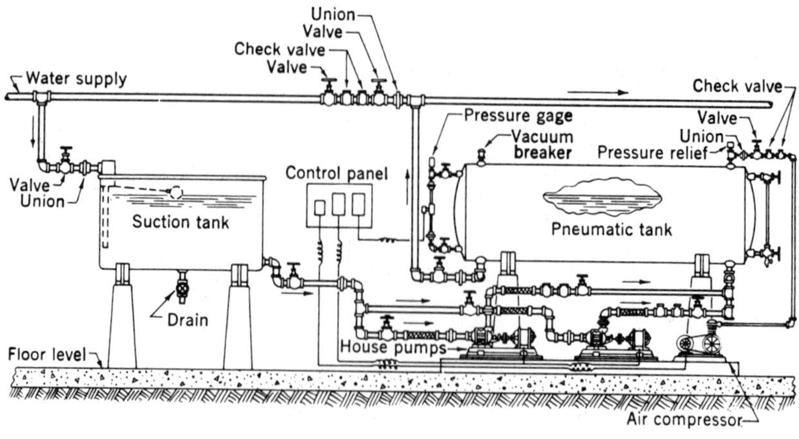

Fig. 4. Piping arrangement for a suction tank used with a pneumatic tank system.

should be large enough to hold at least 1500 gal. Often it is advisable to double the capacity of the suction tank.

In any event, the factors that the engineer must consider in sizing the suction tank are the cycling of the pumping equipment, and the rate at which the water service can replenish the suction tank during the peak demand periods of the day.

An open suction tank has been used in this illustration. The closed or underground tank can be used with equal success. Moreover, the illustration shows the water service line bypassing the suction tank, pumps and pneumatic tank. This arrangement will serve where service water pressure is used to its greatest extent in supplying building needs. If this pressure is not needed, the control valves on this line can be closed which means that the system will be supplied only from the water contained in the suction tank. Here again, the particular arrangement of equipment, valves and devices represents only one of several possible.

Roof Tank

A roof tank system may be installed as shown in Fig. 5. The pumps draw water directly from the suction to the roof tank. So far, the discussion has covered methods whereby a suction tank can be used to safeguard the water supply in buildings from a few stories to 20 stories in height. Systems for taller buildings, and also large industrial and manufacturing plants

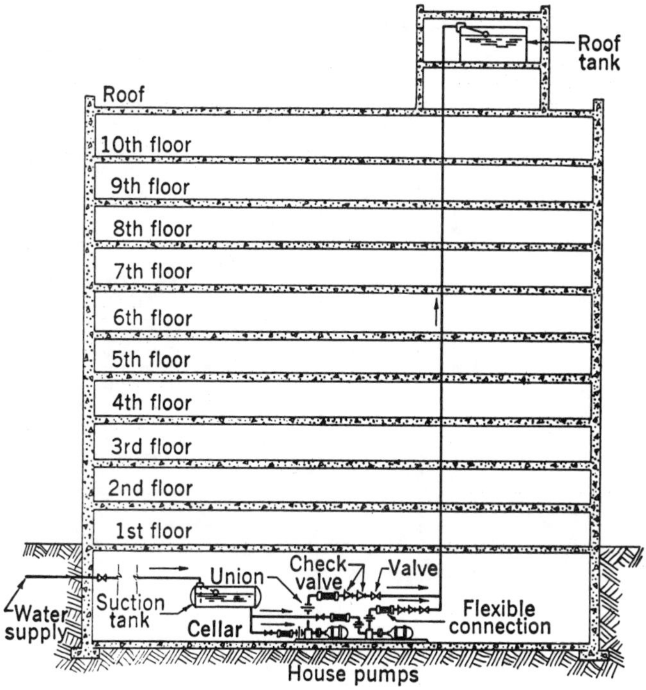

Fig. 5. Suction tank in basement serving a roof tank gravity system.

where enormous quantities of water are needed, will now be discussed. At this point, however, it might be timely to mention that whenever unreliable sources of water exist at a site, it is good practice to check all the available sources with the thought of installing two or more water services into the building.

Dual Water Services

Where inadequate water supply and pressure exists at a site, it is advisable to utilize, when possible, the water resources of two separate supplies as suggested in Fig. 6. This shows a street with buildings on either side, which is typical of many of the larger cities. The various ways that two water services may be utilized are illustrated.

Building *A* is a corner building with the water services supplied from two different street mains. The water services are combined in the building to serve the suction tank and pumping equipment.

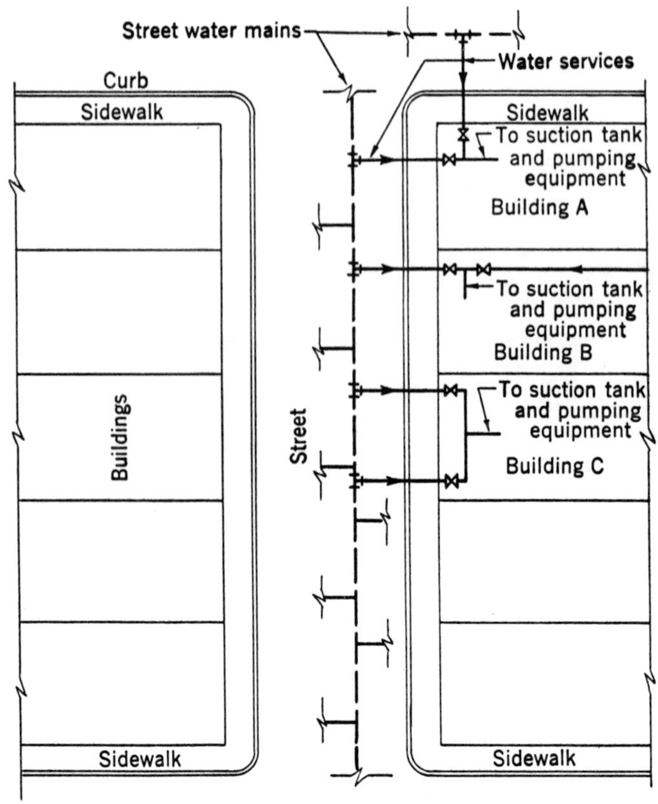

Fig. 6. Typical street with tall buildings, showing various ways for connecting house pumps to water mains in the street.

Building *B* represents a building that extends from one street to the other. Here the water services are taken from each of the two streets. They are also joined inside the building.

Building *C* shows how two water services may be taken from the same street main and joined inside the building.

A large building cannot depend on just one water service, since any failure in one service could disrupt all of the building's activities dependent upon the water supply.

40-Story Buildings

Figure 7 shows how a water distribution system, including a suction tank, may be installed in a 40-story building.

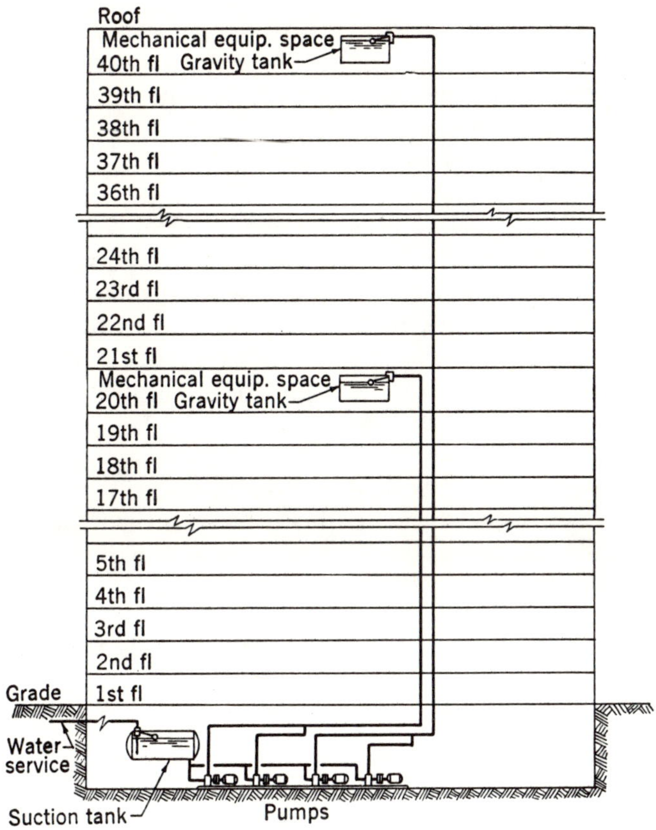

Fig. 7. How a suction tank is utilized with a water distribution system to serve a 40-story building.

While several methods can be used to distribute water in a 40-story building, a method employing two gravity tanks is shown. The tanks are placed at the 20th floor and at the 40th floor. Both of these floors are mechanical equipment floors and are the service floors for other equipment, such as air conditioning, heating, electrical and other apparatus normally located at this level. Each tank is supplied by its own set of duplex pumps. One suction tank is used and each set of pumps draws water from this tank. The tank should be of sufficient size and capacity to supply both sets of pumps during their pumping cycle.

Since each gravity tank holds a water reserve for at least 20-30 minutes or more, the suction tank need be only large enough to supply the full pumping cycle of both pumps operating together for that time. In sizing the suction tank, the engineer must also take into consideration the rate at which the water service can replenish the suction tank to fulfill the demands of the pumping equipment.

Large Industrial Sites

When large industrial, manufacturing, processing or other types of establishments that use considerable quantities of water are involved, it is important that an adequate supply of water be provided. For example, refer to the site shown in Fig. 8. This illustrates how a building may be located on a site and the manner in which the water is supplied to the elevated storage tank and thence to the building.

In many of such industrial plants, located far from the nearest city or town, the water supply may have to come a number of miles to serve the plant. If the water company installs the water line, it generally tries to size it adequately for the demand of the building. Often, however, the plant may have to obtain its water from an existing water main located in a nearby street or highway. It is then quite possible that the water

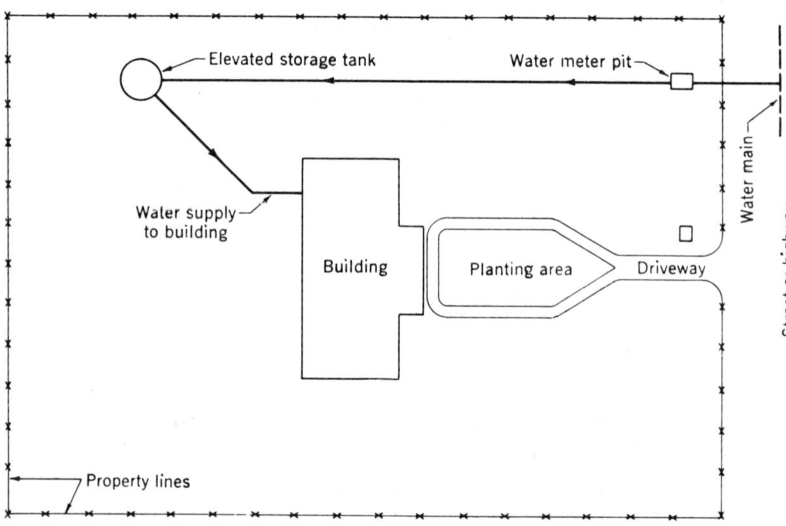

Fig. 8. One way of supplying water to an elevated storage tank serving an industrial building.

line is not of adequate size to supply the water demands of the plant. In this event, it is necessary to provide an elevated storage tank on the site. In addition to the necessary water storage on the site, the elevated storage tank assures a sufficient water pressure.

Referring to Fig. 8, a water service is extended from the street main to the pumping equipment that is located at the base of the elevated storage tank. The pumping equipment replenishes the storage tank. Since the storage tank supplies all the needs of the plant, it must contain an ample quantity of water. It is not unusual therefore, that an elevated storage tank holds 50,000 gal. or more.

Undersized Water Mains

Many times the water main that serves an industrial building is not capable of delivering a dependable supply of water, for this undersized main must also furnish water to nearby plants. The engineer must take every precaution to be certain that his proposed building will receive an adequate supply, even when pumps are drawing water for the nearby plants. There are a number of ways to ensure an adequate supply.

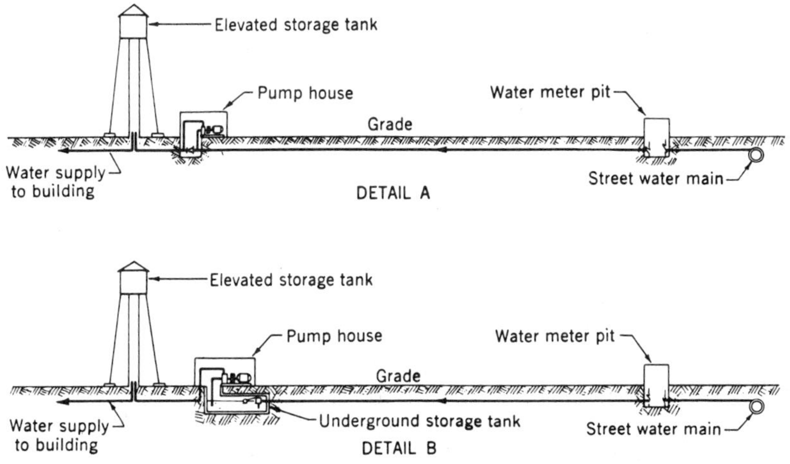

Fig. 9. Two methods that may be used to supply water: elevated storage tank, and combination elevated storage tank and underground storage tank.

Detail *A* in Fig. 9 shows how the water supply is hooked up when the water pressure to a site is adequate. You will note that the water service is run to the pumping equipment at the base of the elevated storage tank where it is raised up to the storage tank. If the water pressure is sufficient, it can directly supply the tank, with the pumps then acting as boosters. A water meter pit containing a water meter and other equipment as may be required by the local water company is installed just inside the property line.

Detail *B* shows how a suction tank may be applied where the water service is unable to fulfill the capacity of the pumping equipment during its cycle of operation. Here, for the purpose of example, the suction tank is located underground. In fact, it is located under the small building that houses the pumping equipment. The size of this underground suction tank is dependent upon a number of factors such as the rate of water flow in the water service line, the rate at which the storage tank must be replenished, and the capacity of the pumping equipment in gallons per minute. All must be taken into consideration when sizing the suction tank. The engineer must determine how much underground water storage is needed.

Underground Storage of Water

Many plants today have enormous underground storage tanks, some holding in excess of one million gallons of water. Such tanks are designed to provide a constant supply of water despite any disruption of water service by a water main break or other cause.

The engineer must size the underground storage tank in accordance with the water requirements of the plant, and the hours or days that the storage tank will be expected to handle the water requirements until normal water supply delivery is resumed in the outside water main.

Even though a large underground storage tank is utilized, an elevated storage tank is still needed. The arrangement of the elevated tank is similar to that shown in Fig. 9, Detail *A* or *B*. The tank supplies the building and the large underground storage serves as a reserve tank. Periodically, the water in the underground storage tank is used for replenishing the elevated storage tank. This design helps to maintain a fresh supply of water in the underground tank.

CHAPTER 42

Salt Water Distribution System

Hotels on the ocean often feature salt water distribution systems. The salt water is reputed to have therapeutic or other benefits.

Obtaining the Salt Water

Two factors must be considered before attempting the design of a salt water distribution system: location of the hotel and the quality of the intended source of supply. The hotel must be located near or adjoining the shore front and permission must be obtained from the local or governing authorities to extend a suction line out into the ocean. The second consideration is the quality of the salt water. If the salt water is polluted or dirty, then the system should not even be considered.

Assuming that the hotel is situated near a clean, wholesome supply of salt water, then the engineer must check with the governing authorities to learn of requirements concerning the installation of piping, how far off shore the intake must be placed, and the type of intake that may be utilized.

The method that may be employed to draw salt water from the ocean is shown in Fig. 1. The distance to which the piping must be extended must conform with the requirements of the governing authorities. Piling is driven down to hard pan and a substantial frame is secured to the piling. The salt water intake consists of a bullheaded tee connection with a diameter twice that of the suction line and set well above the frame to prevent any possibility of sucking sand. A pipe flange anchors the intake pipe rigidly to the frame. The intake connection is protected by a screened cage. The cage is also securely anchored to the frame. Access should be provided at the top of the cage to allow periodic cleaning of its interior. Skin divers accomplish the task of cleaning the interior and exterior of the cage in a very easy manner. Rocks may be placed around the cage if desired. Again, the

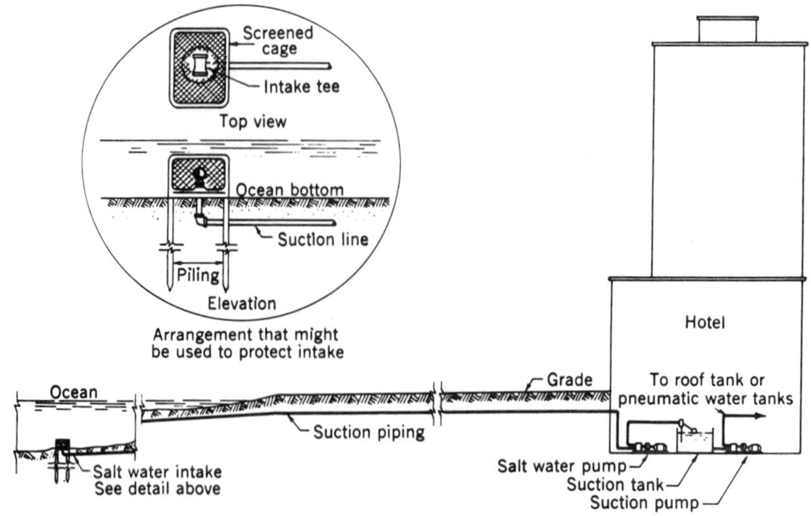

Fig. 1. Arrangement of salt water intake, piping and pumping equipment.

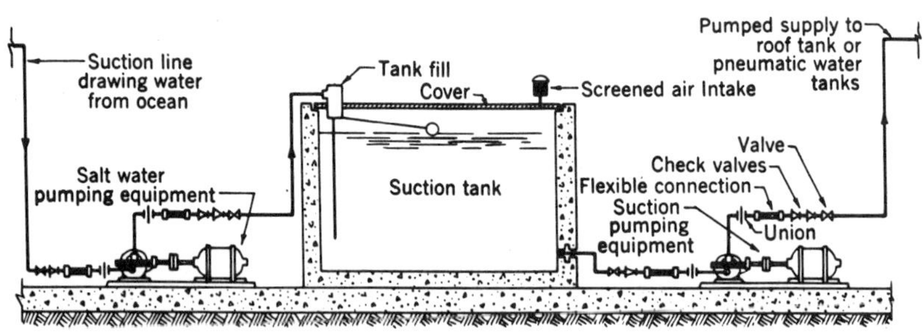

Fig. 2. Arrangement of equipment and apparatus to pump salt water from the ocean to the roof tank or pneumatic tank system.

authorities will usually advise on the specific type of intake arrangement to be constructed. Many other arrangements, including well points, may also be used.

Means of Pumping

One method used to draw water from the ocean is shown in Fig. 2, where a pump draws salt water from the ocean and discharges it into the suction tank. If desired, a strainer may be placed on the suction side of the pump to intercept any foreign objects that might be drawn into the system despite the screened cage. The suction tank can be of waterproofed concrete or other suitable material. The capacity of the suction tank must be determined on the anticipated demand for salt water in the hotel and the particular pumping cycle needed for the roof tank or pneumatic water system. For a 20-story building, for example, the tank should be not less than 10,000 gal. Attention should be given to the feasibility of using a two-compartment tank; this is very practical since it permits one compartment to be cleaned while the other functions. The suction pump delivers salt water to the roof tank or the hydropneumatic tanks.

Although only single pumps are shown in Fig. 2, dual pumping arrangements are preferred. The piping should include

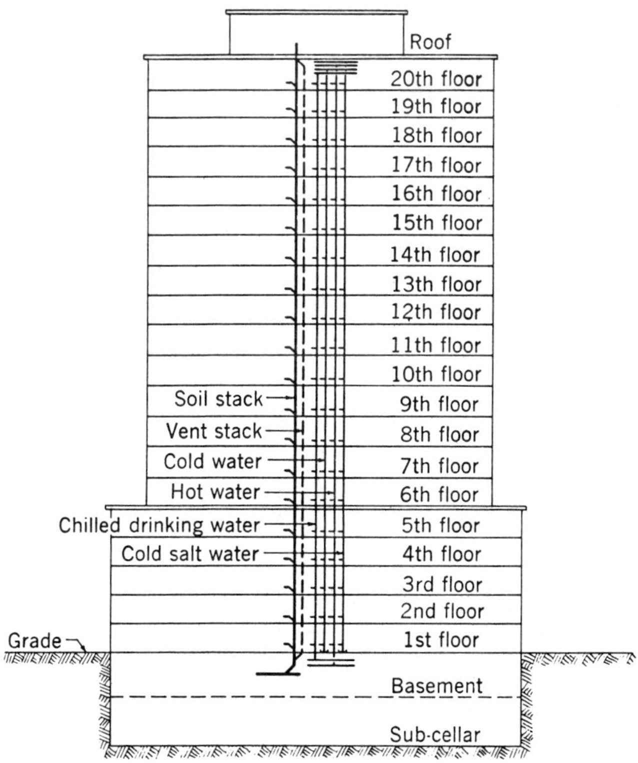

Fig. 3. Single plumbing stack in a 20-story hotel.

flexible connections to prevent the transmission of pumping noises throughout the system. Likewise, the bases of the pumps should be isolated to absorb pumping noises.

Typical Riser Arrangement

Figure 3, shows a typical riser diagram for the 20-story hotel used for example in this article. It illustrates a gravity system served by roof tanks.

This particular riser arrangement is intended to serve the vertical stack of bathrooms. Note that the riser is comprised of a soil stack, vent stack, chilled drinking water, cold water, hot water and cold salt water. A hot salt water distribution system has not been shown or included because of the difficulty encountered in generation and distribution of hot salt water. Hot salt water has a very corrosive effect on all piping materials.

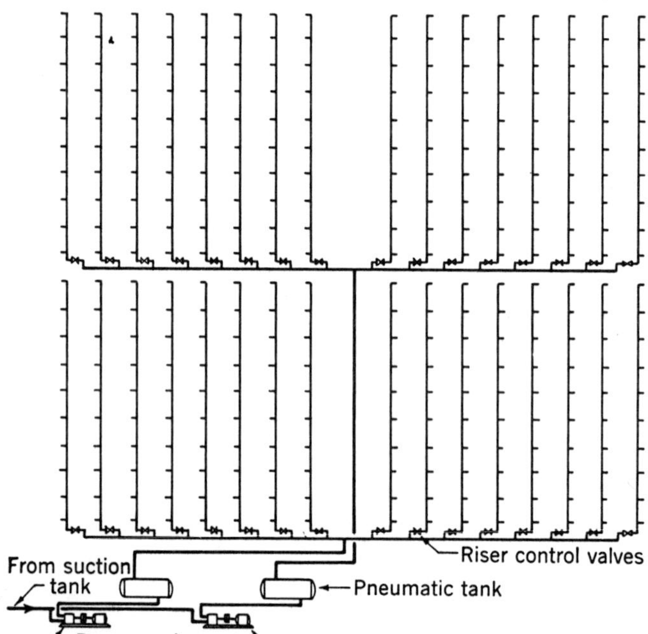

Fig. 4. Riser diagram for a salt water distribution system in a building, served by a pneumatic tank system.

Therefore, unless the hotel owner insists that hot salt water be furnished, it need not be included. A hotel guest can obtain a hot salt water bath merely by filling the tub partially with cold salt water and then adding fresh hot water to gain the desired temperature.

Pneumatic Water System

There are several methods that can be used to distribute cold salt water in the building. One is the pneumatic water system, Fig. 4. Two separate pneumatic water systems are employed. One supplies the upper 10 stories of the building while the other handles the lower 10 stories. Pumps draw water from the suction tank and supply it under pressure to the tanks. An air compressor supplies the quantity of air needed in the tank and furnishes the pressure needed to force the water up to the various floor levels in the building. Other arrangements of the pneumatic water system can be used with equal success.

Roof Tank

Another popular method that can be employed is the roof tank system. This method is illustrated in Fig. 5. The pumping equipment draws the salt water from the suction tank and lifts it up to the roof tank from which the water is distributed by gravity throughout the building. Pressure reducing valves are placed on those branches where the water pressure exceeds any prescribed water pressure that the engineer has previously selected.

Piping Materials

Salt water is rather difficult to handle with any piping material. However, materials are available that are quite favorable for this application. Here again, the piping manufacturers should be consulted for their recommendations. Any specific suggestion on the matter of materials that can be used with salt water has been avoided because it is so important that the engineer carefully check this matter with piping manufacturers. This is the only way in which some reasonable degree of assurance can be gained on the performance of the particular material specified. This same attention must be given to the pumping equipment as well as the valves and other component parts of the building's salt water distribution system.

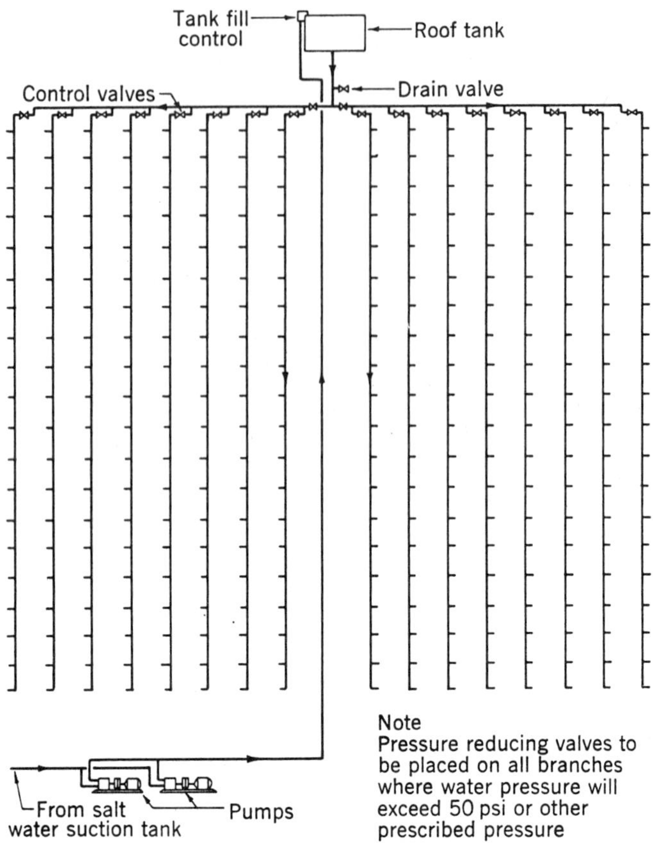

Fig. 5. Riser diagram for a roof tank gravity system.

Typical Hotel Room

For reasons of economy, bathrooms in a hotel are generally placed back to back when practical. A typical arrangement of two hotel rooms is shown in Fig. 6. Here is shown how the salt water piping may be arranged to serve the bathtubs and the water closets.

Detailed Arrangement of Piping

Figure 7 shows a detailed arrangement of all the piping that might be installed to serve the fixtures in a hotel bathroom. Although a singular arrangement of piping has been shown for

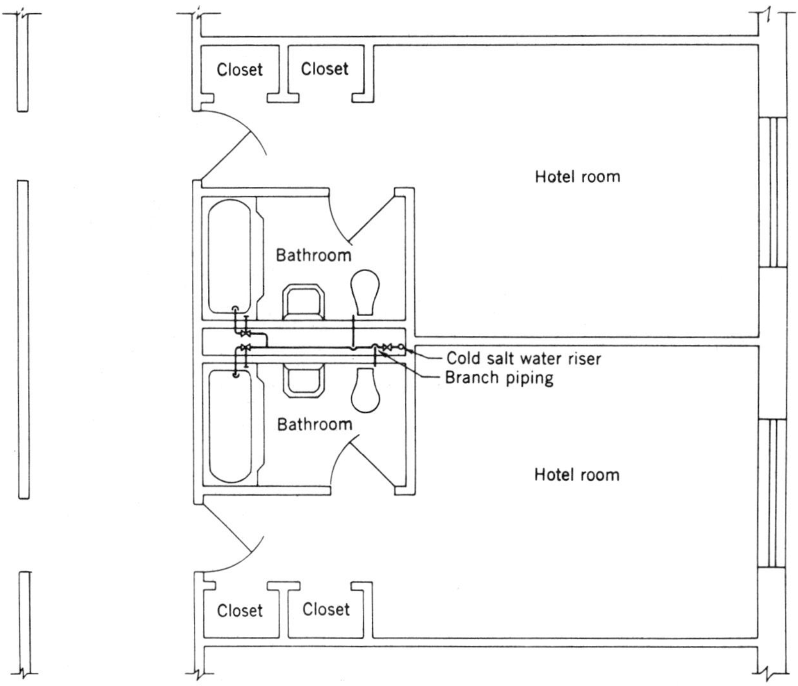

Fig. 6. Typical arrangement of hotel room showing how cold salt water piping is installed.

the sake of clarity, the arrangement for a back-to-back bathroom would be very similar in that branch connections would be extended to supply the back-to-back fixtures.

In this instance, all the plumbing waste, vent and water facilities have been shown to offer a realistic picture of the piping arrangement. Note that cold salt water is supplied to the water closet flush valves. This provision is entirely up to the engineer. This specific application has been included because the salt water is available and certainly reduces the quantity of cold fresh water that would otherwise be wasted for this purpose. Whether salt water is to be used for this function is a factor that must be carefully considered by the engineer. If the hotel must obtain its own fresh water from deep wells, the use of salt water for flushing purposes is a very definite advantage. In any event, the merits of this water saving provision are worthy of attention.

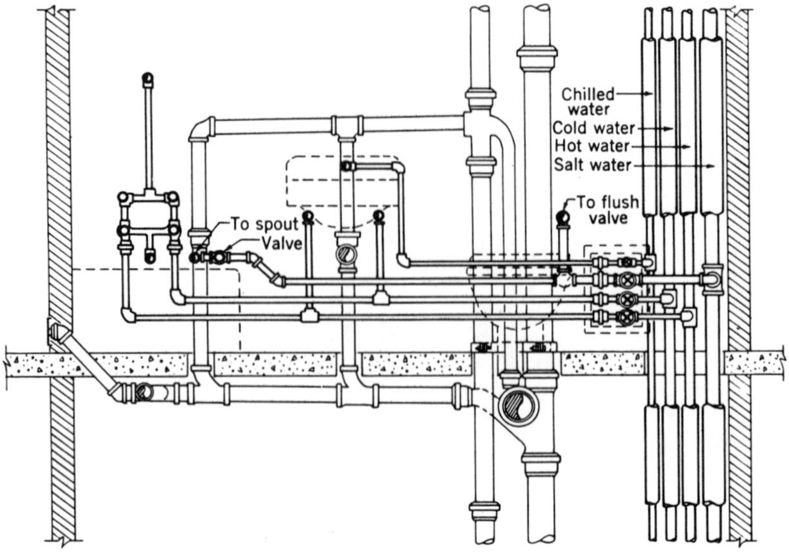

Fig. 7. Arrangement of waste and water facilities, including cold salt water piping.

Salt water is supplied to the bathtub but is controlled by a separate valve and is dispensed into the tub through a separate spout. The reason for this separation is to avoid any cross connection with the fresh water supply. A salt water showering arrangement has not been included since most persons seeking the benefit of salt water fill the tub. However, if a salt water shower is desired it can also be accomplished as a separate arrangement.

Pipe Covering

The salt water riser piping should be covered with an anti-sweat pipe covering to avoid dripping lines. In fact, the same consideration given the fresh cold water piping shall apply to the cold salt water piping.

CHAPTER 43

Pressure Reducing Valves

Excessive pressure in a water distribution system causes a number of disturbing factors in addition to harmful effects on the piping. The passage of water through the piping will generate noises, especially at piping offsets. As the water passes through small diameter pipe, the noise that is associated with its movement is quite audible. Stresses are set up in the pipe and fittings and these can be damaging to the piping. Furthermore, since piping will tend to move with the velocity of the water, a strain is imposed upon piping supports and hangers.

Due to the abrasive action of water under pressure, valves, faucets and equipment served will also suffer. Faucet washers will require more frequent replacement. The innards of gate and globe valves tend to vibrate with the surging of the water. Delicate equipment served by the water supply can rupture as a result of excessive pressure.

Hydrostatic shocks of an intensified nature are always associated with excessive water pressures. Thus, piping will rattle and bang to an annoying extent, and the pressure exerted in the line due to the quick closing of a valve or faucet can actually rupture the pipe. Moreover, systems with excessive pressures waste considerable quantities of water.

Control of Excessive Water Pressures

It is rather difficult for a community to regulate the water pressure in the exterior piping distribution system. In fact, the water pressure will vary in all parts of the community with the result that some sections of the community will have excessive pressures while other sections will have just sufficient pressure. This practice is necessary to ensure that the overall area will obtain adequate pressure. When the community water system serves buildings in addition to fire hydrants, it becomes extremely important that sufficient pressure be maintained.

When an engineer finds that the street water pressure is excessive for a contemplated water distribution system in a building, he must consider means that will reduce and maintain a prescribed pressure for his proposed system. This is accomplished by the use of pressure reducing valves.

There are many types of pressure reducing valves available. Although the basic operation of all the valves is rather similar, certain features incorporated in each design distinguishes the

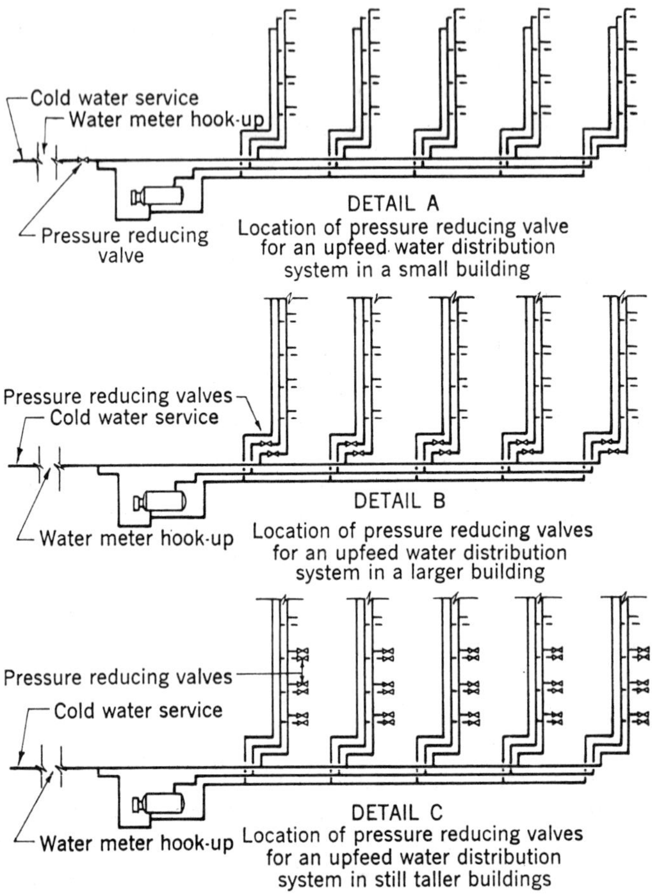

Fig. 1. Locations where pressure reducing valves may be used to reduce and maintain prescribed water pressures. Riser piping consists of cold, hot water supply, and hot water return.

different valve types. In general, a valve can be selected that will handle any specific need in a water distribution system.

Location of Pressure Reducing Valves

Where should pressure reducing valves be placed on a water distribution system to be most effective? Figure 1 offers three basic suggestions.

Detail *A* represents a four-story building. If the water pressure serving this building is excessive, it can be controlled by placing a pressure reducing valve on the house side of the water meter hook-up as shown. In this manner, it is possible to reduce an excessive water pressure to one that is adequate to serve fixtures at the top floor level.

When the building is taller, it may be advisable to locate pressure reducing valves at the base of the cold and hot water risers as shown in Detail *B*. This method provides a flexible control arrangement and allows each riser to be adjusted to a pressure that is best suited for the different types of fixtures served by each set of risers.

For still taller buildings, utilizing an up-feed system of piping, it may be advantageous to locate pressure reducing valves on the water branches serving the first few floors of fixtures as shown in Detail *C*. This method will handle the excessive water pressures at the first few floors and will allow the upper floors to be served by street water pressure without any pressure reduction.

Basement Mains

Figure 2 shows a general arrangement of piping for an up-feed system in a building and denotes where pressure reducing valves can be placed to best serve the different sections of piping.

The building of Fig. 2 is a large one. Illustrated are the different locations where pressure reducing valves can be placed. A reducing valve has been placed on the cold water service main in the building, on the house side of the water meter hook-up. This would be the desired location if the entire water distribution system were to be controlled at this point.

A pressure gage has been placed on either side of the reducing valve to permit an accurate adjustment of the reduced pressure. If it is not desired to place the valve at this location, then the

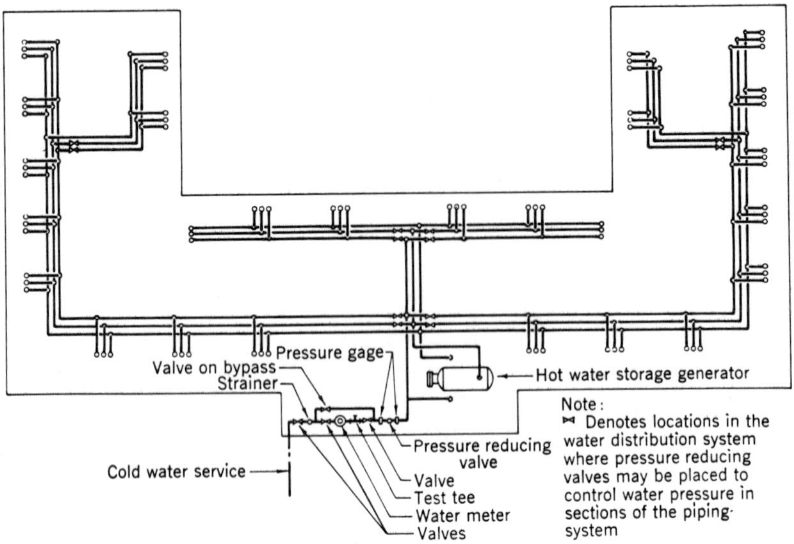

Fig. 2. Suggested arrangement of a water distribution system in a building, indicating where pressure reducing valves may be located.

valves may be placed at alternate locations as suggested in Fig. 2 to control the different sections of piping. Each proposed design requires particular study by the engineer to best satisfy the needs of the building.

Special Conditions

Sometimes, conditions of excessive pressure are encountered in a building which require that special consideration be given to the installation of a pressure reducing valve. Figure 3 shows a single-stage reduction of water pressure. This arrangement consists of a strainer and a pressure reducing valve. If the water pressure is abnormally high, this arrangement may not be sufficient since there will be considerable wear and tear on the operating parts of the valve. Therefore, another method of reduction is advisable.

Figure 3 also shows how a two-stage reduction of water pressure can be accomplished. This arrangement consists of a strainer and two pressure reducing valves. The first valve reduces the abnormally high water pressure to a moderate degree and then the second valve further reduces the pressure to a limit best suited for the needs of the building.

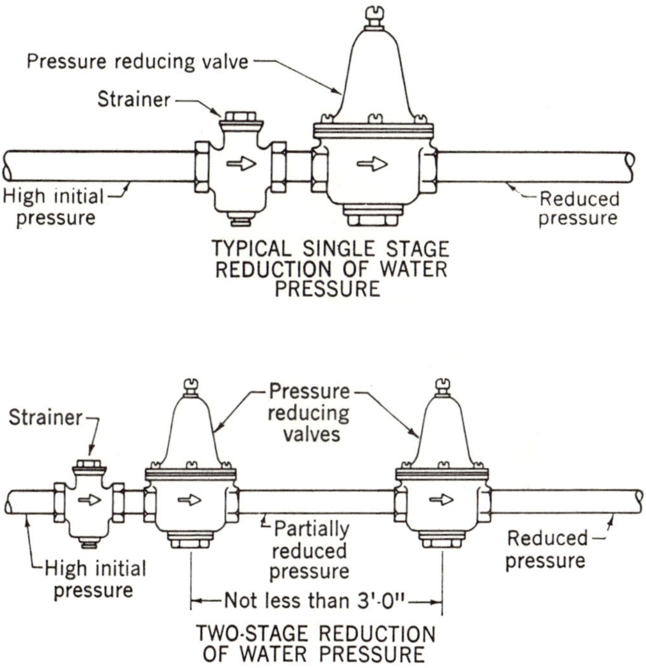

Fig. 3. Two methods for water pressure reduction.

Tall Buildings

In tall buildings down-feed distribution systems are generally installed. A typical example of this type system is shown in Fig. 4.

In this design, the roof tank is set at an elevation that will supply the needed water pressure for the upper stories of the building. However, with this type of system, the water pressure for the lower floors will increase to an excessive degree, due to the height of the piping. Therefore, it is necessary to install pressure reducing valves on branches to the fixtures at the lower floors as shown in Fig. 4. The valves are placed on the cold and hot water branches at the lower floors where the water pressure exceeds a prescribed limit.

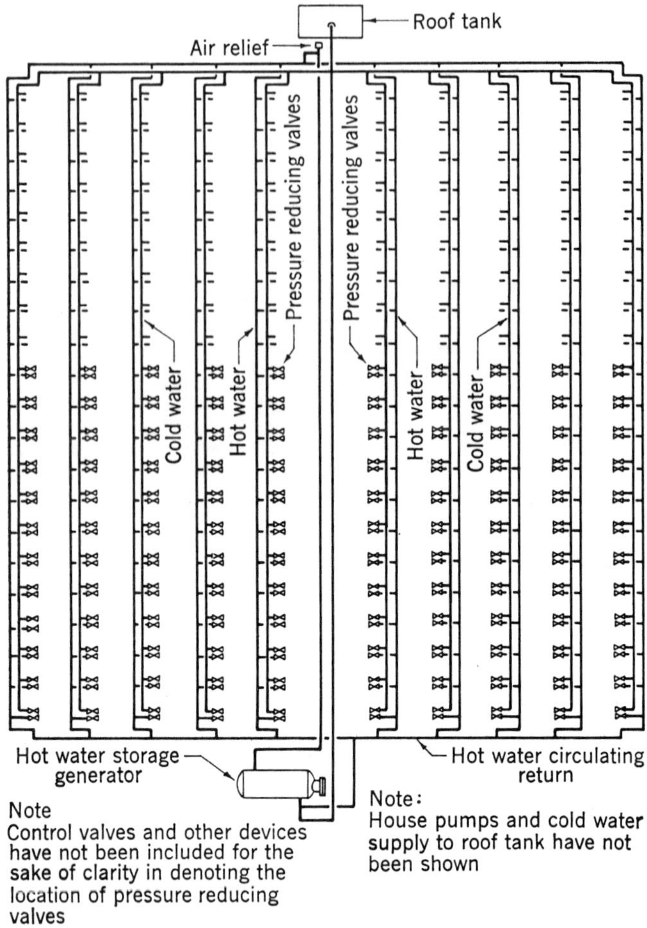

Fig. 4. A typical downfeed water distribution system in a tall building. Pressure reducing valve locations are shown.

Cold and Hot Water Branches

Pressure reduction on typical cold and hot water branches is detailed in Fig. 5. The small diagrammatic arrangement shows how the branches serve a battery of fixtures, consisting of water closets and lavatories. The enlarged detail shows a placement of the pressure reducing valves to provide a regulated pressure to the fixtures. This same arrangement is typical for the pressure reducing valves shown in Detail *C* of Fig. 1.

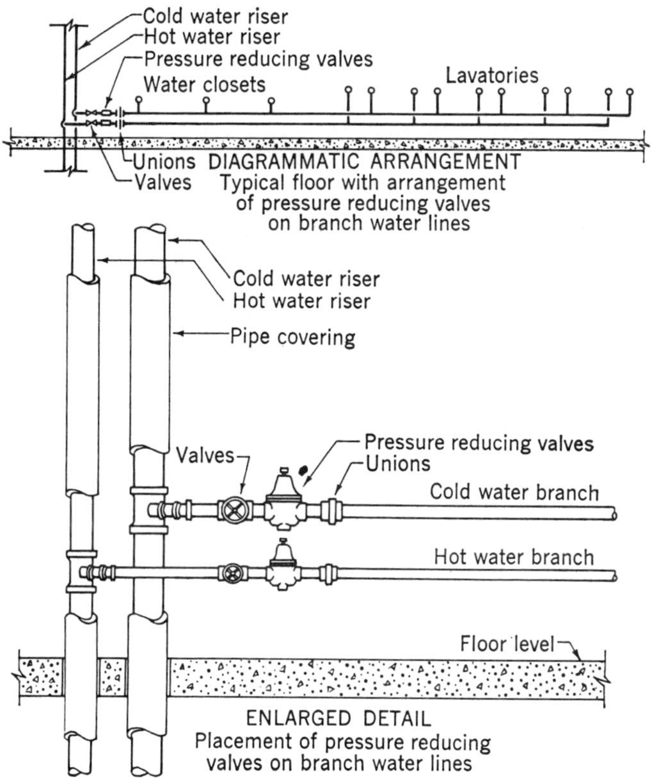

Fig. 5. How pressure reducing valves are placed on branch water lines which serve a battery of fixtures.

Types of Pressure Reducing Valves

Many types of pressure reducing valves are available. A few of the different types produced by one manufacturer are detailed in Fig. 6.

A illustrates a small type of reducing valve that can be placed on supply lines to drinking fountains, kitchen equipment, laboratory equipment, distilled water producers, and many other types of apparatus where a regulated water supply is needed.

B shows a pressure reducing valve with a built-in strainer. Strainers protect the valve from grit and foreign objects that could damage the working parts.

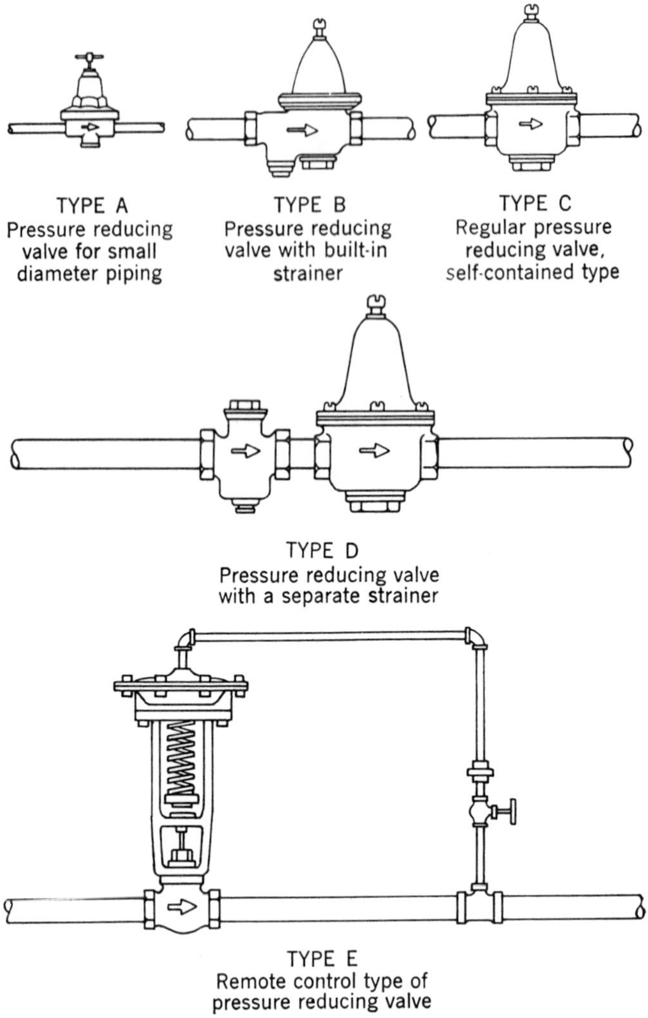

Fig. 6. Several different types of pressure reducing valves produced by one manufacturer.

C shows the type valve that is most popular in application. This particular valve has been used in many of the illustrations in this article.

D is the same reducing valve with the exception that a separate strainer is placed in front of the valve.

E is the type valve employed when it is desired to maintain very accurately controlled pressures. The valve is a remote control type.

The capacity of a pressure reducing valve is usually shown as the maximum flow obtained through the valve when it is wide open and when varying pressures are maintained by the incoming and outgoing pressure at the valve connections. This is not a realistic rating and should be used only for comparison of the various types and makes of valves.

The capacity to use for estimating will vary from 50 to 75% of its maximum rating and therefore a reasonably safe estimate would be to use 60% of the maximum rating.

CHAPTER 44

Vacuum Breakers

While pressure in a supply line is adequate, water freely discharges into an open vessel. Should a drop in water pressure occur, the water will drop in the supply line. This situation is called backflow since the water reverses its direction of flow. Suction created in the line can draw water from the fixture into

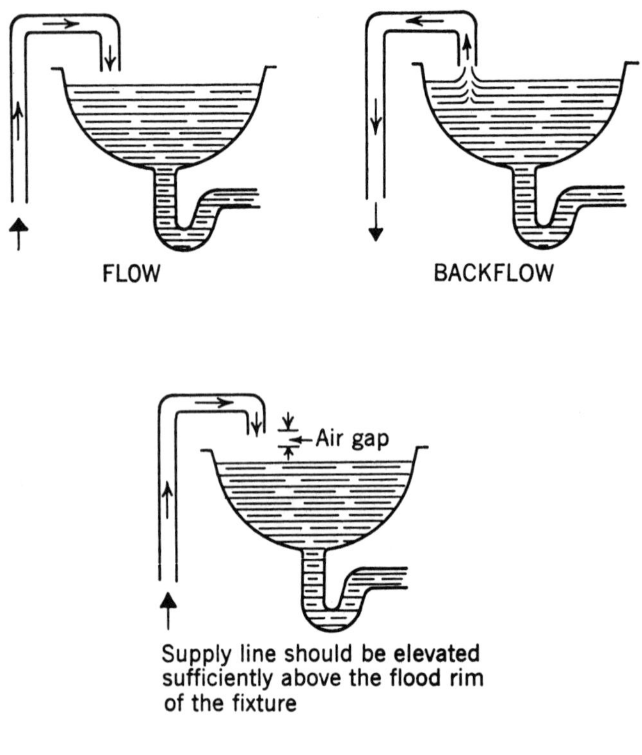

Fig. 1. How an air gap prevents back siphonage.

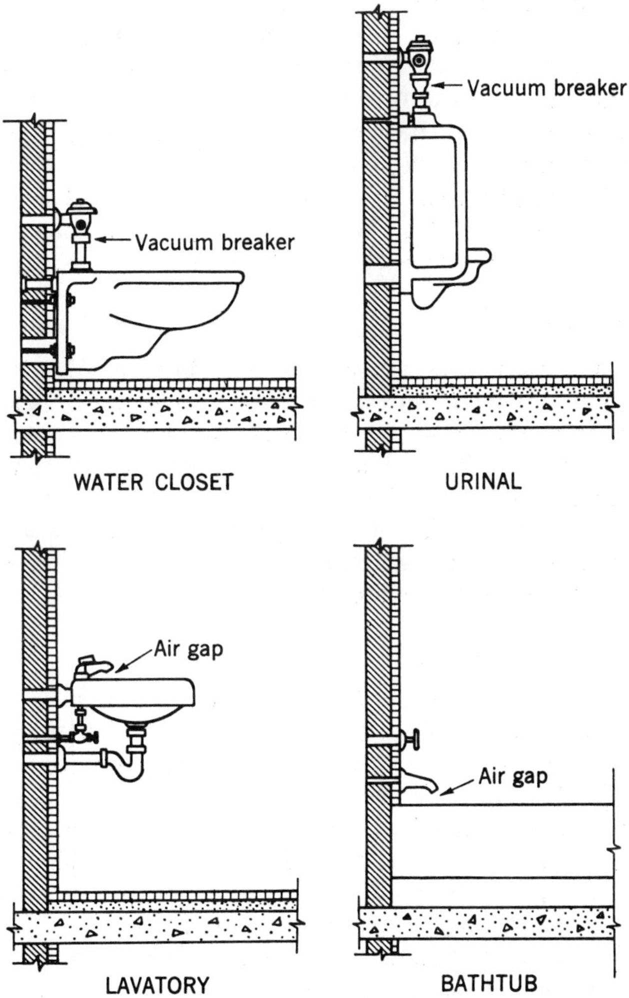

Fig. 2. How fixtures are protected against back siphonage.

the supply line. In this way dangerous waste material can be drawn into the water distribution system.

To safeguard the water supply against this hazard, it is necessary to provide an air gap between the supply pipe and the flood rim level of the vessel, as shown in Fig. 1.

When backflow occurs and the proper air gap has been provided, the supply line will draw in air and not water. With reference to an air gap, there are certain prescribed distances that must be maintained between the supply outlet and the flood rim level of the fixture. Most fixture trim is now designed to incorporate the minimum air gap requirement for the particular trim. In general, the minimum air gap should be at least twice the diameter of the effective opening in the spout of the trim or the supply line. To be positive on this matter for those critical installations which are encountered from time to time, it is suggested that the air gap be exceeded.

Cause of Backflow

There are numerous causes for backflow in a water supply line. Some of the common ones are: an inadequately sized water distribution system, heavy demand of water at lower floor levels, a momentary drop in pressure at the pumping equipment, and a break in a street or building main.

When street fire hydrants are supplied from a street water main, there is a danger of backflow when one or several of the hydrants are operated in the event of a fire.

Vacuum Breaker Devices

Water Closet. A wall or floor mounted water closet supplied by a flush valve can siphon water out of the closet under certain conditions, unless the supply line is protected with a vacuum breaker. In this instance, the vacuum breaker is installed at the flush valve and in the event of backflow, closes tight and allows air to enter the air vent ports. The vacuum breaker must be installed in a prescribed manner dependent upon the particular arrangement of piping which may be utilized.

Urinal. The water supply to a urinal also requires the protection of a vacuum breaker. Here again, the vacuum breaker may be installed as illustrated or must be properly placed in accord with the specific arrangement of piping used.

Lavatory. To avoid siphonic action at the supply line serving a lavatory, the fixture trim spout should be elevated to afford a prescribed air gap between the spout and the flood rim level of the fixture. Most trim manufacturers have incorporated the proper air gap into the design of their trim.

Bathtub. The water spout of the bathtub trim should be elevated sufficiently above the flood level rim of the bathtub. Since the contractor installs the spout, he should make certain that the proper air gap is provided. The engineer should indicate in his specifications that the minimum requirements concerning air gaps be strictly followed.

Figure 3 illustrates the parts of a flush valve vacuum breaker. A rubber insert fits into a tailpiece which connects to the underside of the flush valve. The bottom of the tailpiece slips into and is connected to the inlet spud at the water closet. The lower portion of the rubber insert is designed with a slit which will permit water to pass through but seals tight in the event of backflow. Air enters the air vents and breaks the suction caused by siphonic action. The vacuum breaker thus safeguards the water supply serving the fixture.

An alternate type of vacuum breaker is the shallow unit shown above the tailpiece in Fig. 3. This is also an approved type device and functions in a manner similar to that of the larger insert. The smaller rubber insert is equipped with a criss-cross slit at

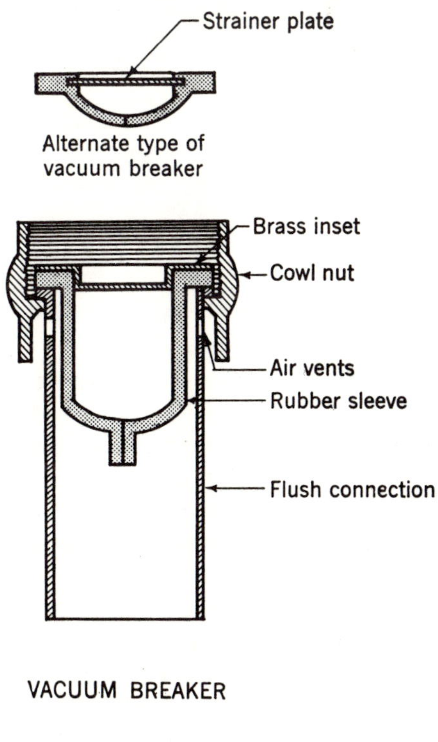

Fig. 3. Detail of vacuum breaker used with flush valves.

its bottom. Other types of vacuum breakers are also available for use with flush valves depending on the specific location of the vacuum breaker and the type featured by the particular flush valve manufacturer.

The service sink and mop sink must also have vacuum breaker protection, as illustrated in Fig. 4.

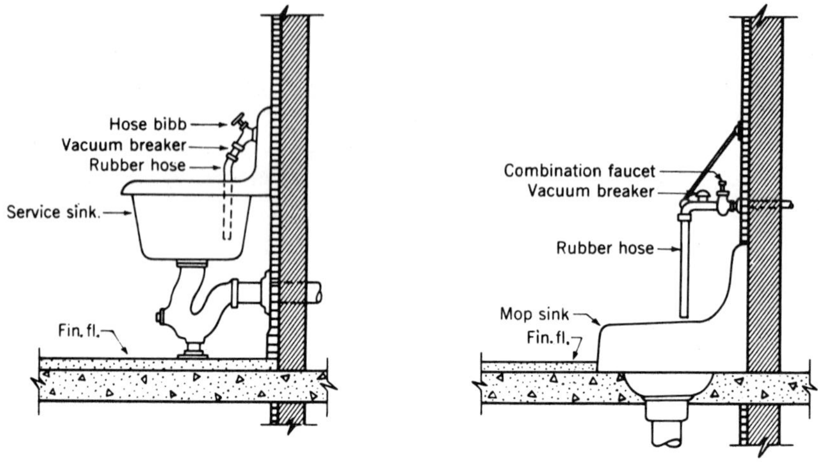

Application of a vacuum breaker on the water supply to a service sink.

Application of a vacuum breaker on the water supply to a mop sink.

Fig. 4. Vacuum breakers on service and mop sinks safeguard water supply.

Although the faucets or combination faucet on a service sink is elevated well above the flood rim level of the service sink, it is common practice to apply a short piece of rubber hose to the spout of the faucet, as shown. The hose is used to facilitate filling pails. Nevertheless, since this hose extends below the flood rim level of the fixture, it poses a potential hazard.

If the sink contains a quantity of waste water and the hose is submerged in it, it would be possible to siphon out the contents of the sink when backflow occurs. This can be a most dangerous situation, in view of the varied mixtures of contaminants which are poured in the service sink. It is therefore recommended that the rubber hose be omitted. If it must be used, then it is recommended that a vacuum breaker be placed between the spout of the faucet and the rubber hose. There are several types and styles of approved vacuum breakers for this purpose.

With reference to the mop sink, the rubber hose is generally required with this installation to facilitate the filling of pails in mop carts. In this instance, a vacuum breaker is an absolute requirement, especially since the rubber hose often extends into the mop sink. The vacuum breaker should be placed between the faucet spout and the rubber hose.

Some combination faucets are equipped with a vacuum breaker on the outlet spout, as in Fig. 4.

In hospitals, there are many special fixtures to care for the needs of the patients. The supply lines to all of these fixtures must be protected by the proper air gap or vacuum breaker devices. In a hospital, the water distribution system, unless properly protected could easily convey pathogenic germs.

Figure 5 shows how a bed pan washer is equipped with a flush valve which includes a vacuum breaker. The bed pan washer is the semi-recessed type with a concealment panel. Siphoning the contents of a bed pan washer could impose a highly potential

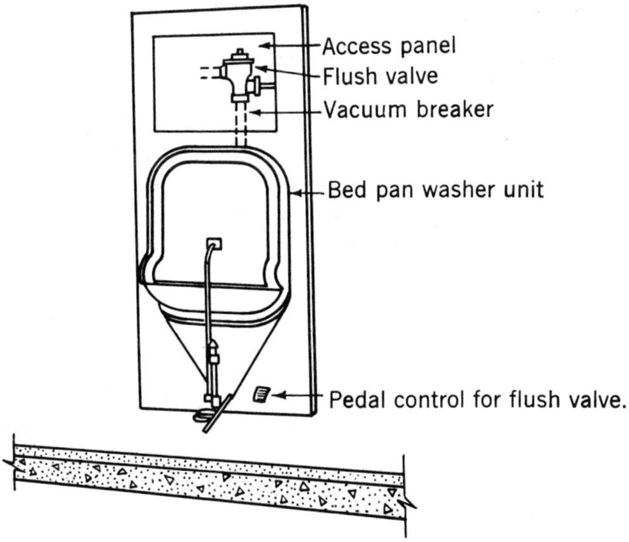

Fig. 5. Vacuum breaker installed on flush valve serving semi-recessed bed pan washer.

hazard with regard to contaminating the water distribution system.

Detail *A* of Fig. 6 shows what happens when siphonic action occurs in a tank to which a water supply line is connected. The contents are drawn out and into the water distribution system.

Detail *B* shows a vacuum breaker on the water supply to the tank. When siphonic action takes place, the vacuum breaker closes to the backflow and allows air to enter the inlet portion of the valve. This breaks the vacuum created by siphonic action and safeguards the connected water supply line.

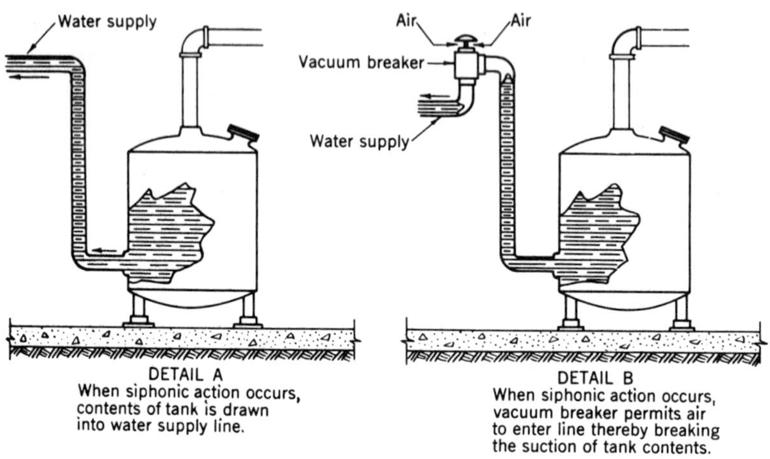

Fig. 6. How vacuum breaker safeguards water supply system against siphonic action.

Approved Vacuum Breakers

There are many types of vacuum breakers expressly manufactured to protect the water supply line to apparatus against the hazards of backflow and contamination. Figure 7 shows the details of one approved type vacuum breaker.

Under flow conditions, the piston of the vacuum breaker rises immediately, sealing the air port under pressure to prevent water leakage. The plunger rises next to form the flow orifice, permitting the water to enter through the inlet, whence it flows up and around the plunger and out through the outlet. The piston and plunger are both up in the operating position.

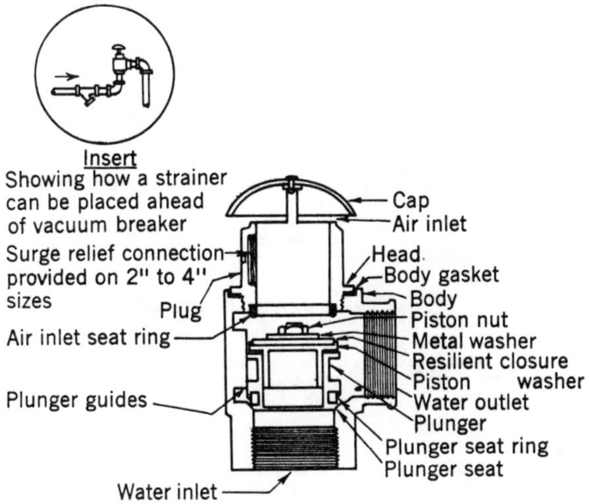

Fig. 7. Operating parts of an approved vacuum breaker.

Under backflow or unbalanced pressure conditions, that is, when the flow on the inlet side of the unit is suddenly cut off and reversed as happens when a partial vacuum develops, the plunger instantly drops to the lower seat, sealing the water inlet and acting as a check against backflow from the outlet piping. The piston drops at the same time, admitting air and breaking the effect of the vacuum. With piston and plunger in the down position, back siphonage is impossible.

The small detail of Fig. 7 shows how a strainer is often placed ahead of the vacuum breaker to prevent foreign objects from lodging in the working parts of the vacuum breaker. If the water in a given area is known to contain impurities, it is always good practice to install a strainer on the water service in the building. A check with the local water department can quickly determine the needs.

Typical Vacuum Breaker Installations

Figure 8 shows several typical applications for vacuum breakers.

Laundry Machines. Both cold and hot water supply to laundry machines should be protected against contamination from the wash water contained in the laundry machines. Figure 8 shows

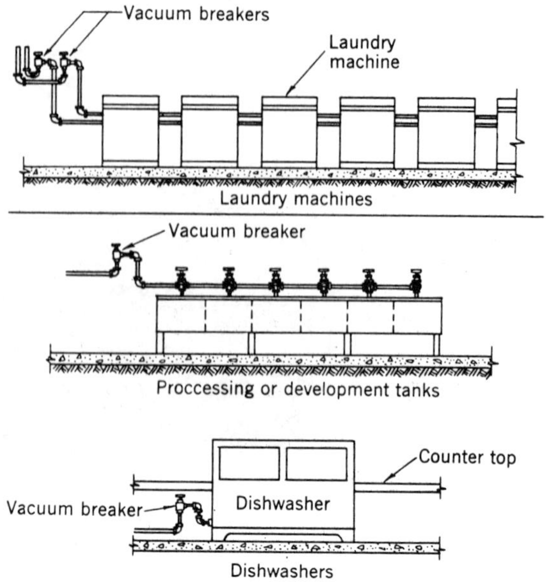

Fig. 8. Installation of vacuum breakers for special equipment.

one of the several different piping arrangements which may be used to serve the machines. Note the placement of the vacuum breakers.

Tanks. A water supply is needed for many different types of processing and developing tanks. It is always highly recommended that the water supply to such tanks terminate above the tank with a prescribed air gap between the water outlet and the tank. However, this provision is often neglected and the water outlet then terminates below the flood rim level on the tank or is even submerged in the tank. Therefore, it is of utmost importance that the water supply be equipped with an approved type vacuum breaker which is installed in a manner similar to the arrangement shown.

Dishwashers. The water supply to dishwashers are also subject to the hazard of siphonic action due to conditions of backflow. One suggested placement of vacuum breaker is indicated in the detail. However, the engineer should design an arrangement of vacuum breaker and piping that best satisfies his specific needs.

Other Vacuum Breaker Applications

There are very many needs for vacuum breakers, and the few applications described here are intended only to indicate the varied scope of equipment and apparatus for which they are needed.

Hot Water Generator Equipment. It is important that a vacuum breaker be utilized on the water supply to a hot water storage generator, for two reasons: (1) The vacuum breaker prevents the possibility of drawing the contents of the hot water storage generator into the water distribution system, thereby avoiding the risk of contamination, and (2) the vacuum breaker prevents the possibility of drawing the hot water into the cold water distribution system, where it can create a hazard. In this instance, a person opening a cold water faucet might find that hot water is dispensed rather than cold. This could be serious when very hot water is maintained in the storage generator.

Space Heating Plant. It certainly is not desired to siphon the contents of the space heating plant into the water distribution system. Yet, backflow can cause siphonic action if it occurs at the instant that water is being added to the space heating plant. The water contained in the space heating plant can be quite harmful for human consumption when you consider the various mixtures that are added to the boiler water for sealing leaks or to inhibit the formation of rust scale. A vacuum breaker is of extreme importance in this application.

Lawn Irrigation Supply. The water supply to wall hydrants and lawn irrigation systems should most definately be protected with a vacuum breaker. Take the wall hydrant for example. This hydrant is used for many gardening reasons including:

1. A hose from the hydrant is often directly connected to a jar containing a chemical mixture which is sprayed on the grass, shrubs or flowers. Such mixtures can be most harmful when consumed by humans and especially children.

2. The hydrant is often used to provide pressure for paint spray apparatus. It is needless to mention the potential hazard to humans when paint is siphoned into the water distribution system.

3. The lawn irrigation system also presents an extreme hazard. Here again, dangerous mixtures are in solution at the grass surface when the lawn sprinkler system is in operation. The solution is easily siphoned into the water distribution piping when conditions of backflow occur.

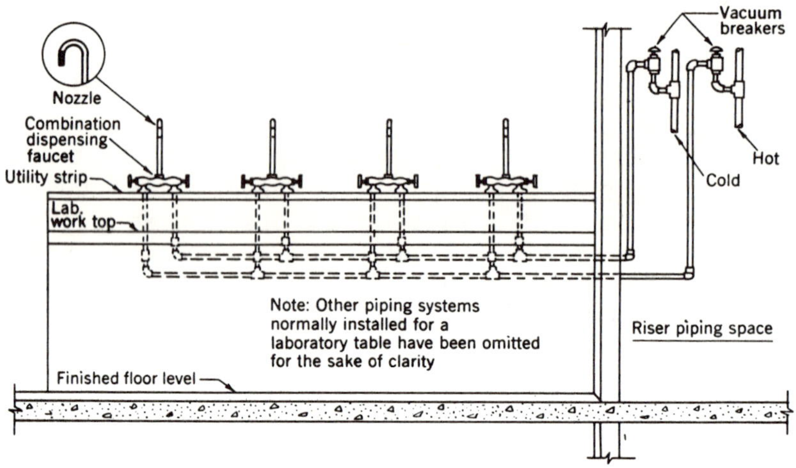

Fig. 9. Installation of vacuum breakers on cold and hot water lines serving laboratory table.

One possible arrangement of laboratory piping and vacuum breakers is shown in Fig. 9. The combination faucets shown are typical of the type used in laboratory research. When used only for dispensing water into an open jar or vessel there is little danger involved. However, the nozzles are equipped with serrated tips to permit the use of a hose when certain experiments are performed. Should a condition of backflow occur during one of the experiments, the entire contents of any apparatus to which the hose is connected could easily be siphoned into the water distribution system. Therefore, it is of the utmost importance that the cold and hot water piping be protected by approved type vacuum breakers. These same considerations apply to the varied equipment and apparatus installed in hospitals. With the high degree of sanitary cleanliness maintained in a hospital, the water supply is the only medium that can convey contamination from one area to another unless the proper safeguards in the form of vacuum breakers are applied to the water supply piping.

Care of Vacuum Breaker

The vacuum breaker should always be in an exposed location where it can be visually inspected at periodic intervals. If it is concealed, an access door or other means must be provided to facilitate inspection and replacement. When, in time, the vacuum breaker shows indications of failure, it should be immediately replaced.

Extra Precautionary Measures

Very often, the engineer will indicate the placement of two check valves on a water supply to certain equipment or apparatus. The check valves are in addition to the vacuum breaker. Figure 10 shows this installation.

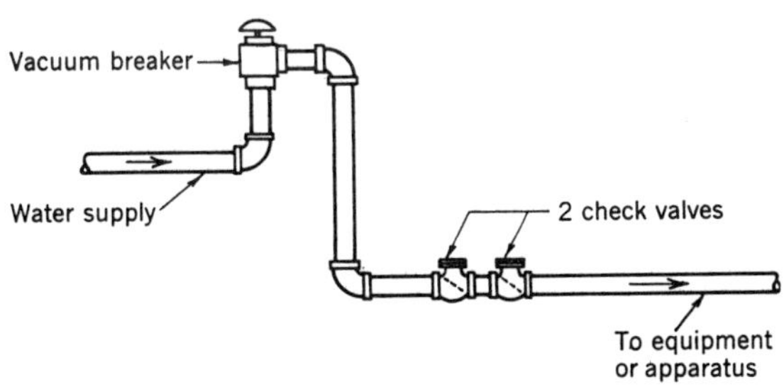

Fig. 10. Added protection is provided by two check valves.

The check valves cannot replace the protection of a vacuum breaker but help in the prevention of backflow. Should backflow occur, the flappers in the check valves will close and the vacuum breaker is there to handle any action that manages to pass the check valves.

Part 4
Building Fire Protection

CHAPTER 45

First Aid Fire Protection

A fire standpipe system is required in almost all types of buildings. This system serves not only as the first initial means of combatting a fire by the building occupants but it also serves the fire department. The design of a required fire standpipe system is governed by the regulations of the National Board of Fire Underwriters as well as the local plumbing or fire department. The rules and regulations of the National Board of Fire Underwriters are available in booklet form and are more or less general for the entire nation.

A building owner may desire fire standpipe protection in that type of a building where it is not normally required, or he may request additional fire protection facilities in a building in which the proposed fire standpipe system meets approved minimum rules and regulations. When this type of a request is encountered, the engineer must give consideration to systems designated as "first aid fire protection." Some of the systems used for this purpose are described in this chapter.

Occupancy Systems

The first type system normally associated with first aid fire protection is that which is installed at the owner's request but which is not required by the national or local fire authorities. A private home is an example of the type of a building in which this system might be placed. Another good example is an apartment house. Many apartment houses do not generally require fire standpipe protection because they are of fireproof construction with metalclad doors and have stairways serving as a means of fire escape exit.

However, although this system is not a compulsory one, certain rules and regulations must necessarily govern its installation. These rules and regulations are necessary to maintain some jurisdiction over the type of systems installed and some control over the proposed use of the water.

Figure 1, shows a typical system that might be installed in a 5-story apartment house.

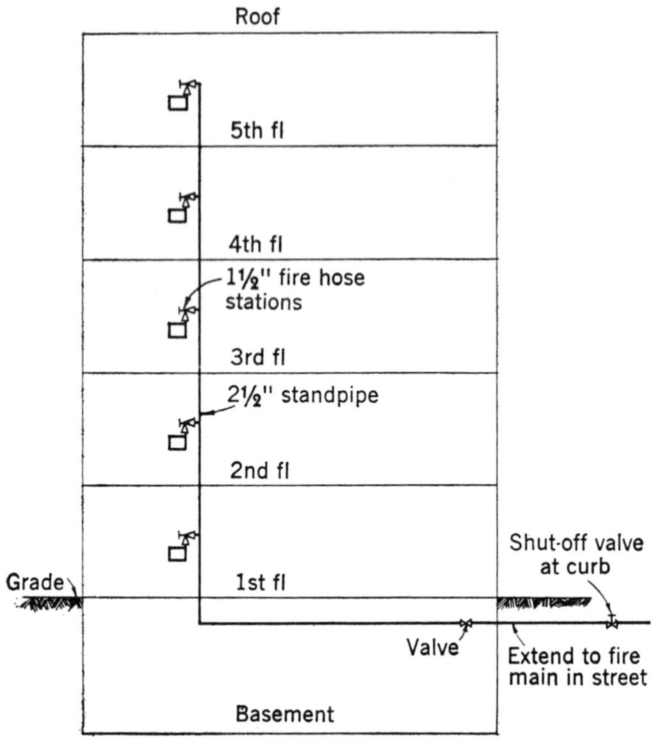

Fig. 1. Typical installation of a fire standpipe system for specific use by building occupants.

Assuming that the local authorities do not regard the building as one that must be equipped with a regulation type system, then the system may be classified in the occupancy category. The system illustrated in Fig. 1, has been designed to meet the rules and regulations governing its installation. Therefore, since the building is more than four stories high and exceeds 50 ft. in height, the fire standpipe riser must be 2½ inches in diameter. The fire hose is 1½ in. diameter with a ⅜ or ½ in. nozzle.

If the building does not exceed four stories or is less than 50 ft. in height, a 2 in. diameter fire standpipe riser could be used. Since the system is classified as an occupancy type, only one approved source of water is needed. If a fire main exists in the street, it may be used for this purpose; otherwise the cold water service in the building may be utilized.

When the fire main is used, a control valve and curb box are generally placed near the curb. A post indicator valve might also be considered for the function. Valves offer a means for shutting off the water supply in the event that the building suffers a severe fire with resulting damage to water lines which could waste considerable water. The water company or local authorities may require that a water flow meter be installed on the fire main in the building to prevent the misuse of the fire line.

If the fire line is to be connected to the cold water main inside the building, it is generally connected as close as possible to the water service entry into the building. Just how this supply line should be connected to the service main and the component piping that must be included, can only be ascertained in each instance by consulting the local plumbing, fire or water departments. When this system is properly installed, it can be counted upon to provide the necessary fire protection sought by the building occupants.

First Aid Fire Protection

An owner or administrator of a building often requests additional facilities to the fire standpipe system that exceeds the minimum requirements of the national and local fire authorities.

Examples are certain types of office buildings, industrial plants, research laboratories, warehouses and other types of buildings where for extra safety, small stream hose stations are placed near the interior office or working space in the proximity of the fire exit to the stairway escape. Such stations normally have an approved fire standpipe system with 2½ in. hose. The additional small stream hose station is for use by occupants in the event of a fire and permits some control of the fire until all the occupants are safely out of the area. Figure 2 represents such a system.

Referring to Fig. 2 note that the basic fire standpipe system is of the approved type that meets the requirements of the national and local authorities. The only modification is the addition of the small stream first aid hose stations in the office or work areas. Figure 3 illustrates in detail, the suggested hook-up for the arrangement of hose racks on a typical floor level. The standpipe riser in this figure is 4 in. diameter; the hose in the stairway is 2½ in. diameter which is the regulation type for use by the fire department. To supply the remote small stream hose stations, a connection is made to the 4 in. standpipe as shown. The 1½ in. gate valve is the means by which the firemen can shut off the water supply to the small stream hose station when the 2½ in. hose station is to be used.

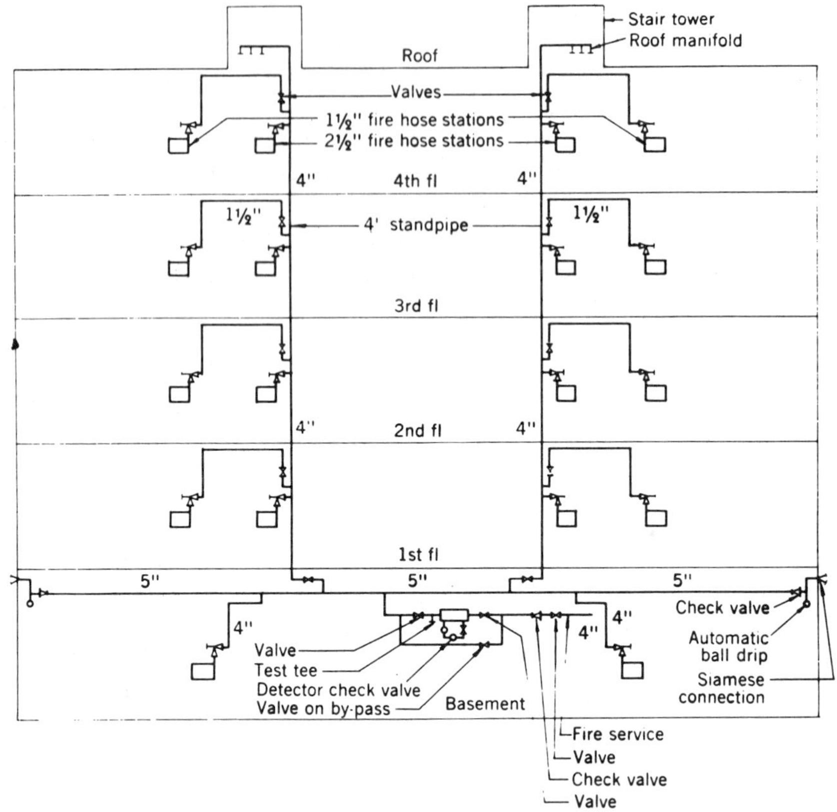

Fig. 2. Arrangement of approved fire standpipe system with extra small stream house station for use by building occupants.

Occupants of a building can usually handle the small diameter fire hose but it takes trained firemen to handle the 2½ in. regulation hose. When full water pressure is turned on in a 2½ in. hose, it can present a serious hazard to the untrained fire-fighters who attempt to handle it. Such a line out of control endangers both life and property. It is for this reason that the smaller hose is recommended.

Illustrated in Fig. 3 is an approved way of furnishing first aid fire protection in a building. In every instance, the national and local

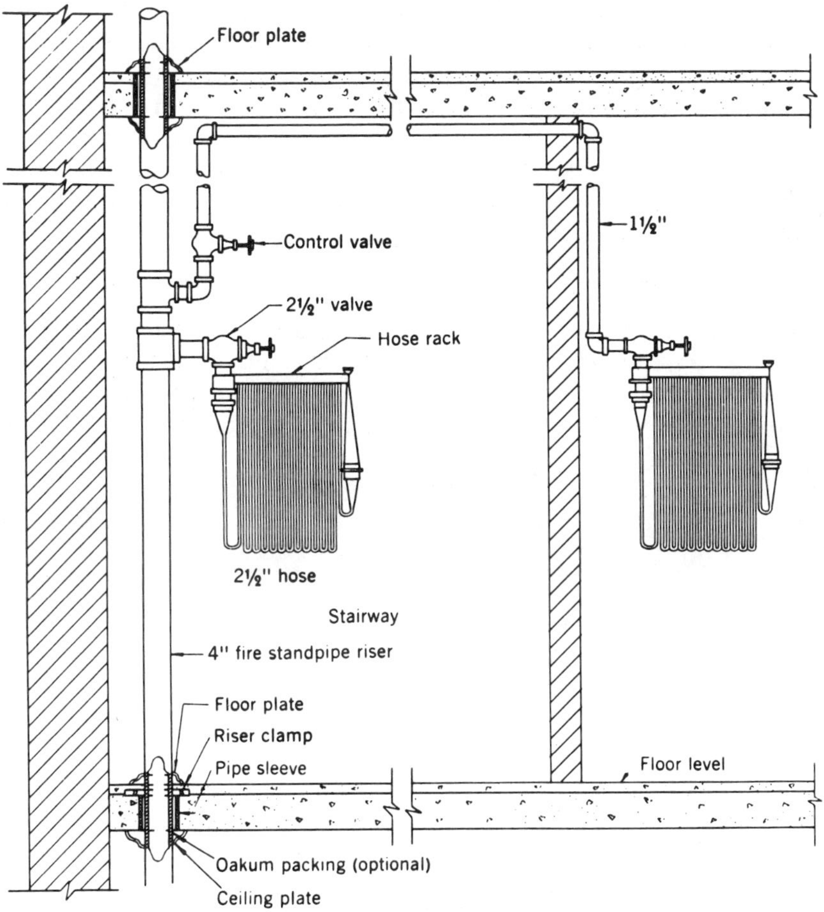

Fig. 3. Arrangement of approved fire hose station and small stream hose station, including valves and piping.

authorities should be informed of your proposed design. It is important for the local fire department to have in their files a print of the complete fire protection system installed in the building.

First aid fire protection can be provided in the basement. However, it is general practice to run a 4 in. diameter branch in the basement over to the proposed hose station and then reduce to 2½ in. for the valve and hose connection. The hose would be of the regulation 2½ in. size. The reason why 2½ in. hose is used in this case is because the building maintenance help are usually qualified to handle the larger hose. If desired, 1½ in. hose can be used by fitting a 2½ in. x 1½ in. adapter to the 2½ in. hose valve. Here again, if the hose station is located in a remote area, a valve of the OS&Y type is required on the supply line to the hose station. The valve is generally located in an exit or escape area where it is accessible to the firemen. The valve permits firemen to shut off the water supply to the hazardous area, when necessary.

Combination Hose Rack

First aid fire protection is a current consideration in many new buildings now under construction. One way of furnishing first aid protection and yet permit the use of 2½ in. hose is by the method shown in Fig. 4. The standpipe is 4 inches in diameter. Regulation hose of course would be 2½ in. diameter. However, with the speed that fire departments respond to fires today, relatively few minutes elapse between the time a signal is received and the time it takes for the equipment to arrive at the scene.

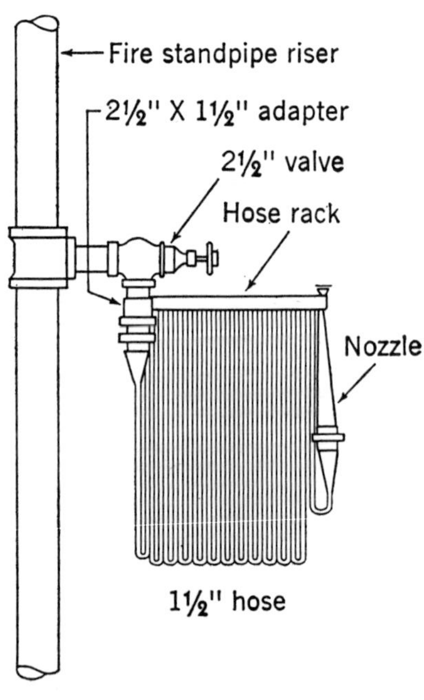

Fig. 4. Detail of approved hose station equipped with 1½ in. fire hose.

By installing a 2½ in. x 1½ in. adapter at the 2½ in. valve, it is possible to use 1½ in. hose for the benefit of the building occupants. In the event of a large fire, the firemen can easily disconnect the 1½ in. hose and connect their own 2½ in. hose.

Firemen normally prefer to use their own hose instead of the 2½ in. hose that may be on the rack. Since this hose remains unused for a period of time, the fire department prefers to use its own hose of known quality.

CHAPTER 46

Fire Protection for Apartment Houses

Modern apartment houses, by reason of advanced trends in their design and function, have become exceedingly popular with the result that they are now found in small towns, in the suburbs, as well as in large cities. Moreover, because of their neat appearance, planted zones, play areas and parking facilities, they have become valuable assets to the local community. Since the modern apartment house is considered home by a community of perhaps over 100 families living under the same roof, careful consideration must be given to achieve adequate fire protection.

In most instances, the apartment house because of its construction is classified as a fireproof building. However, even fireproof buildings burn as the records of fire reports show, and therefore some physical means of fire protection must be considered for the building. The system that affords this protection is called a fire standpipe system and is comprised of an arrangement of piping and hose racks, installed in the escape stairways of the building and in other hazardous areas for the purpose of controlling and extinguishing the outbreak of a fire at its source.

The function then, of the fire hose station located at each floor level is to provide the tenants with a means of first aid protection in the event of a fire. More important, however, it furnishes connections of adequate size to which the fire department may attach its hose and fight the fire at any floor level.

Figure 1 is a riser diagram of a typical fire standpipe system for a building we shall assume to be a 6-story apartment house. The various parts of the fire standpipe system are illustrated and labeled. In this chapter we shall discuss the component parts that are described and the necessary consideration that should be given each part during the design stages of the proposed system.

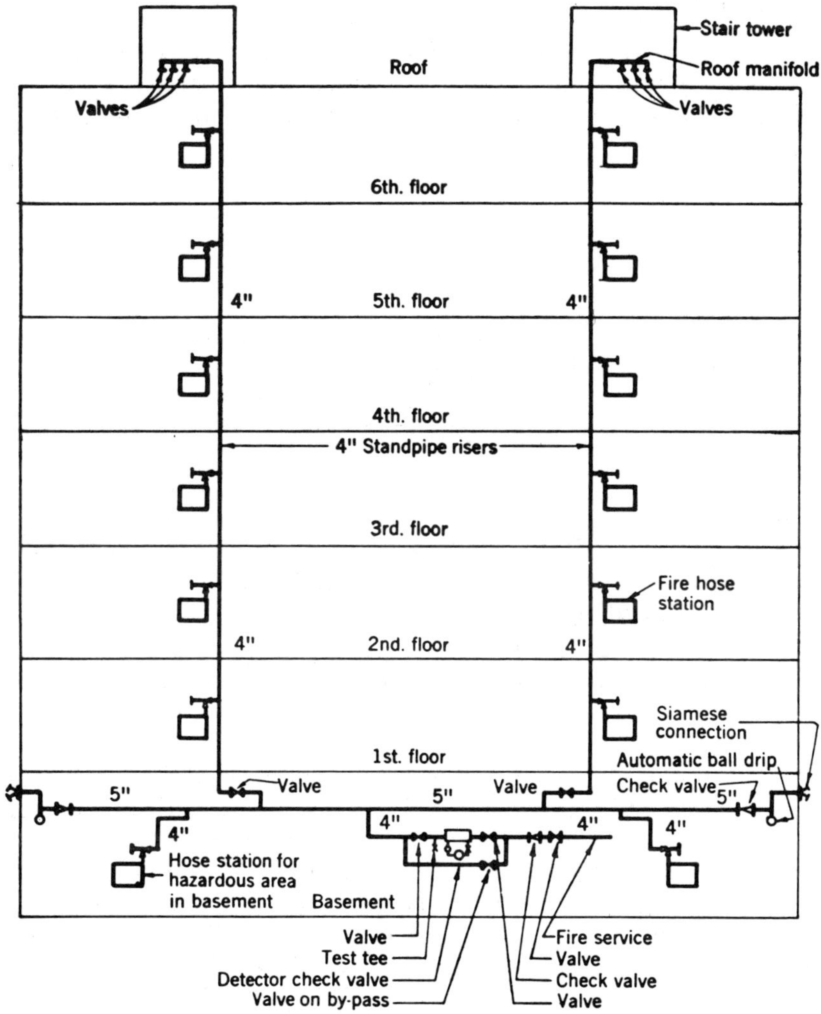

Fig. 1. A typical riser diagram for a 6-story apartment house. These risers serve the fire hose stations on each floor and the roof manifold.

Omitted are the various requirements of the fire underwriters or local authorities that determine the type or scope of system to be used. Therefore, it must be assumed, that the water pressure is adequate and that the system shown, including the pipe sizes and the arrangement of piping, meet all the necessary requirements of the governing departments.

Basement Piping

The first part of the fire standpipe system to be discussed is the basement piping. This part of the system may best be explained with the help of Fig. 2, which shows a typical arrangement of such piping.

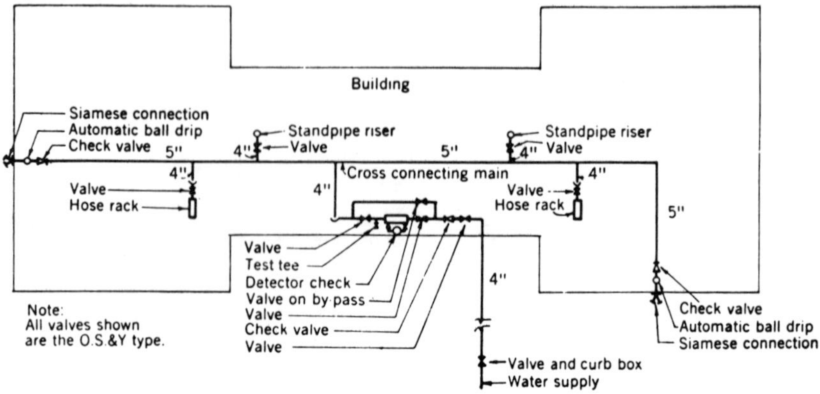

Fig. 2. Arrangement of piping in the basement that is part of the building's fire standpipe system. Plan shows the location of the various valves.

As shown in Fig. 2, all the piping and equipment, either necessary or required by code has been indicated. For a better understanding of this part of the system, the purpose of each item will be covered.

Water supply. The 4-inch diameter water supply shown is considered the primary source of water because it is capable of furnishing at least 20 psi flowing pressure at the highest hose outlet, excluding the roof outlet. The exact pressure requirements for each locality should be ascertained from the local authorities.

The supply starts with a connection to a street water main or a separate street fire main. A valve and curb box or other means of control is installed on this line at some point usually designated by the local water company. The principal purpose of this valve is to shut off the water supply to the building in the event of a fire serious enough to endanger the interior piping. This practice prevents the loss of water or water pressure, should part of the standpipe system be broken or fractured by the heat of the fire.

On the inside of the building, the water supply is equipped with an OS&Y (outside stem and yolk) valve and a check valve. The

OS&Y valve is the interior control for the entire system. The check valve is installed to block the entry of water from the secondary source of water supply. During a fire, the fire department, by means of its pumping equipment and through the siamese connections, maintains a high water pressure in the standpipe system. Since this pressure is greater than the available pressure in the primary source of supply, a check valve is needed to prevent the loss of pressure or the short circuiting of the pumped water through the primary source of water supply.

Water meter. Included on the interior water supply piping is a water meter with its recommended hook-up as shown in Fig. 2. The meter indicated is of the detector check valve type and although it is not required in all localities, it has been shown to cover the requirements of many water companies. The purpose of this unit is to meter the passage of water without restricting the flow and to guard against the leakage or misuse of water intended for fire protection purposes. Such meters are tested frequently.

Siamese connections. These are fire department connections and are considered the secondary source of water supply. In the event of fire, the fire department can, by means of its pumping equipment, pipe water from the nearest fire hydrants to the siamese connection to serve and maintain an adequate water pressure within the standpipe system. Two siamese connections are shown in Fig. 2, and have been so located, that each connection faces a different street, thus assuring accessibility to the fire department. Siamese connections are joined together by means of a 5-inch cross connection line, which assures an ample flow of water to one or both standpipe risers when needed. When specifying the proposed siamese connection, it is important to mention that the hose threads be of the same type used by the local fire department.

An automatic ball drip and a check valve are installed in the location shown in Fig. 2. The automatic ball drip, drains the water out of the line between the siamese connection and the check valve, after the connection has been used or tested. This automatic drainage is necessary to prevent freezing in cold weather climates. The check valve prevents water from the primary source from entering the drained line.

Fire standpipe risers. The two 4-inch standpipe risers shown, are supplied from the 5-inch cross-connecting main. The OS&Y valves, installed in accessible locations are for use by the fire department.

The valve is closed if, during a fire the riser can no longer be used or is in danger of fracturing from the heat of the fire. The base of the riser is usually supported by means of a base elbow resting on a pipe stand or on a brick or concrete pier; or else by means of a clevis type hanger, the piping is supported from the first floor slab. Both methods are satisfactory.

Hose Stations

The important part of the fire standpipe system is the hose station. It contains equipment that is actually used to extinguish a fire and a detail of a typical hose station is illustrated in Fig. 3.

Although the hose rack may be enclosed in a cabinet, and several different arrangements of hose racks and equipment are available for use, the arrangement shown has been used for example in this particular case. As shown in Fig. 3, the hose station consists of a number of various items, all necessary for the satisfactory operation of the unit in the event of a fire. The purpose of each of these items is best explained in the following paragraphs. Design data are included.

Hose. The hose contained on the rack is 1½-inch unlined linen hose used because it is easier for the tenants to handle than 2½-inch hose, and is therefore classified as first aid fire protection. By means of the 2½ in. x 1½ in. adapter shown, the 1½-inch hose may be quickly removed by the fire department to permit the use of its 2½-inch hose. The length of the hose required is determined by the most distant point in the apartments to be reached by the hose. Hose of 50 ft. to 75 ft. lengths are generally used, depending on the particular application requirements.

Pressure reducing device. A pressure reducing device is used when the water pressure at the hose station exceeds the prescribed pressure recommended by the fire underwriters or local authorities. Often this pressure is set at 50 psi. There are several approved types of reducing devices available, but the one intended for use in this case is the pressure reducing disc, that may be adjusted for the desired pressure. This disc is inserted into the hose coupling and may be quickly removed during a fire.

Hose valve. The 2½-inch hose valve shown is set at a distance of approximately 5 ft. to 5½ ft. above the floor as this has been found to be the most desirable height. Here again it is important to mention that the hose threads should be of the same type used by the fire department.

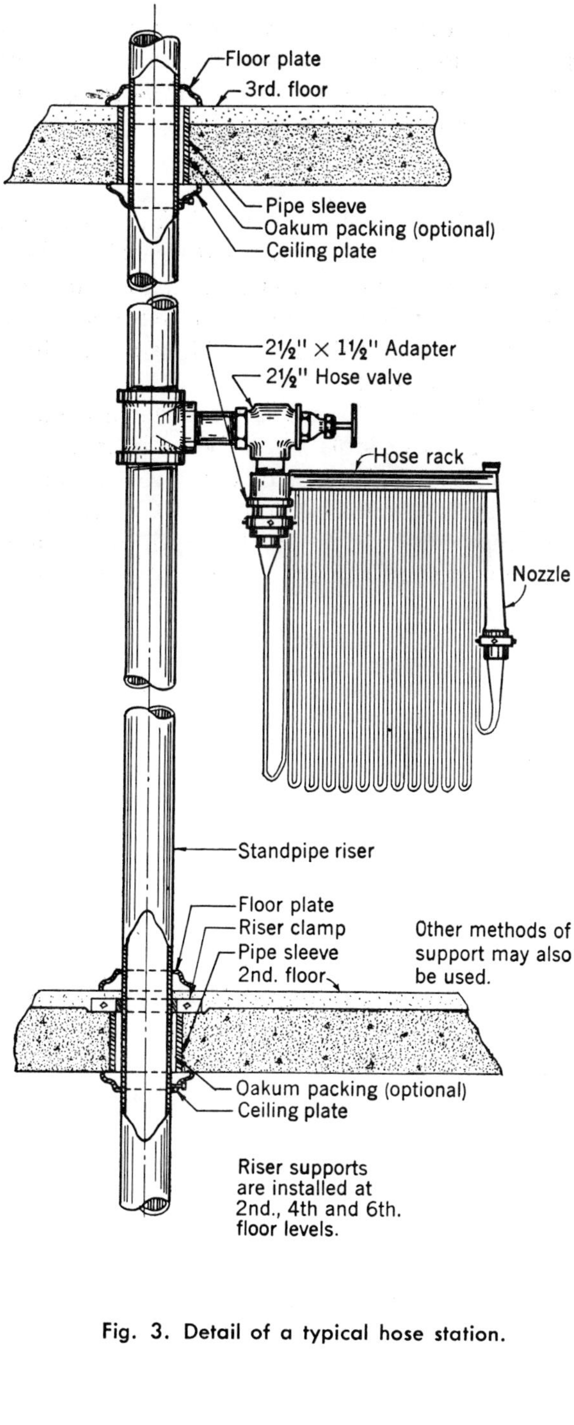

Fig. 3. Detail of a typical hose station.

Riser Support

A typical riser support has been included in Fig. 3, to show the manner in which the piping is supported. Although, there are several methods that may be used for the support of the risers, the method shown is for the purpose of illustration only.

Roof Manifold

A roof manifold which consists of an OS&Y valve and two or more $2\frac{1}{2}$-inch hose connections is installed in each of the two stair towers at the roof level to enable firemen to fight a fire on the roof. It also makes it possible to protect the apartment house in the event of a fire in an adjoining building, by cooling down the roof and exposed side of apartment house.

The unit is shown in Fig. 1, and is installed in a heated space. If a heated space is not available, then the valve must be installed below the roof, in a heated space, with the valve handle so located, to be easily opened by the fireman on the roof. This practice prevents freezing in cold weather. A 50 to 75 ft. length of $2\frac{1}{2}$-inch rubber lined hose is placed on a rack near the manifold for use by the fire department.

CHAPTER 47

Fire Protection in Tall Buildings

Some specific fire protection means is required in every new building being erected. This may consist of sprinklers, fire standpipe or other systems required by the insurance underwriters, local fire department or even the building owners. Of all available systems, the fire standpipe is most generally employed.

To cover design details, this chapter will consider the installation of a fire standpipe system in a 25-story office building. This size building is used for an example because (1) it represents a tall building and (2) because it is the tallest building allowed in many of our cities.

Fire Standpipe Systems

The particular arrangement of piping and equipment that may be utilized in a building of this size is best illustrated by Fig. 1. Two standpipe risers are used in the structure illustrated. The number of standpipe risers is determined by the total area of each floor serviced by the hose stations. Since only 75 ft. of hose are generally allowed at each hose station, sufficient risers must be considered to fulfill the dimensional requirements. Some areas of course, allow 100 and 125 ft. of hose which is quite helpful in a good number of cases. The standpipe risers and hose stations are usually contained in the stairways of the building.

A brief explanation of how the system functions is as follows: Water is supplied to the roof tank by the house pumps which have not been shown. The roof tank holds water for both the domestic and fire reserve and the minimum fire reserve for this installation is 5000 gal. The tank is set at an elevation above the roof that will supply domestic water at a prescribed pressure to the toilet fixtures on the 25th floor. The required pressure at this point ranges from 15 to 25 psi, depending upon the particular type of water closet.

In Fig. 1, the supply from the roof tank is shown as being connected to both standpipe risers. Often this line may be connected to only one of the risers and it will supply the other riser in an upfeed manner by means of the cross connecting line in the subcellar. The fire reserve furnishes the standpipe system with the water it needs for fire protection. On the lower floors where the pressure will exceed 80 psi at the hose station, pressure reducing devices are employed to maintain prescribed pressures. The type of device used here, for example, is the orifice type, which fits into the hose connection at the $1\frac{1}{2}$ in. valve.

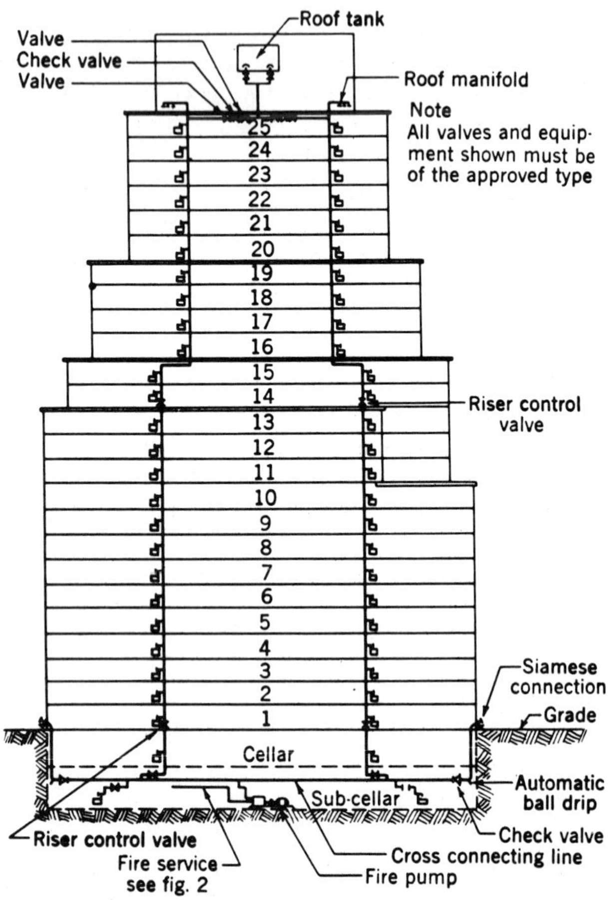

Fig. 1. Typical arrangement of a fire standpipe system in a 25-story office building.

The base of the standpipe risers are connected to a cross connecting line which terminates at the two siamese connections shown. In this building a fire pump is also required and this unit is placed in the sub-cellar in a room that adjoins one of the stairways. Two fire service mains are installed to supply this pump with the water needed to combat a serious fire.

Component parts of the fire standpipe system are now covered, starting with the piping and equipment located in the sub-cellar.

Sub-cellar Floor Level

Figure 2 shows a typical arrangement of the piping and equipment as may be found at this floor level in the building.

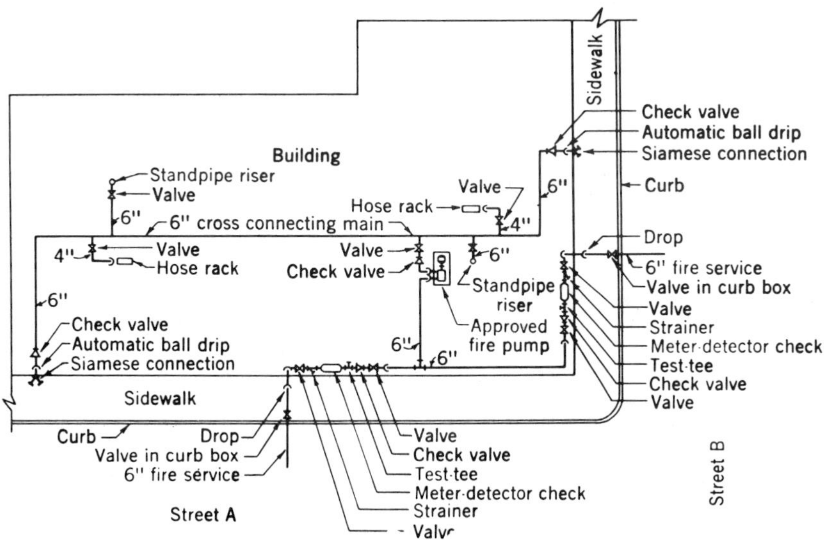

Fig. 2. Arrangement of fire standpipe piping in sub-cellar of building discussed.

The 6-inch fire mains equipped with valves and curb boxes enter the building from both streets and pass through the valved and meter arrangement before it connects with the fire pump. The fire pump must be of the approved type, capable of maintaining the pressure required in the fire standpipe system. The size of this pump is easily calculated but it is always good practice to consult with the pump manufacturers representative to ascertain the performance characteristics of the pump selected. It is also well to check with the local authorities for their requirements regarding the fire pump.

As previously mentioned, the fire pump room should be located adjacent to one of the stairways, since this room must be easily accessible to firemen entering from the street. Several signs indicating the way to the fire pump room should also be placed both inside the building and out, to point the way to this room. It is always advisable to check both the location and requirements of this room with the local fire authorities.

A cross connection line with a siamese connection on two different streets is required. A check valve and automatic ball drip are placed on this line inside the building. A drop has been shown on the fire service line before it enters the building. The fire service line is installed near the surface of the grade. The drop is required to allow the line to be low enough for entering the sub-cellar.

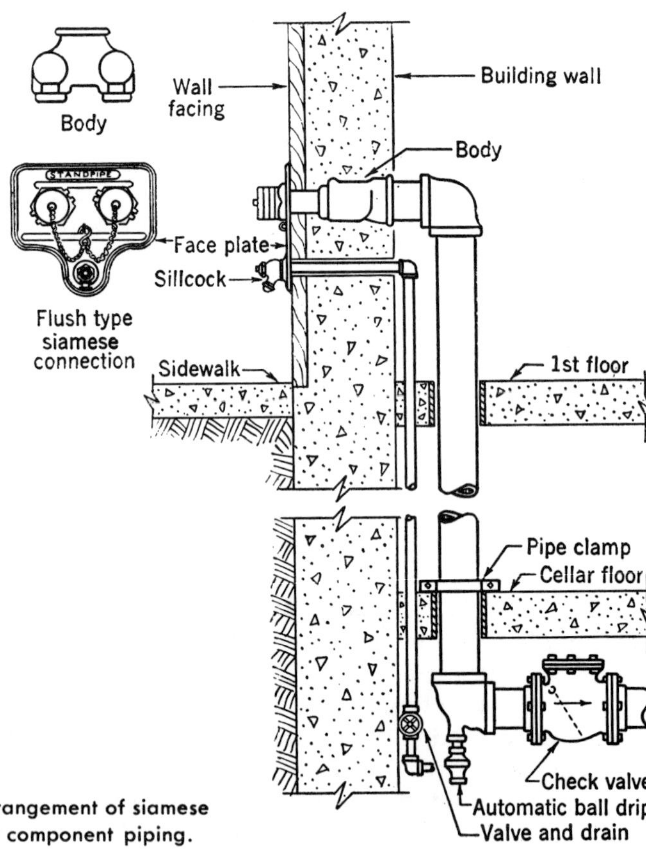

Fig. 3. Typical arrangement of siamese connection and component piping.

Siamese Connection

There are several types of siamese connections available for installation on the ouside face of the building. The type shown in Fig. 3, is for example only, and is a flush type siamese connection equipped with the number of hose connections required by the fire department or local authorities. It is very important to check with the local fire department for the specific type of hose threads that should be specified. In fact, this consideration applies to all threaded equipment in the building. Unless the hose threads conform to local fire department standards, it will be impossible for the fire department to connect their equipment in the event of a serious fire.

A check valve and an automatic ball drip must be placed on the cross connection line as shown. The check valve presents interior water pressures from reaching the siamese connection while the automatic ball drip drains the line between the check valve and the siamese connection of any water that may accumulate.

Sillcocks, other than the type shown, may be utilized. Of these, the non-freeze type is very popular, especially in temperate climates.

Fire Pump

A detail of the fire pump is illustrated in Fig. 4. Here the fire pump is set on an adequate concrete base. The supply and discharge

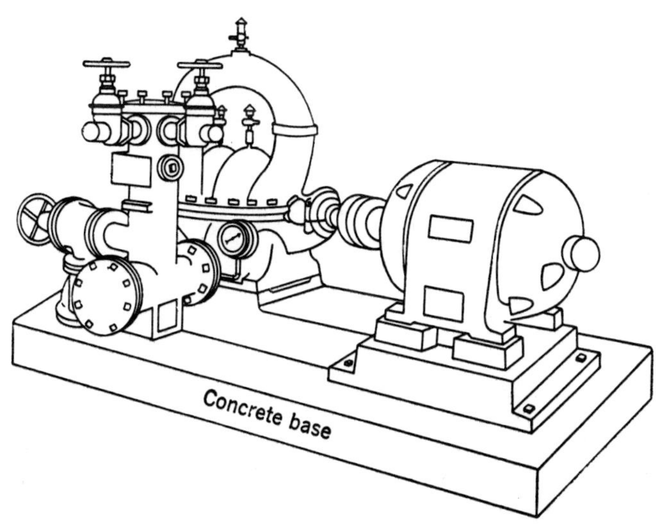

Fig. 4. Type of pump installed for fire service in a building.

connections to the pump are equipped with flexible connections to prevent piping vibration and clatter when the pump operates. Since the pump is tested periodically, these are worthy provisions. All the other control devices normally installed with the pump have been omitted for the sake of clarity.

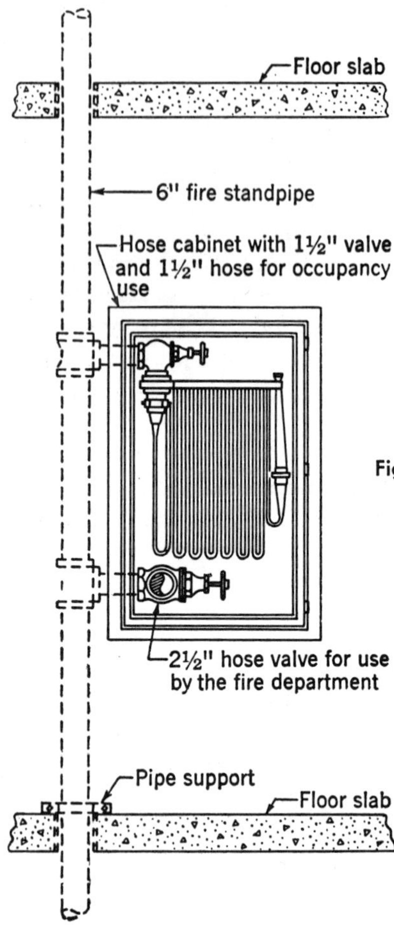

Fig. 5. Detail of a fire hose station.

Base of Riser Supports

It is highly recommended that adequate supports be placed at the base of the standpipe risers. Even though the risers themselves are held firmly in place with pipe clamps at intervals of every two or three floors, a support at the base prevents any possibility of the heavy riser slipping.

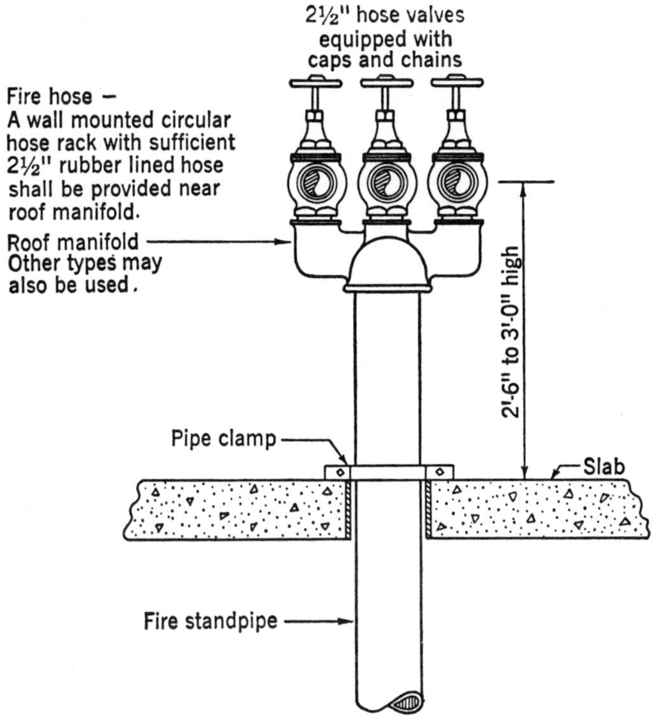

Fig. 6. Detail of roof manifold installed in a heated stair tower.

The supports usually consist of a special elbow resting on a brick or masonry pier. If there is not sufficient space for a pier, then an adequate hanger support should be given some attention.

Hose Stations

In this case the hose station is a fire hose cabinet as shown in Fig. 5. The hose cabinet houses a 1½-inch valve and 75 ft. of unlined linen hose. It is common practice to utilize 1½-inch hose because hose of this diameter is more easily handled by the occupants of the building than 2½-inch hose, which is the size used by the fire department. A 2½-inch hose valve is placed in the lower portion of the hose cabinet for fire department use. The fire department will attach their own regulation 2½-inch hose to this valve.

Where, because of the height of the riser, the water pressure will exceed prescribed limits, pressure reducing devices can be employed.

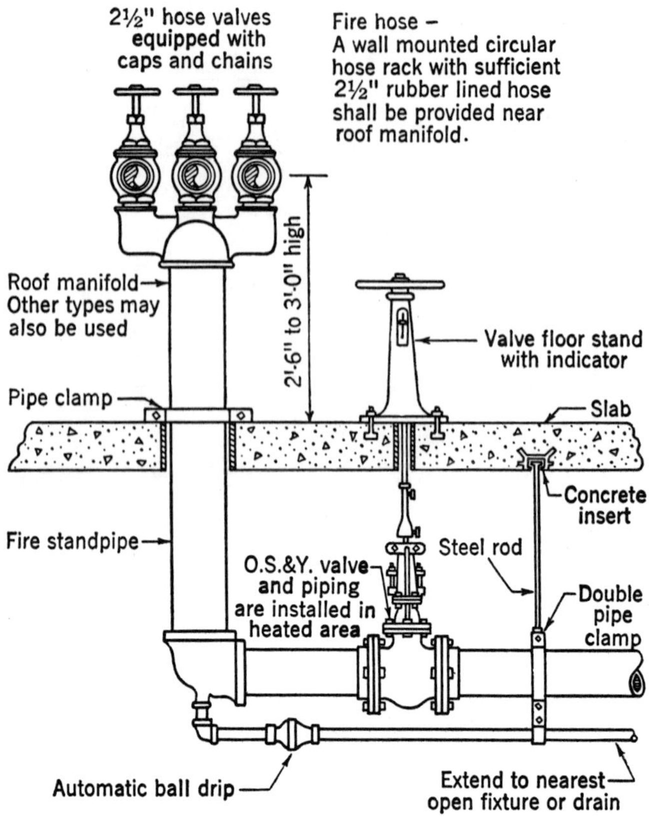

Fig. 7. Detail of roof manifold installed in an unheated stair tower.

When these devices are used, they are placed only on the 1½-inch hose rack. Several manufactured devices are available for this purpose. The local fire department should again be consulted for any requirements they may have for the hose cabinet and the equipment it should contain.

The standpipe riser is generally supported at every second or third floor by means of the pipe clamp shown.

Roof Manifold

A roof manifold is required at the roof of the building to combat fires that develop at this level or just for wetting down this area in the event of fires at adjoining structures. A typical roof manifold is detailed in Fig. 6.

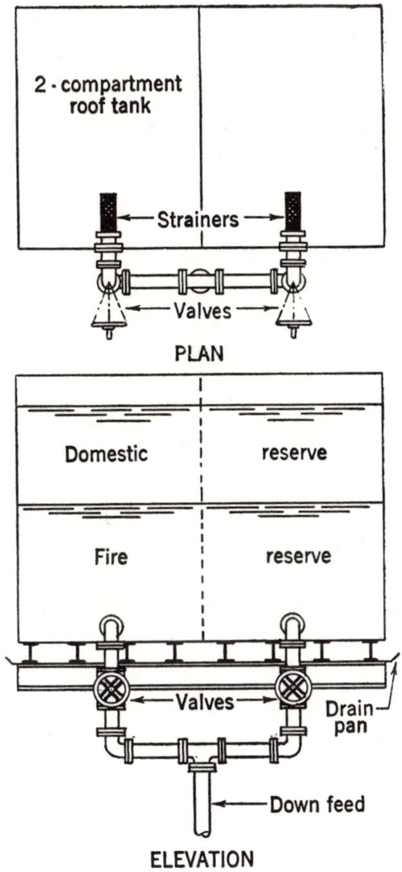

Fig. 8. Arrangement of piping connection to roof tank.

This unit illustrated consists of a patented manifold with three capped hose outlets. The hose used at this area is rubber lined to withstand dampness or the elements that tend to affect it.

If the space where the roof manifold is to be located is not a heated area, then a special arrangement must be utilized as given in Fig. 7.

Here the manifold and hose is located in the unheated space while the OS&Y (outside stem and yoke), valve controlling the water supply is placed in or at the ceiling below. The valve handle is extended through the roof in the manner shown. An automatic ball drip removes any water that remains in the line between the OS&Y valve and the manifold after the unit has been used or tested. This device prevents the possibility of water freezing in the manifold. It is positive assurance against damage in the event that the OS&Y valve leaks or is accidentally left open. Roof manifolds have an additional use in many new government buildings, in that, they are used to wash down roofs should a heavy atomic fallout occur.

Roof Tank

The roof tank is installed either indoors or out. When it is installed indoors there is no danger of the water in the tank freezing. However, when it is installed outdoors, the tank should be equipped with heaters which may consist of steam coils or electric heating elements. Office buildings are subject to long week-ends when the building is unoccupied. It is at these times that freezing will occur. Figure 8 shows the general arrangement of the piping at the roof tank.

The roof tank illustrated is constructed in two sections or compartments. This type of construction permits the cleaning of one compartment while the other still furnishes a safe reserve of water for fire protection purposes.

The connections for the fire lines is taken at the bottom of the tank with enough clearance from the bottom of the tank to avoid the sediment that accumulates here. Strainers are placed on these lines in the tank to prevent any foreign objects from entering the lines. A drainage pan is installed under the tank to catch and drain any condensation that will form and run down the sides of the tank.

Only the fire standpipe lines have been shown in Fig. 8; all other piping, including the tank controls, have not been illustrated. Although roof tanks may also be constructed of wood, the tank shown in Fig. 8, is of steel construction.

Part 5
Plumbing Fixtures

CHAPTER 48

Access Space for Plumbing Fixtures

Over the past number of years, architects and engineers have shown an increasing tendency to provide an access space behind or between rows of plumbing fixtures. Basically, an access space is a narrow corridor behind a single row of fixtures or between a double row of fixtures through which a man may pass for inspection and maintenance purposes. Acceptance of access space in design work, even though it uses up some floor area, is understandable because of the ease that this arrangement provides for inspection and maintenance, and the better appearance that results in toilet rooms when all valves are moved back of the fixture-supporting wall, into the access space.

In certain institutions, as prisons or mental hospitals, the access space often ceases to be optional and becomes a necessity because of the requirement that fixture trim of any type or form must be concealed to prevent inmates from inflicting injury upon themselves with such trim.

Access spaces are widely used in various types of buildings and they are particularly desirable because it is possible to use push button or oscillating handle controls in the toilet rooms and to conceal the flush valves. Since the flush valves, other types of valves, and all piping are installed in the access space, and only the fixtures are exposed in the toilet room, a high standard of sanitary conditions can be achieved.

The decision of whether or not an access space should be provided, depends on the desires of the owner or the architect, the particular nature of the building, and the proposed location of the toilet rooms. If numerous fixtures are required in the toilet rooms, the access space would be a definite advantage.

As floor area is an important item to both the architect and the owner, a major consideration is the amount of floor area required for the access space. Two conditions will be considered — the minimum size of an access space for a single battery of fixtures and the minimum size access space for a double battery of fixtures that are back to back, together with certain considerations relating to the piping installed in these spaces.

Single Battery of Fixtures

As an example for a single battery of fixtures, we shall assume one consisting of water closets, urinals and lavatories as would be found in a men's toilet room. Figure 1 is a typical arrangement of such a condition.

In Fig. 1, only the water closets have been indicated since such roughing dimensions determine the required extra space needed for access. All piping, including soil, waste, vent and water piping, is installed as close as possible to the wall on which the fixtures are supported. The 2 ft. access width is minimum and yet will allow a man to move through the space for maintenance and repair purposes. Since the flush valves are also located in the pipe space, they can be inspected and adjusted with ease.

Conventional fittings are used in Fig. 1 to show the overall piping dimensions. Patented closet fittings of various types could be used

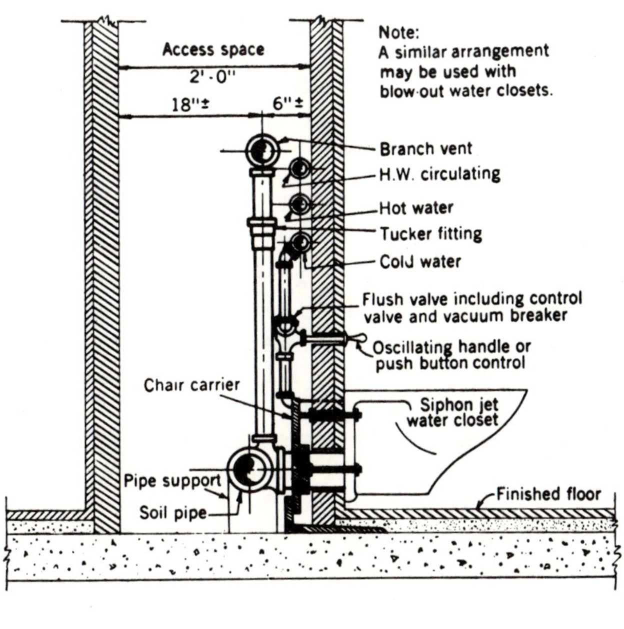

Fig. 1. Access space for a single battery of fixtures.

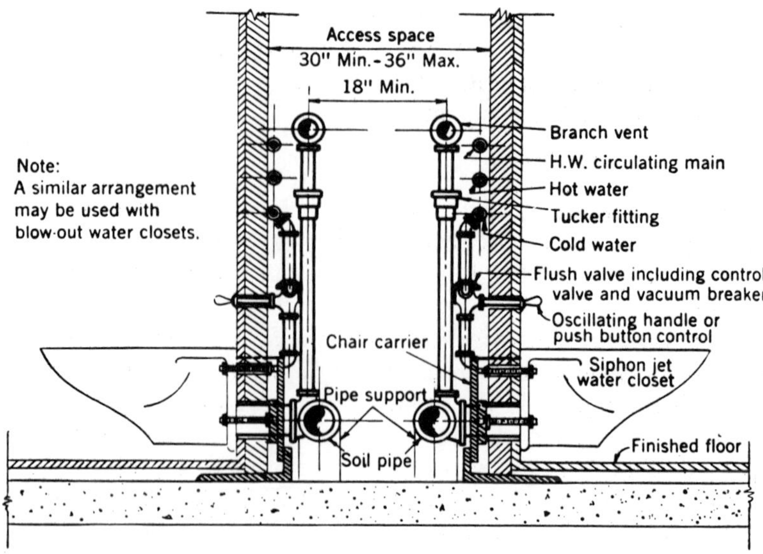

Fig. 2. Access space for a double battery of fixtures.

to better advantage than the fittings shown because they are more compact. But in this example, the conventional type fittings are used to illustrate the condition.

Entrance to this back-wall space is provided by an access door placed at one end of the space or in the wall, not supporting the fixtures. The architect normally chooses the door location.

Double Battery of Fixtures

As an example of a double battery of fixtures, we shall assume two batteries of fixtures, back to back. One battery has the water closets, urinals and lavatories used in a men's toilet room, and the other battery consists of water closets and lavatories as installed in a women's toilet room. Figure 2 is a typical arrangement.

In Fig. 2, again, only the water closets have been shown since their roughing dimensions determine the required extra space needed for access. Although patented closet fittings could be used to best advantage in this case, conventional type fittings were used for the purpose of illustration.

In an access space of this kind, the important consideration is to provide a clear passage through which a man may freely move and operate for maintenance and repair. To accomplish this, the soil, waste, vent and water piping must be installed close against each wall. Soil lines, especially, should not interfere with clear passage through the space and should, therefore, be installed along each wall. The installation of the piping, as shown, will be more costly than regular roughing but it is necessary in order to keep the space clear.

Entrance to this space is usually provided by means of a full size access door at the end of the space. If this is not possible, then a smaller access door must be provided at some convenient location preferably in the men's toilet.

With these considerations, it will be possible to provide an access space of minimum, yet adequate size for easy inspection, and repair of the valves and piping.

Advantages of an Access Space

The width of access space shown in Figs. 1 and 2 are sufficient for any type of plumbing fixtures used. This includes floor type water closets or any other combination of plumbing fixtures normally installed in the toilet rooms.

By means of the access space, the necessary maintenance and repair may be performed without disturbing persons using the toilet facilities. Access space illumination can be furnished by a fixed electric light and an electric utility outlet into which an extension cord may be plugged.

Water flow in the piping must be cushioned to avoid the objectionable noise that water hammer can cause. This cushioning effect is usually obtained by individual pipe air chambers at each fixture supply take-off, or by single pipe air chambers at the end of the cold and hot water branches. Mechanical type air chambers may also be used at the end of these branches. When pipe air chambers are used, the air cushion requires frequent air replenishment to make up for air that is gradually absorbed by the water. To replenish this air, it is necessary to shut off the valves controlling the water branches and to open the fixture faucets or flush valves. This will allow the water in the pipe to drain and air to enter. When the water is turned on, a sufficient air cushion will again be available in the chambers. This type of air chamber is not very efficient and the introduction of air must be performed at such time when the fixtures are not in use. It is time consuming.

Since the access space is available, it is of much greater benefit to install pipe air chambers at the end of the branches, properly supported and equipped with a pet cock, control valve and a drain valve. This type of a chamber, Fig. 3, because of its greater air volume will cushion the water more effectively. When it is necessary to replenish the air, the control valve is closed and the pet cock and drain valves are opened, without disrupting the fixture usage. The water in the chamber will drain out and is replaced with air. After this change

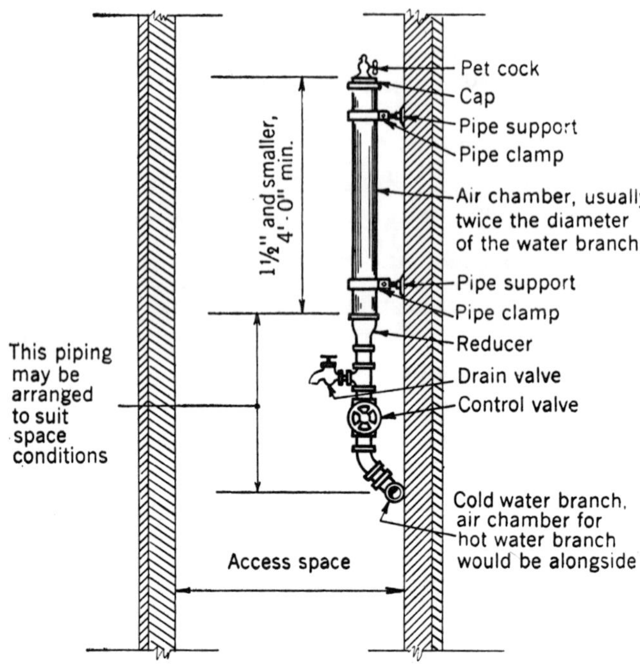

Fig. 3. Typical air chamber at end of cold and hot water branches.

has taken place, the chamber is once again ready for operation. Mechanical air chambers are also advantageous, but they should be properly supported to prevent strain on the piping.

For two batteries of fixtures, served by hot and cold water lines, one air chamber can protect the two cold water lines and similarly, one air chamber can safeguard the two hot water lines.

Aside from the advantages of an access space that were mentioned, there is the point that such space simplifies toilet ventilation. Instead of taking away part of the toilet space for ducts and for air outlets, the supply duct is run in the upper portion of the access space. It thus uses space that would normally be wasted, and it is also out of the way of the men who may have to work back of the pipes.

With regard to the amount of floor area utilized by the access space, the space required for a single battery of fixtures is just about twice that required for a regular pipe space for the fixture roughing without access. For a double battery of fixtures, the space is again just about twice that of a regular pipe space, the conditions in both cases being equal. Therefore, if on a proposed project, an access space is being considered for the fixtures by both architect and engineer, the advantages derived from the space far exceed the loss of some additional floor area.

In addition, if it is possible to locate the toilet rooms so that the fixtures will be back to back as shown in Fig. 2, there will be greater savings in space than if two separate toilet rooms in different locations on the same floor are planned. To clarify this point, two separate access spaces for single batteries of fixtures would use up about four lineal feet, whereas one access space for a double battery of fixtures requires about 30 inches as indicated in Fig. 2.

Although many favorable points have been made for access spaces, they were motivated by the thought that certain considerations should be noted if access spaces are contemplated. It is not the intention of this chapter to show a decided preference for access over regular pipe spaces.

The points presented should be kept in mind during the design stages of the project. In this manner it will be possible to have not only an access space of minimum dimensions, but also one of adequate size. The chief advantage is that the finished walls in the toilet rooms are free of everything except the fixtures themselves. In addition, care in the selection and installation of valves, air chambers and equipment will provide for a system that is trouble-free in operation.

CHAPTER 49

Fixture Roughing for Low Rent Housing Projects

There will always be a demand for low rent housing projects of the type now being erected in great numbers throughout the country. This class of apartment house has significant importance because it provides modern living quarters for families in the low-income bracket. As a result, many slum areas have been cleared and in their place, now stand, handsome structures with spacious grounds. The benefits gained from this type of project are many, including a higher standard in living conditions, playgrounds, trees and most of all, a finer sense of community spirit.

Since low rent housing projects are financed for the greater part by federal funds, economy in construction is a necessary consideration. Economy, however, is a broad term and should receive careful interpretation. Economy does not mean that cheap and often inferior materials or equipment can be installed. Rather, economy, means that the best possible plumbing systems be provided for the least possible cost. This requires careful planning for the project.

Various features of design are described that should be included in the minimum requirements of good plumbing systems.

Typical Layout of Toilet and Kitchen Fixtures

Usable floor space is the prime factor in each apartment. Therefore, the bathroom, although it may be limited in size, should contain all the necessary facilities. A typical arrangement for such fixtures is shown in Fig. 1. As illustrated in this figure, the kitchen and toilet room fixtures are installed on either side of a common pipe space with the minimum of spacing between the fixtures. The gas or electric range, depending upon which one is most economical to operate in the particular area, is generally installed next to the sink. The range as well as the service connections have not been shown because this subject merits more thorough discussion.

Minimum Pipe Space Requirements

Referring to Fig. 1, it will be noted that all the vertical and horizontal piping is contained within a 10 in. pipe space. This is the amount of space generally required for the installation of the piping. If it is possible, a 12 in. pipe space should be provided for ease of installing the roughing.

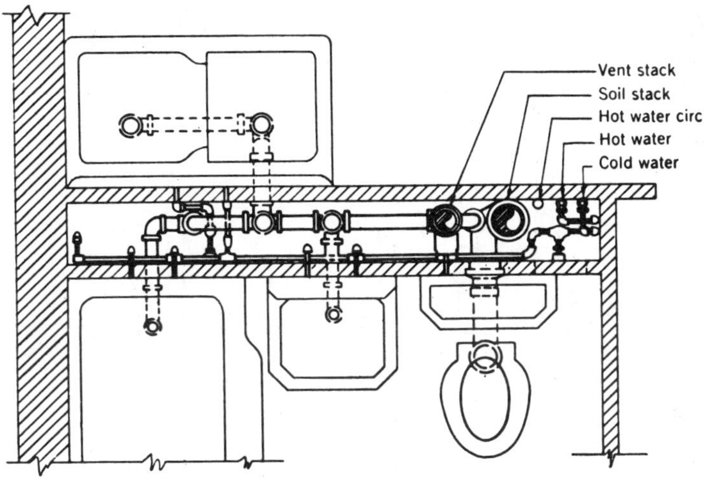

Fig. 1. Plan of typical arrangement of fixtures and piping.

Typical Elevation of Required Roughing

The roughing for the kitchen and toilet room fixtures can be installed in a number of different ways and of varied materials depending upon the preference of the engineer and contractor and the requirements of the local code. A system offered only for the purpose of example is shown in Fig. 2.

The soil and vent stacks consist of a patented system of piping and fittings as manufactured by the Central Foundry Company. However, conventional pipe and fittings of the caulking or threaded type can be used equally well. Other patented systems of pipe and fittings may also be used. Galvanized wrought iron or galvanized steel of the threaded type has been shown for the branch waste and vent piping. Again, other selection of materials are also acceptable. The water piping to be installed is usually specified of material that best withstands the corrosive action of water.

The pipe supports shown in Fig. 2 are quite important and have been placed at specific locations to prevent sagging of lines thereby eliminating the danger of leakage and escape of foul odors. The branch water lines are offset near the connection to the vertical risers. The offsets provide flexibility to compensate for the expansion and contraction of the vertical risers.

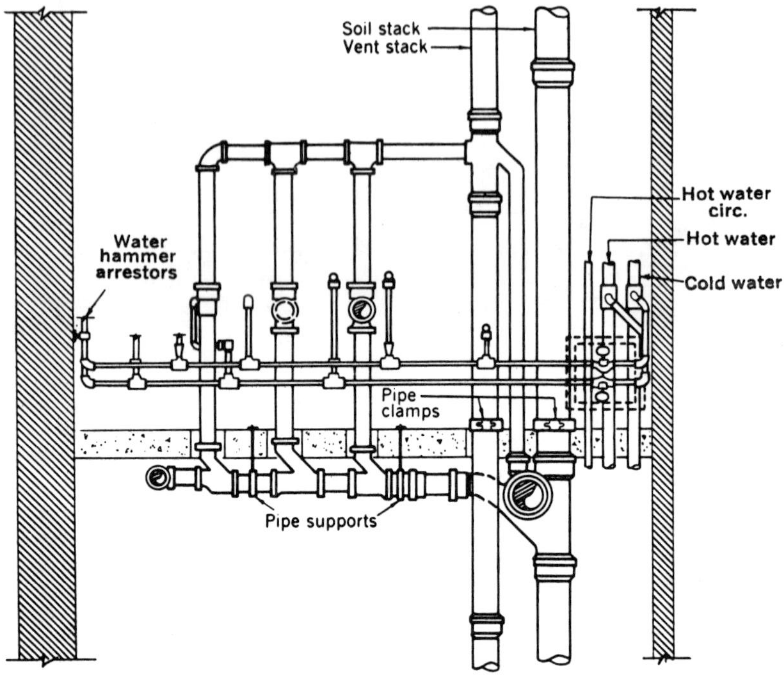

Fig. 2. Elevation showing the roughing for fixtures.

Water Control Valves

Valves are needed to control the water branches. It is always economical to install quality valves at this point rather than to replace valves of lesser quality, after a few short years. Often, control valves are installed in the pipe space with just the stem and valve handle projecting through the finished wall. This is not good practice because all parts of the valve should be accessible for inspection and repair. As a result, if the need would arise, a hole would have to be chopped in the wall to reach the rest of the valve. It is better practice to furnish a small access door and frame of say 8 in. by 8 in. to ensure the ease of inspection or repair.

Cuts in Floor Slab

In Fig. 2, it will be noted that cuts have been provided in the slab for the installation of the vertical piping. Whether or not a full cut will be provided in the slab within the confines of the pipe space, or cuts just large enough for the passage of the piping, is a condition that should always be discussed with the structural engineer. Normally, just cuts are furnished by the structural engineer to maintain strength of slab construction.

Typical Sections of Fixture Roughing

In order to point out various considerations related to the roughing for the water closet, lavatory, sink and bathtub, the following illustrations are used.

Figure 3 illustrates the floor type water closets which are generally

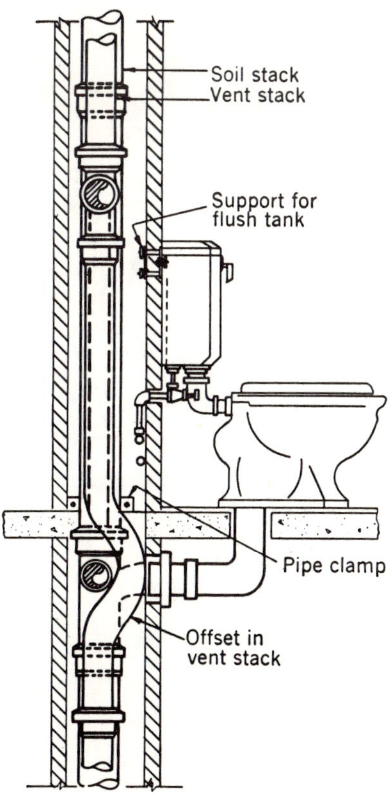

Fig. 3. Roughing for a water closet.

installed in low rent housing projects because they are considered more economical. The flush tank should be carefully supported because the finished wall of these structures are of lightweight construction. Therefore, steel plates bolted together on either side of the wall support the tank. The steel plates distribute the weight load over a larger area rather than at a concentrated point.

The conventional lead bend or other patented device is installed under the slab but is exposed in the toilet room below. This installation does not lend to a neat appearance or add to the degree of sanitary cleanliness in the toilet room below. A partially furred ceiling would hide the exposed roughing but because of the additional expense involved, is not usually included. This is a point where economy could be expanded a bit.

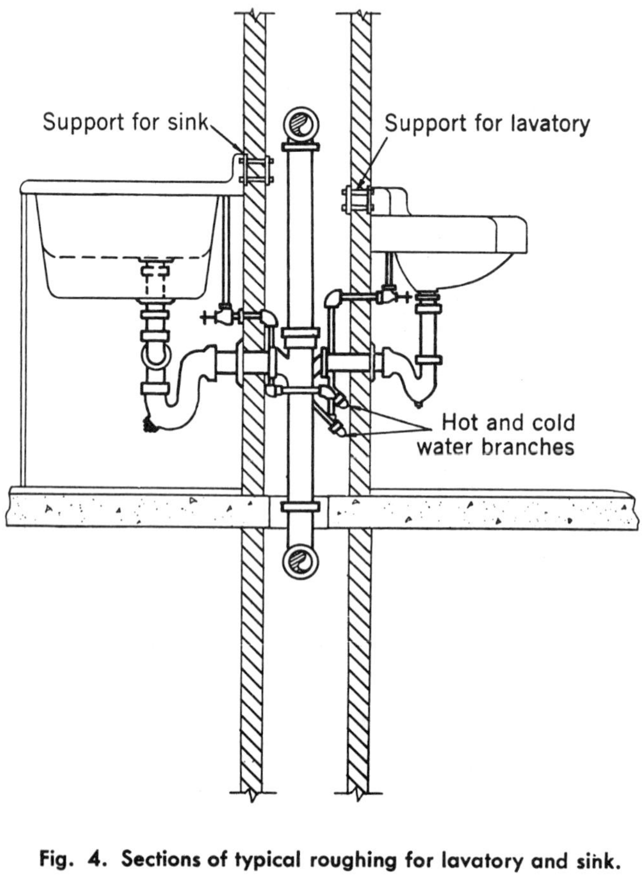

Fig. 4. Sections of typical roughing for lavatory and sink.

Use of Valves

It will be noted that steel plates are also utilized for the support of the lavatory and the kitchen sink, as illustrated in Fig. 4. Control valves have been included on the hot and cold water supplies to each fixture. These valves are often omitted. However, they do save considerable maintenance time when faucet washers are replaced and are quite helpful for adjusting the proper amount of water desired at each fixture. Such valves bear a little further consideration since repair or maintenance time is also a form of economy in the operation of a low rent housing project.

Roughing for the bathtub is shown in Fig. 5. The P-trap is installed below the slab and is exposed in the toilet room below. Moreover, since the cleanout plug is located at the bottom of the trap, it becomes necessary to clear any stoppages from the apartment below. If the bathtub waste piping and all the other piping for that matter is carefully reamed before it is assembled there will be much less possibility of stoppage.

Projected Economy

The various factors of economy in construction have been covered. Certain features have been added or recommended which add to the cost of construction. These added features, however, will also reduce the required maintenance time. Economy, therefore, cannot be measured in terms of the cost of construction alone for cost of maintenance time must also be considered. When both considerations are weighed, a favorable balance should result. This balance represents the true picture of economy. It must be remembered that all items added or deducted in low rent housing projects are multiplied in cost by 100 to several thousand units depending upon the size of the project.

Other Housing Projects

Many privately owned projects are being erected with kitchen and toilet fixtures installed in a similar manner to that shown for the low rent housing projects. The illustrations shown for the low rent housing projects will therefore also apply to the privately owned projects. If wall hung water closets are desired, the required roughing would be similar to that illustrated in Fig. 6.

As detailed in Fig. 6, the soil and vent stacks have been modified to include the necessary chair carrier for the siphon jet water closet. In addition, the diameter of the cold water supply lines will have to

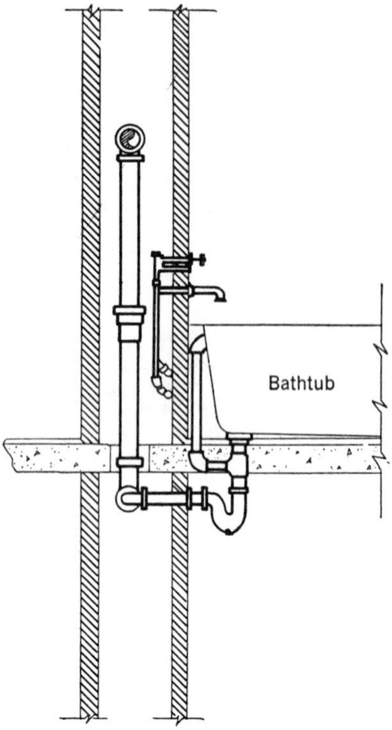

Fig. 5. Section of roughing for a bathtub.

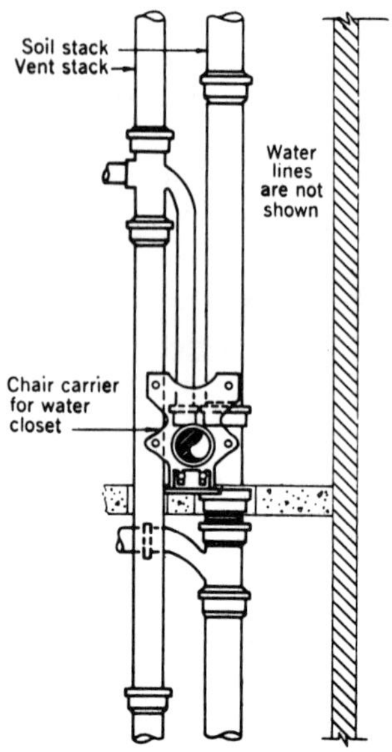

Fig. 6. Typical roughing for a wall-hung water closet.

be increased proportionately in order to furnish ample water for the flush valves.

Many variations in arrangement or type of fixtures is possible with the basic roughing system. The important consideration, however, is that the systems be durable, trouble-free in operation and offer ease of maintenance.

CHAPTER 50

Floor Mounted Water Closet Installations

One subject that has always caused consternation among architects and engineers is the amount of pipe space required for the roughing of floor mounted water closets. It is the *net* floor space from which the owner derives his income.

Naturally, much thought must be given to the careful placement of systems and equipment for best utilization of floor space. Unfortunately, there is a tendency to reduce space needed for the systems and equipment to gain a few inches of useable floor space and this can lead to many systems and equipment installation complications.

There is often a tendency to minimize the pipe chases for installing the roughing for plumbing fixtures and equipment. Certain limitations must be maintained. These minimum dimensions will be discussed.

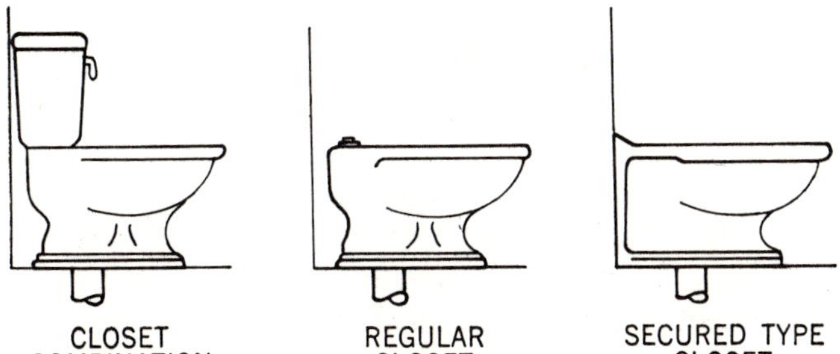

Fig. 1. Three basic types of floor mounted water closets.

Types of Water Closets

This chapter is confined to floor mounted water closets. The three basic types are the closet combination, regular closet and the secured type closet. All are illustrated in Fig. 1.

The closet combination comes in several types and is primarily used in private residences. It can also be used in other types of buildings. The closet tank is supplied by a small diameter water line.

The regular closet is also available in a number of styles and shapes. A flush valve is employed for dispensing water. This type of closet is used in quantity in all types of buildings, including private residences, where the water piping is of ample size.

The secured type closet is generally used in institutions or for any special application where it is desired to secure the closet to both the floor and the wall to prevent it being torn loose. This closet is equipped with a concealed flush valve.

Closet Combinations

The closet combination is generally installed in private residences where the pipe space or wall construction is narrow.

Figure 2 shows a typical installation of this closet. The roughing is installed in a wall made of 2 x 6-inch stud framing, sheetrock finish on one wall, and a sheetrock finish and tile on the other wall in the bathroom. The soil stack is generally 4-inch diameter piping, although 3-inch diameter piping is acceptable in many areas of the country. Although cast iron hub and spigot piping has been indicated, galvanized steel, galvanized wrought iron or copper tubing can be used. Cast iron hub and spigot piping has been detailed throughout the chapter because it is of larger physical proportions than all of the other materials mentioned. As such, it serves as the extreme for the amount of space needed for roughing purposes.

The 2 x 6-inch studding generally measures somewhat less than 6 inches. The extra heavy cast iron soil piping hub has an overall diameter of $6\frac{3}{16}$ inches, and therefore it is necessary to gouge some of the sheetrock wall out so that the piping will fit into the space. This is a tight fit but it can be accomplished in the space allotted. Piping materials of smaller proportions will fit a little easier.

When 3-inch diameter hub and spigot piping is used, the pipe

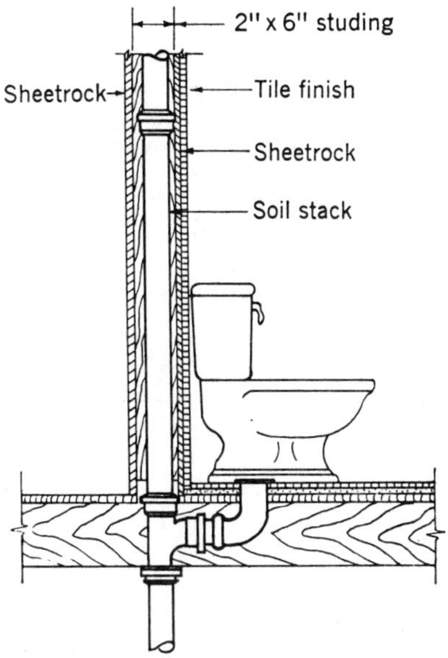

Fig. 2. Typical installation of a closet combination in a residence.

space should not be lessened. It should be 2 x 6-inch stud construction for ease of installation; 2 x 4-inch construction will not provide sufficient space. Another reason for the 2 x 6-inch construction is that another closet combination is often located on the floor above, thereby necessitating space for the proper venting of the lower floor closet. The water supply piping poses no problem to the engineer because it is of small diameter.

Back-to-Back Installations

Quite often, closet combinations are installed back-to-back, as illustrated in Fig. 3.

For this type installation, the 2 x 6-inch stud construction may again be utilized, as shown. A number of different types of closet bends, as the lead bends shown, are available.

Regular Closets

The regular closet is of prime importance because of the vast quantities which are installed in all types of buildings. Many such installations include a battery of closets as suggested by

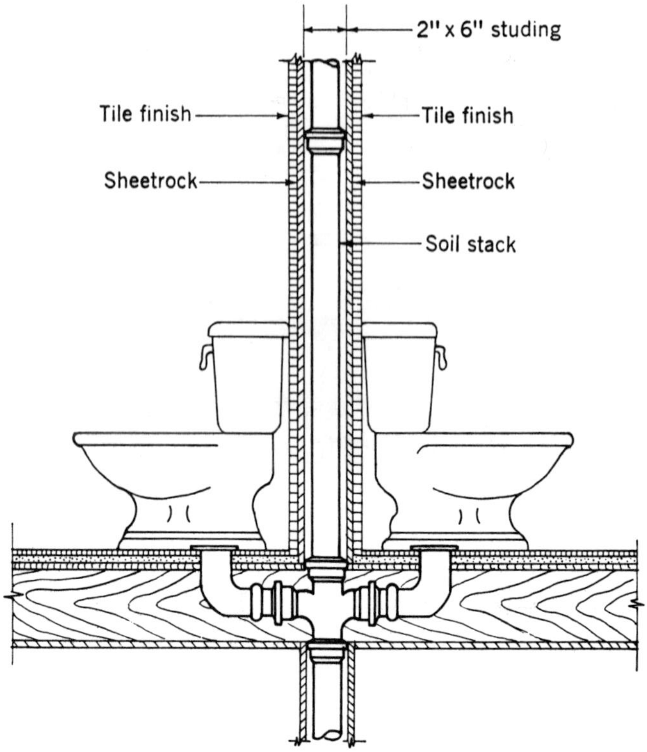

Fig. 3. Back-to-back closet combination installation in a residence.

Fig. 4. A typical method of roughing is illustrated.

In this instance, a cast iron closet bend is illustrated. Each bend is vented in the manner shown. The pipe space will thus contain a soil stack, vent stack, cold water riser, in addition to the branch vent and water piping.

The elevation shown in Fig. 5, section A-A of Fig. 4, shows the roughing which will be contained in the pipe space above the floor level. The drainage piping is installed in the pipe space under the floor. The required pipe space for the fixture roughing for the situation shown is 12 inches.

Minimum Requirements

Refer to Fig. 5 and note the tight grouping of the vertical soil stack, vent stack and cold water riser. The required pipe space is

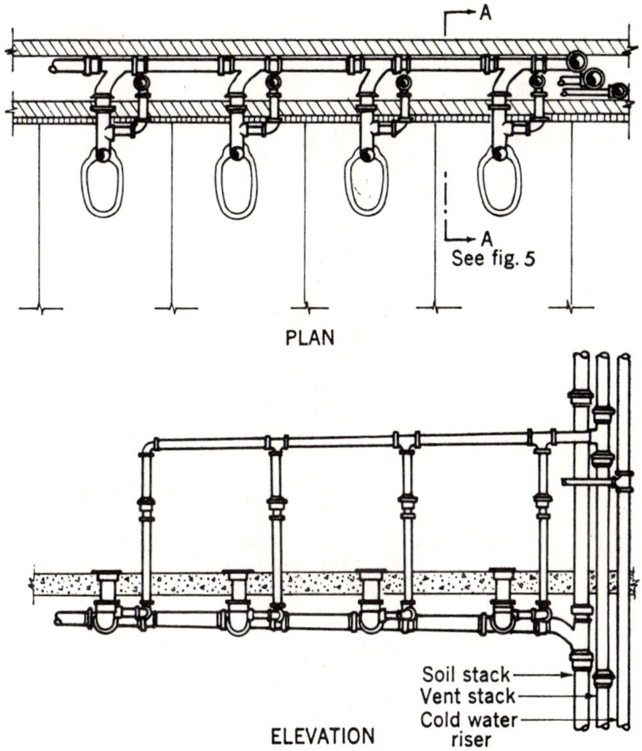

Fig. 4. Arrangement of roughing for a battery of water closets.

12 inches for the ease of installing this fixture roughing. Although the minimum space that might be used is 10 inches, this means there may be some complicated placement of the cold water riser.

Another consideration is the exhaust ducts that are required for interior toilet rooms. Since these ducts are generally placed in the upper portion of the pipe space, sufficient room must be allotted for them. A 12-inch pipe space is usually ample for this. It is always good practice to consult with the engineer designing the ventilation system to learn his exact requirements.

Many persons may believe that they can install the roughing in smaller pipe spaces. This is entirely possible. However, it is recommended that the proposed installation be viewed from the aspect of design and the ease of installation, as well as from the desire for standardization before attempting a design for one

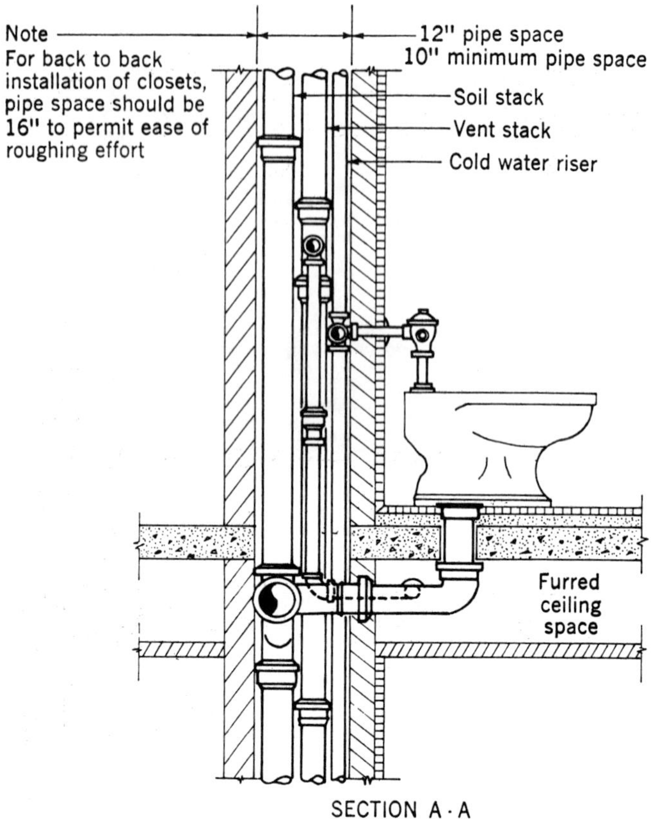

Fig. 5. Section of Fig. 4 showing arrangement of plumbing risers and roughing for fixtures.

project that may not be suitable for another. In addition to the particular arrangement of roughing shown in Fig. 4, several other methods may also be employed, some of which are quite cumbersome. It is better practice for the industry to standardize on a pipe space that is adequate.

Access Space for a Single Battery of Closets

In many buildings there is the desire to use an access space for a single battery of regular water closets. A typical installation is shown in Fig. 6.

An access space permits maintenance of concealed flush valves without disturbing the occupants of the toilet room. If an access

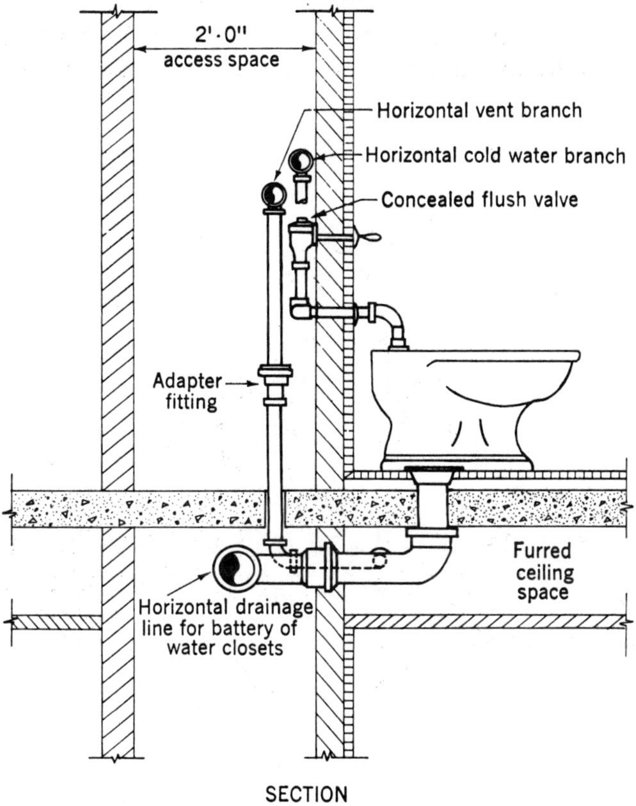

Fig. 6. Typical roughing in an access space for a battery of regular water closets.

space is proposed, it should be of ample proportions for the purpose intended. Very often, inadequate access spaces are provided and this results in limited movement for the maintenance man. Unless the space is sufficient, it does not serve a useful function. The width of the pipe space shown in Fig. 6 is 2 ft. This assures a space of 18 inches between the wall and the vertical vent piping and this is sufficient to allow the maintenance man freedom of movement. Furthermore, electric lighting should be provided in the space to avoid the need for a flashlight. Generally, no problems are experienced in installing an exhaust ductwork in an access space.

Access Space for a Double Battery of Closets

When two batteries of water closets are involved and an access space is contemplated, a typical installation might be that shown in Fig. 7.

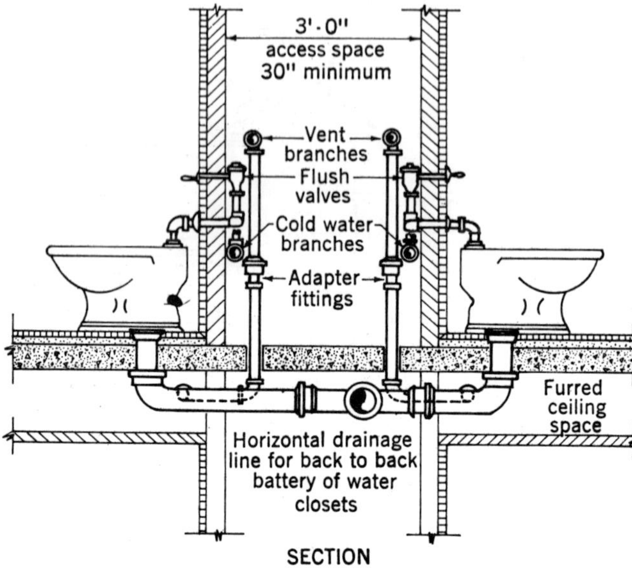

Fig. 7. Roughing in an access space for a double battery of regular water closets.

Here separate vent and water branches are considered for each battery of fixtures. This system is just as economical as to install one vent and water branch and to supply the opposite fixtures with takeoffs from the single branch. Furthermore, this practice would congest the access corridor. A 36-inch access corridor has been shown in this case to provide ample room for the maintenance man. A minimum space of 30 inches can be utilized if the roughing is installed close to the wall as illustrated.

The arrangement of flush valves was chosen arbitrarily. The cold water branch in Fig. 6 is shown above the flush valves, in Fig. 7, below.

Cast iron closet bends, of which there are a number of types, have been used for this example. The connections to the horizontal

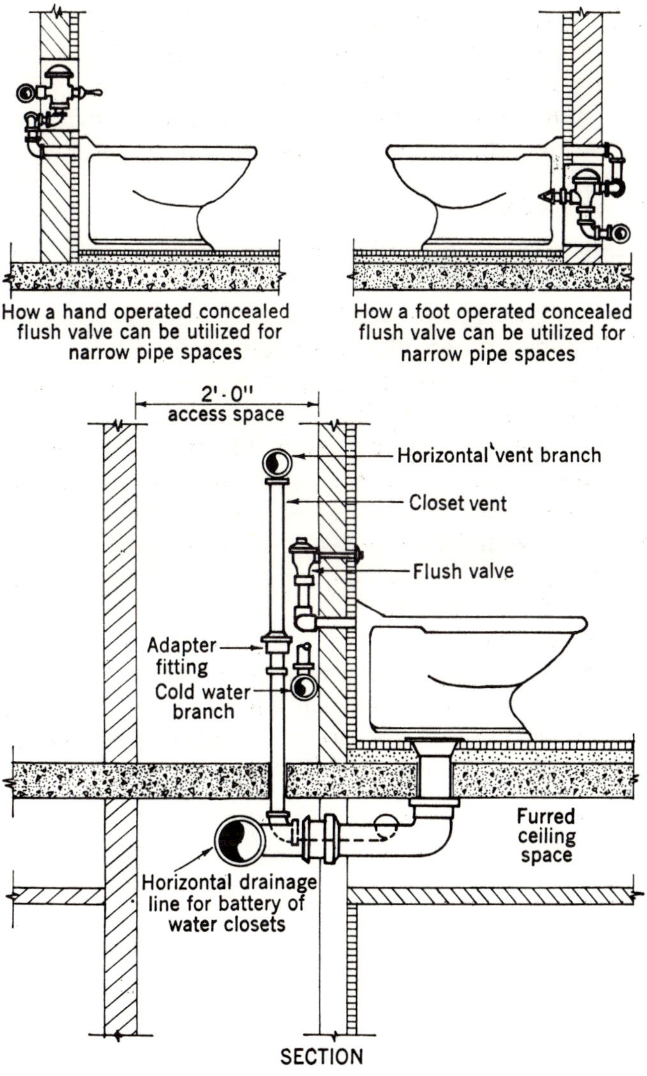

Fig. 8. Roughing in an access space for a battery of secured type closets.

drainage line can be accomplished in a manner favored by the engineer or one that conforms with local code requirements. Individual venting for the closets has been followed throughout this chapter and reflects maximum requirements. The installation of the roughing should always conform to the local code.

For the regular closets in these several installations, the flush valve piping is connected to the top spud of the closet. Closets with back spud connections are also available and fall under the general heading of regular closets, in this chapter.

Secured Type Closets

The secured type closet is most frequently used in applications where rough treatment is expected. Thus, the closets are secured to the floor and wall to prevent being torn loose. A typical installation of the secured type closet is shown in Fig. 8.

When this type closet is used, it is always equipped with a concealed flush valve. The flush valve shown in the main detail features push-button type operation. The flush valve can be maintained by means of the access corridor shown. If an access corridor is required for a back-to-back battery of water closets, the installation would be similar to that shown in Fig. 7. Access corridors are generally desired for secured type closets to offer a tamper-proof installation.

If an access corridor is not required, the roughing for a battery of secured type closets can be installed in a 12-inch pipe space similar to that illustrated in Fig. 5. When the 12-inch pipe space is used, the flush valves can be installed in a manner similar to that shown by the two small details in Fig. 8.

CHAPTER 51

Wall Mounted Water Closet Installations

In the case of wall mounted closet installations, as with floor mounted installations, much thought should be given to the placement of rest rooms. Economy in occupied floor space can be effectively gained when rest rooms are placed back-to-back rather than at separate locations. Furthermore, savings may be realized when a common set of risers serves a back-to-back fixture installation rather than supplying separate risers for individual locations.

Rest room location is a detail the engineer must work out with the architect.

Types of Water Closets

The three basic styles of water closet discussed in the previous chapter are shown in their wall-mounted versions in **Fig. 1**.

The closet combination or residential closet is available in several different types, some of which include an exposed tank, as shown, or a tank that is concealed in the wall construction.

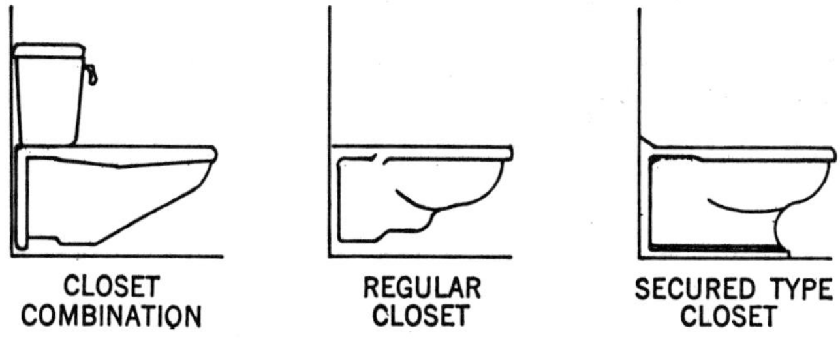

Fig. 1. Three basic types of wall mounted water closets.

This closet is intended primarily for the private residence and the water for the tank is supplied by a small diameter water line.

The regular closet is also available in a number of different styles and shapes, with either blow-out or siphon-jet flushing action. The regular closet is used in all types of buildings. The siphon-jet, because of its quiet action, is preferred in buildings

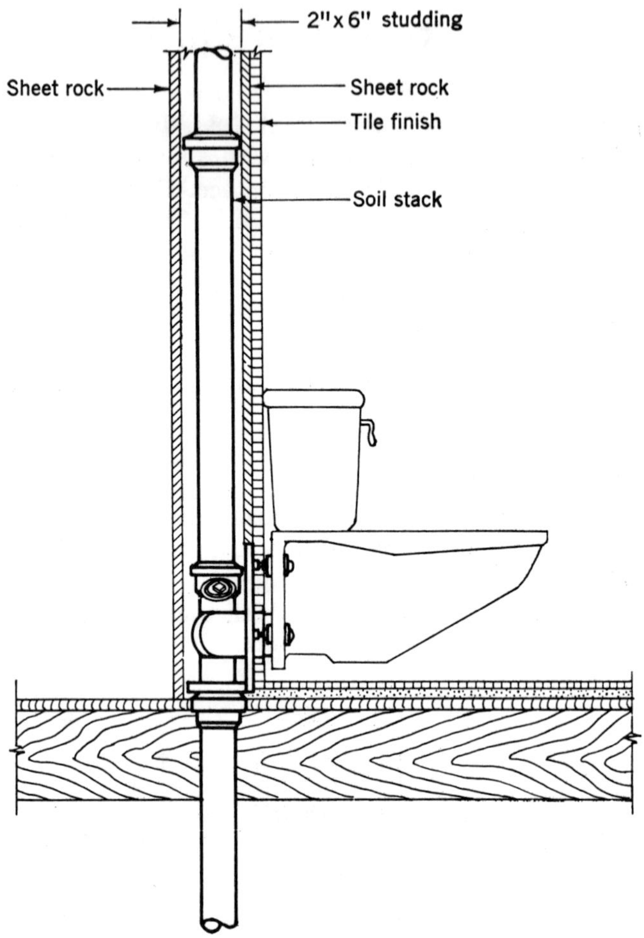

Fig. 2. Typical installation of a wall hung closet combination in a residence.

where sound has to be held down. The blow-out closet is utilized in areas where sound is not an important factor and a strong flushing action is desired.

The secured type closet is generally installed in institutions or where it is desired to secure the closet to both the floor and wall to prevent its being torn loose. This closet is equipped with a concealed flush valve. The secured closet in this case differs from the secured closets discussed under floor mounted closet installations in that the discharge outlet is extended through the wall.

Closet Combinations

The closet combination or residential closet is generally installed in private homes where the pipe space or wall construction is of narrow proportions as shown in Fig. 2. The reader is referred to the sub-section on closet combinations in the previous chapter for further details.

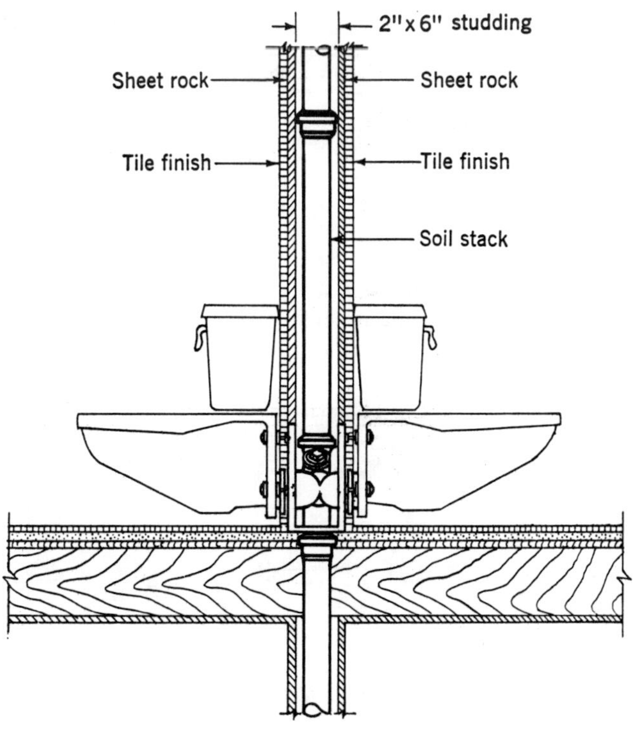

Fig. 3. Typical installation of back to back closets in a residence.

Back to Back Installations

Quite often, residential closets are installed back to back as shown in Fig. 3. For this type installation, the 2 x 6-inch stud construction may again be used. A wide range of manufactured carrier fittings is available. Back-to-back installations are highly favored for motels because of the installation economy.

Regular Closets

The regular closet is installed in vast quantities in all types of buildings. Many applications include batteries of closets as suggested in Fig. 4 which indicates a typical method of roughing for the closets. The pipe space contains a soil stack, vent stack and cold water riser. Figure 4 shows how the horizontal drainage piping is included above the floor level, along with the branch vent and water lines.

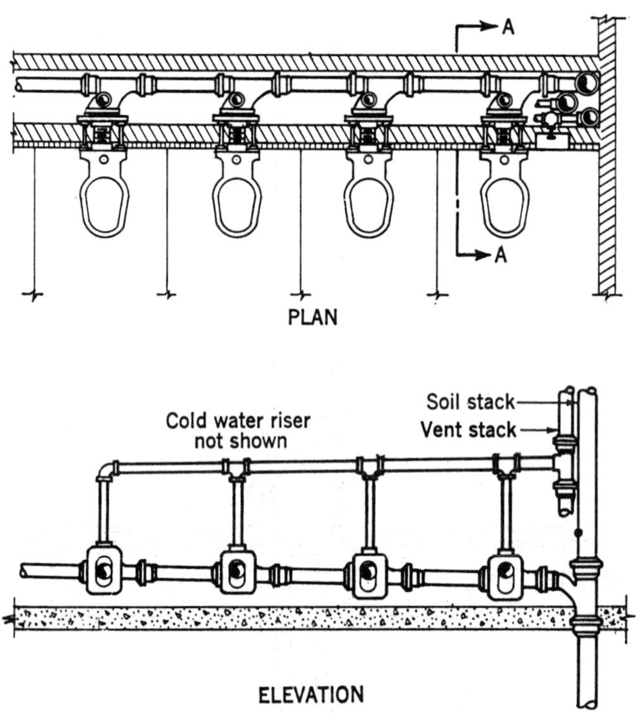

Fig. 4. Roughing for a battery of regular wall mounted water closets.

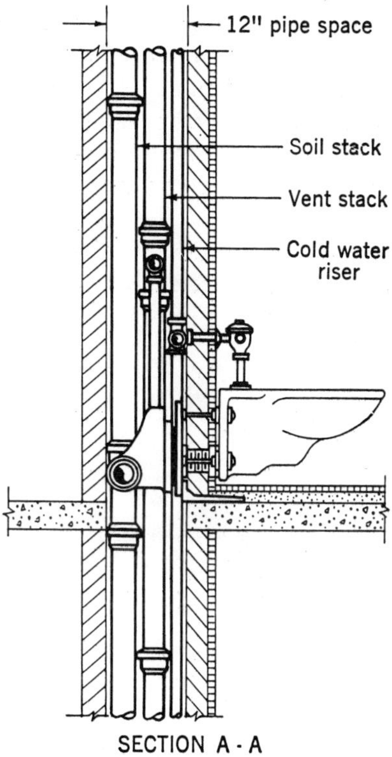

Fig. 5. Section showing arrangement of plumbing risers and roughing for fixtures.

Manufactured carrier fittings are used to receive the discharge from the closets in addition to supporting the closets. The pipe space allotted for the fixture roughing in this instance is 12 inches.

Minimum Requirements

It is the soil stack, vent stack and cold water riser that prescribes the amount of space needed for the fixture roughing. Figure 5 is Section A-A of Fig. 4. It shows the tight grouping of the vertical soil stack, vent stack and cold water riser. The required space is 12 inches for the ease of installing this fixture roughing. This is the minimum space for containing the carrier fitting. When the installation of such fittings is somewhat difficult, the fitting can be installed partially in the wall.

Exhaust ducts are usually required for interior located rest rooms. Since these ducts are generally placed in the upper portion of the pipe space, sufficient room must thus be allotted for this facility. A 12-inch space is usually ample. It is always good practice to consult with the engineer designing the ventilation system for his exact requirements. This will avoid many difficulties.

The 12-inch space is the minimum dimension to contain the roughing for wall mounted water closets. It is not practical to

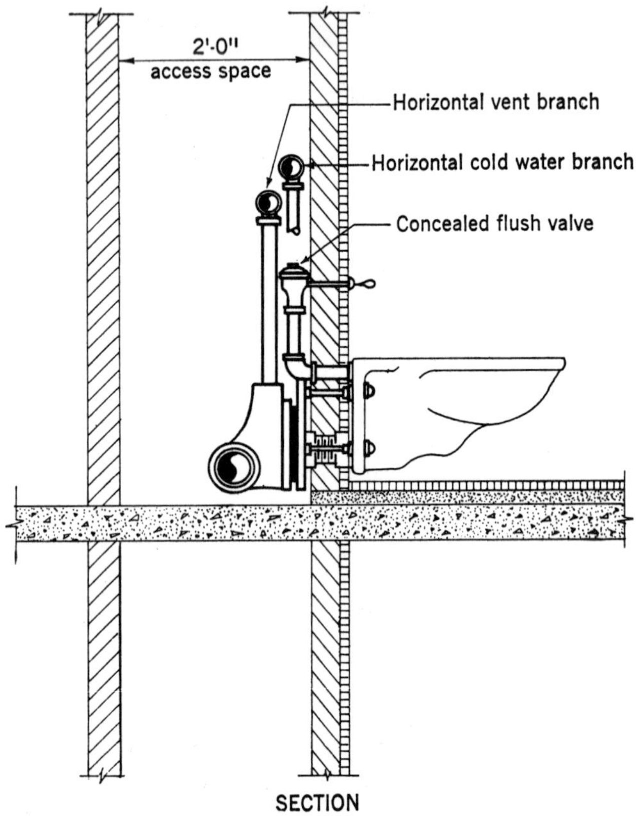

Fig. 6. Installation of roughing in an access space for a battery of wall mounted water closets.

consider a smaller space. The arrangement of roughing shown is generally typical for all manufactured carrier fittings.

Access Space

In many instances there is a definite need for an access corridor to serve a single battery of regular wall mounted closets. A typical example of this provision is shown in Fig. 6.

When two batteries of water closets are involved and an access corridor is contemplated, the installation might be typical to that offered in Fig. 7.

The reader is referred to the previous chapter on floor-mounted closets. The discussion there on access space applies equally here.

Secured Type Closets

A typical installation of the secured type closet is shown in Fig. 8. One detail denotes a concealed flush valve with a handle. Access to the flush valve is gained by means of the access box shown. The detail denotes a concealed flush valve with a foot

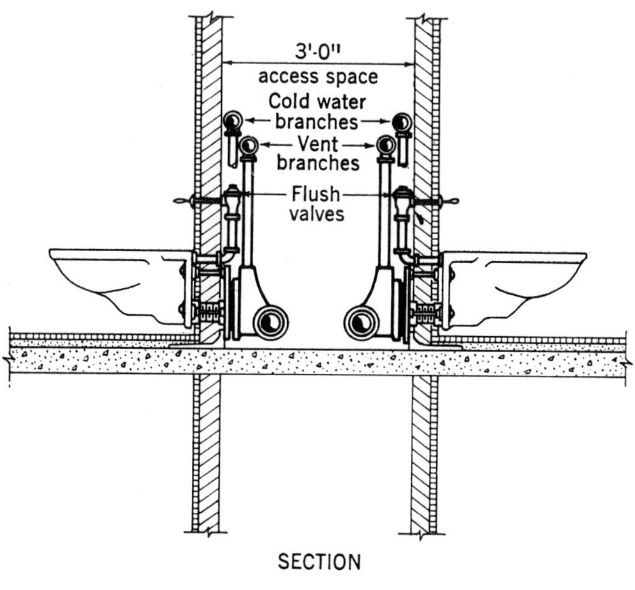

Fig. 7. Roughing in an access space for a double battery of regular wall hung water closets.

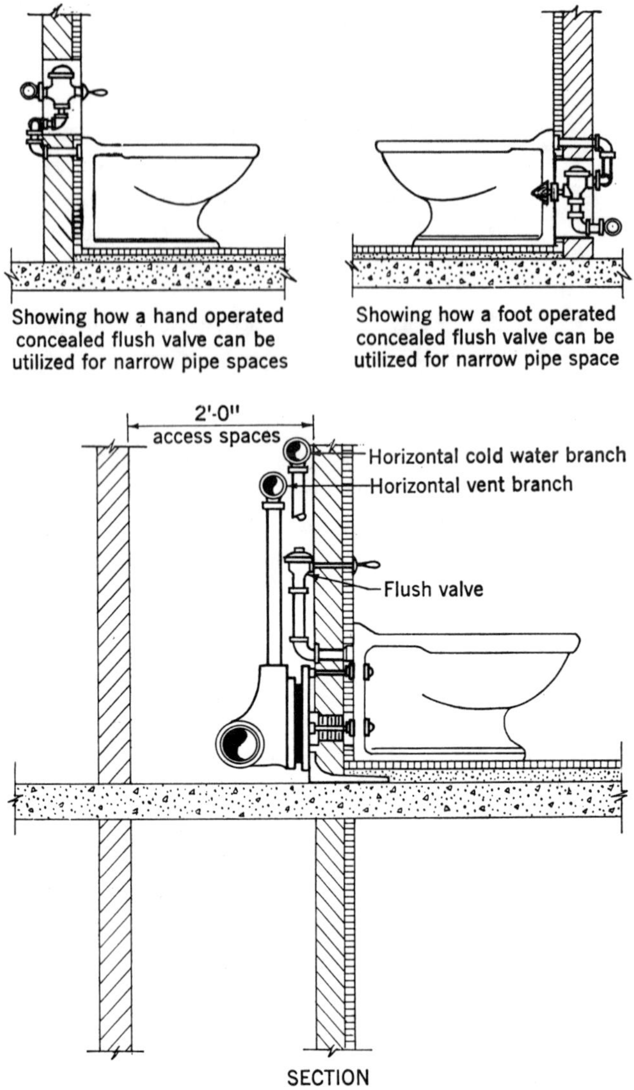

Fig. 8. Roughing in an access space for a battery of secured type closets.

pedal. Here again, access is gained by means of the access box. The access boxes are the tamper proof type.

The remarks on this type closet in the previous chapter apply equally here.

CHAPTER 52

Floor Mounted Urinal Installations

Urinals are installed in all types of buildings, and as with water closets, a prescribed amount of space is needed for the roughing for the fixtures.

Since floor space in every type of building is a prime consideration, it is of utmost importance that minimum dimensions be followed for the required roughing. It is of equal importance that the space is not reduced to below limits, which would hamper the ease of installing the roughing.

Types of Urinals

There are many styles and shapes of floor type urinals and each fixture has a distinct feature in performance which makes one urinal a little bit different than the other. The three basic types of urinals commonly used are: Stall type urinal, pedestal type urinal, and floor mounted women's urinal. These are illustrated in Fig. 1.

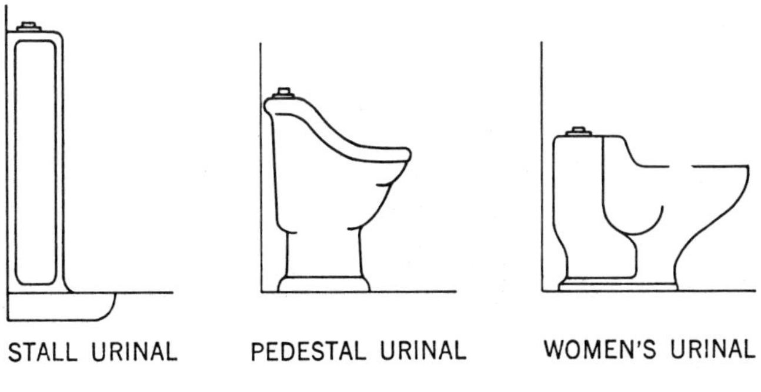

Fig. 1. Three basic types of floor mounted urinals.

The stall type urinal is available in several types and the water is supplied by either a flush tank in an elevated position, or a concealed or exposed flush valve. In this chapter, flush valves are shown rather than elevated flush tanks.

The pedestal type urinal is also a popular type urinal for many installations. It is a free standing fixture and is equipped with either an exposed or concealed flush valve.

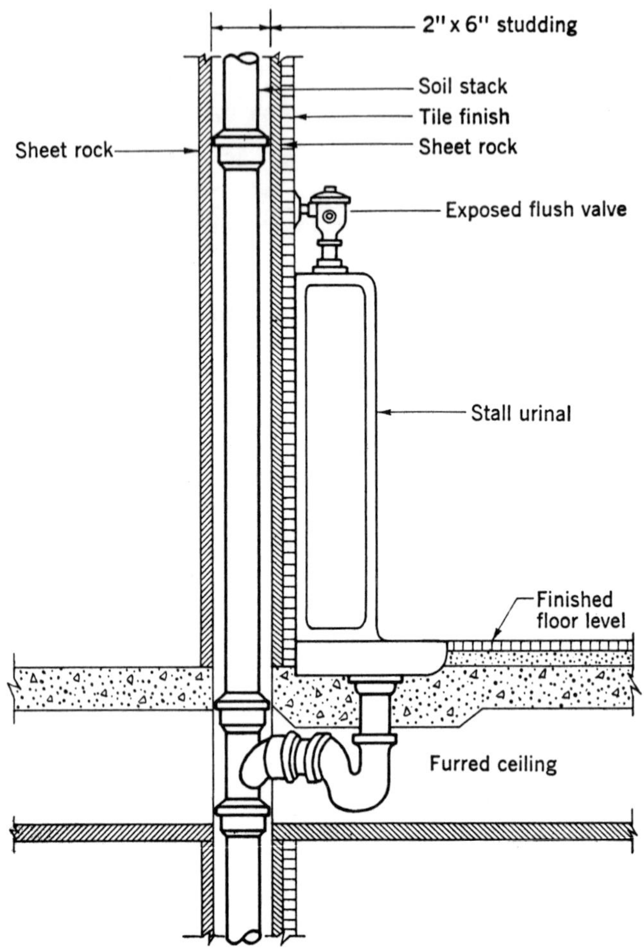

Fig. 2. Installation of a stall type urinal with the roughing contained in a minimum size pipe space.

The women's urinal has become more and more popular in recent years. This, too, is a free standing fixture and is installed in all types of buildings. This particular urinal can also be equipped with either an exposed or concealed type flush valve.

Stall Urinal

The stall urinal can be installed as a unit or as a battery of fixtures. A typical single fixture installation is shown in Fig. 2.

Roughing is installed in a wall constructed of 2 by 6-inch stud framing with a sheet rock and tile finish on one wall and a sheet rock finish on the other wall. The stall urinal is generally placed in a room with one or more water closets and lavatories so that the roughing for the urinal ties in with the same roughing for the other fixtures. In Fig. 2 the roughing consists of a 3-inch soil stack. The 3-inch stack can be increased to 4 inches in its passage through the roof. In areas where a 4-inch stack is required, it also can be fitted into the space shown.

Referring to Fig. 2, the waste line from the urinal may be 2 or 3 inches and since the distance between the stack and the centerline of the fixture outlet is not sufficient to make the necessary connection, the stack has been offset to the side of the urinal to offer sufficient room to make a connection.

Flush valve piping is easily installed in the pipe space provided. Should another urinal be placed on a floor above, venting must be accomplished in accordance with local code requirements.

An advantage of the stall urinal is that it serves as a floor drain. Since the rim of the urinal is flush with the finished floor, with the floor generally sloped to the urinal, it will intercept any water that is spilled on the floor.

Pedestal Urinal

The pedestal urinal can be installed singly or in a battery arrangement. This urinal, too, is usually installed in conjunction with one or more water closets and one or more lavatories, and roughing is tied in with the roughing of the other fixtures. Figure 3 illustrates the roughing when the fixture is installed as a single unit.

A 4-inch soil stack has been indicated for this installation since the outlet from the fixture is 4 inches in diameter. A closet flange engages the horn of the urinal. A lead bend has been shown in this instance to provide for ease of installation. Other devices

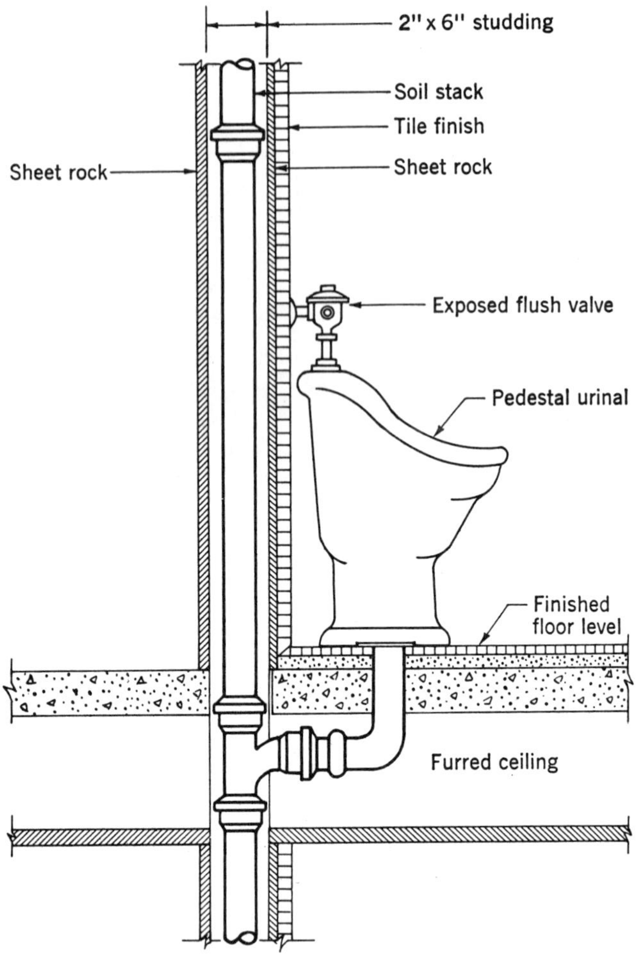

Fig. 3. Installation of a pedestal type urinal. Roughing is shown in a minimum size pipe space.

that fit the roughing space may be used with equal success. An exposed type flush valve has been utilized and its supply piping is easily installed in the narrow confines of the pipe space.

The pipe space is similar to that of Fig. 2. This is the minimum space in which the roughing for a pedestal type urinal can be installed.

Women's Urinal

The women's urinal can be installed singly or in a battery arrangement. Figure 4 shows the minimum roughing requirements when the fixture is installed as a single unit.

A closet flange engages the horn of the urinal and a 4-inch lead bend offers the final connection to the vertical 4-inch soil stack. Water piping is contained within the narrow confines of the pipe space. The soil stack is easily installed in the 2 by 4-inch stud

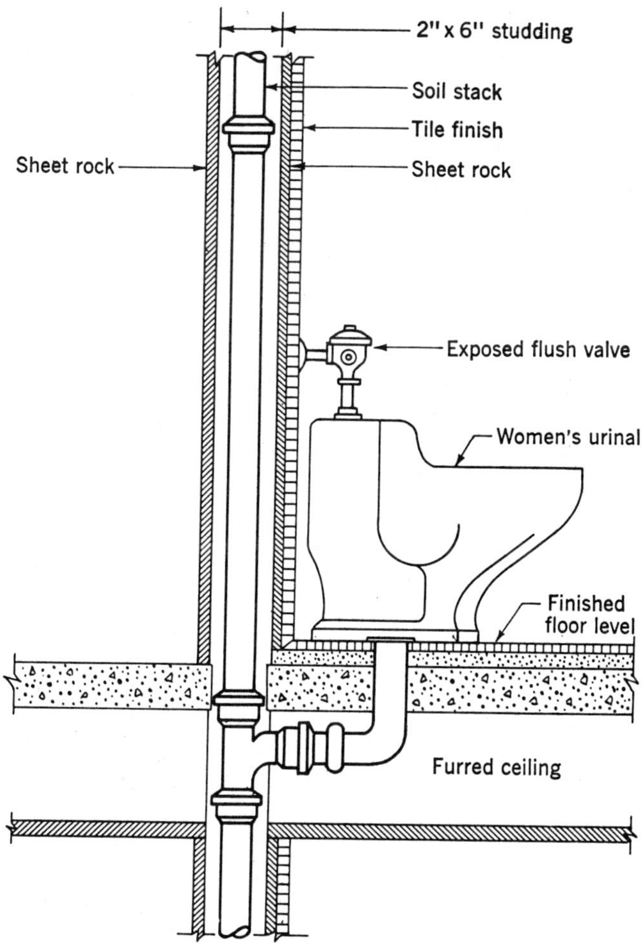

Fig. 4. Roughing and installation of a women's type urinal.

construction and represents the minimum dimensions for a fixture of this type.

Cast iron hub and spigot piping has been shown for the roughing in Figs. 2, 3 and 4 for it represents the maximum in physical dimensions of all the types of piping that could be used. Therefore, galvanized steel, galvanized wrought iron or copper tubing can easily fit into the space allotted. The cast iron pipe has thus been used only for the purpose of example. Fixtures detailed are for illustrative purposes only; several fixture manufacturers offer a variety of fixtures of the same type.

Back-to-back urinals have not been considered since it is rare that urinals are installed that way. The usual practice is to provide a men's and women's toilet room, back-to-back.

Battery of Stall Urinals

When stall urinals are installed in battery arrangements, the installation would be similar to that shown in Fig. 5. The minimum space requirement is 12 inches for a battery of stall urinals.

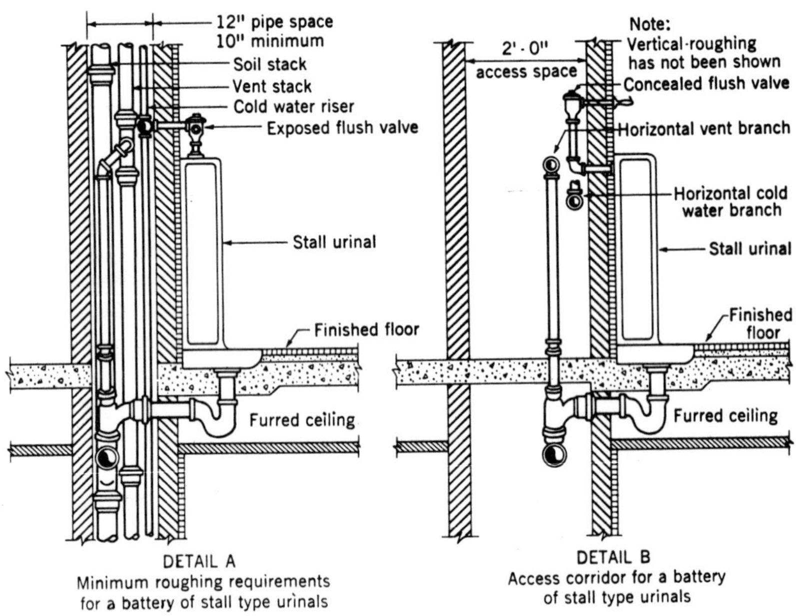

Fig. 5. Two different installations for stall urinals.

The conditions that prescribe the space limitations is the arrangement of vertical roughing as shown. The vertical roughing consists of a soil stack, vent stack and cold water riser. The trapped waste from the urinal is connected to the horizontal waste piping as shown. The horizontal waste piping in turn connects with the vertical soil stack. It is very possible that the vertical roughing can be installed in a pipe space of less than 10 inches but this would require a careful arrangement of the roughing, and the use of offset vent piping to reduce the width of the pipe space. The piping material shown is cast iron hub and spigot piping, for this requires the maximum dimensions. Therefore, for ease in installation, it is recommended that the pipe space be 12-inches minimum for a battery of stall urinals. If necessary, roughing can be arranged in a confined manner to fit the limits of a 10-inch pipe space.

When an access corridor is desired, the space should be 24 inches, as shown. This provides ample room for the maintenance man. When an access corridor is contemplated, the roughing should be so arranged to offer maximum room for passage. The space should also be well lighted and ventilated.

Battery of Pedestal Urinals

The roughing for a battery of pedestal urinals would be similar to that shown in Fig. 6. The arrangement of vertical roughing is similar to that shown in Fig. 5, and again, cast iron hub and spigot piping has been used. Roughing from the urinal consists of a closet bend which is connected to the horizontal waste piping. The closet bend is vented in the manner shown by the detail.

The recommended roughing space for this arrangement is 12 inches. Once again, the space can be reduced to 10 inches when some extra careful attention is paid to the roughing.

When an access corridor is contemplated, the space should be 24 inches, as shown. Although cast iron closet bends have been shown, other types of closet bends can be used with equal success.

Battery of Women's Urinals

For a battery of women's urinals, the arrangement of roughing would be similar to that shown in Fig. 7.

A cast iron closet bend connects the urinal to the horizontal waste piping. Other materials besides the cast iron closet bend can be used with equal success. It is the vertical roughing that

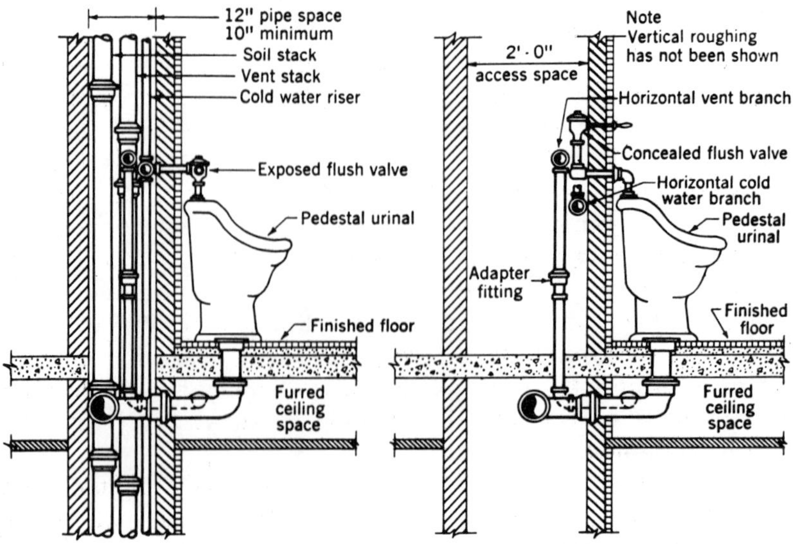

DETAIL A
Minimum roughing requirements for a battery of pedestal type urinals

DETAIL B
Access corridor for a battery of pedestal type urinals

Fig. 6. Two installations for pedestal type urinals.

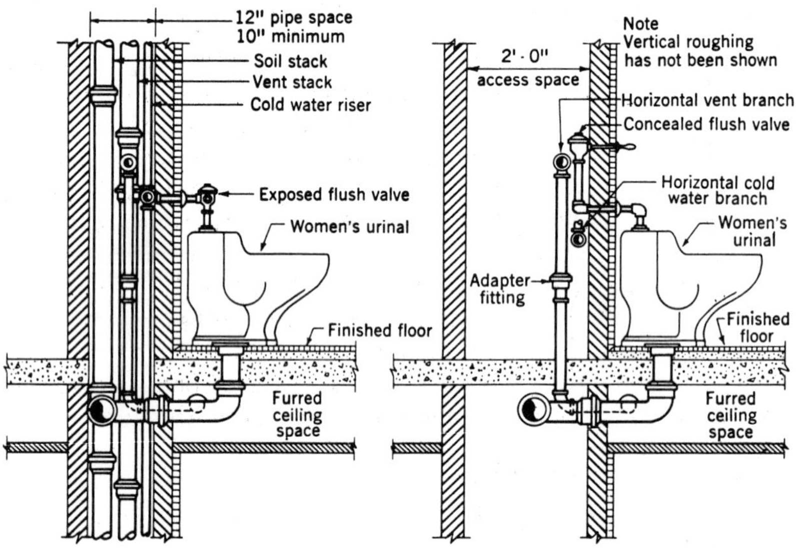

DETAIL A
Minimum roughing requirements for a battery of women's type urinals

DETAIL B
Access corridor for a battery of women's type urinals

Fig. 7. Two installations for women's type urinals.

prescribes the amount of space needed. Here it consists of a soil stack, vent stack and cold water riser. If absolutely necessary, this space can be reduced with a tight arrangement of roughing.

The minimum roughing space for a battery of women's urinals is thus suggested as 12-inches with a further reduction to 10-inches only when actually required.

Roughing Arrangements

Figure 8 shows typical arrangements of roughing for the urinals included in this chapter.

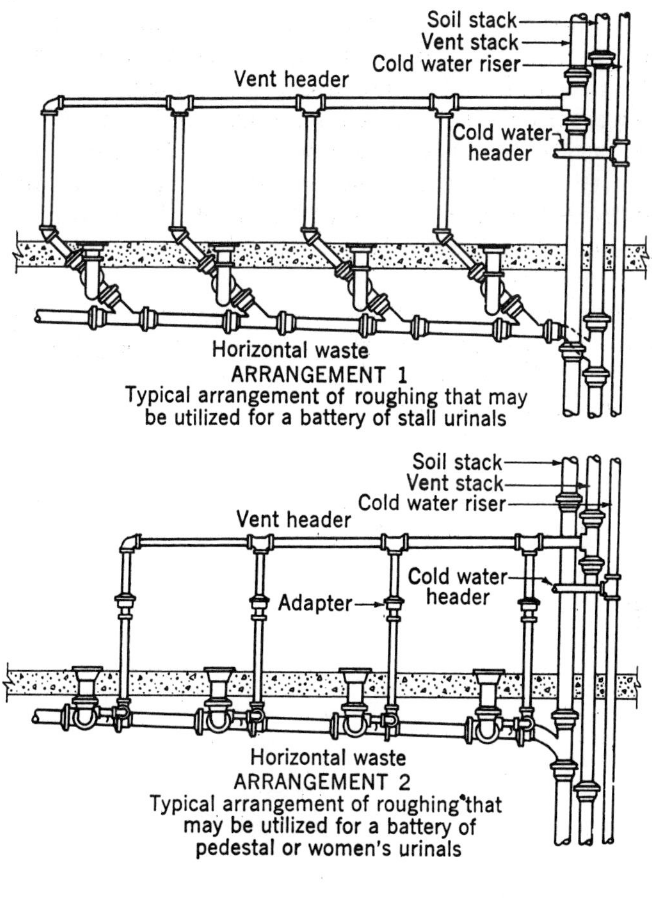

Fig. 8. Roughing arrangements that may be used for floor mounted urinals.

Arrangement 1 shows the roughing employed for the stall type urinals illustrated in Fig. 5, and Arrangement 2, the roughing that might be utilized for the pedestal and women's urinals, shown in Figs. 6 and 7.

For the stall type urinals, Arrangement 1, the trapped connection may be 2 or 3-inches as desired. The horizontal waste line is 3-inches and connects to a 4-inch soil stack. The size of the vertical soil and vent stacks are determined in each case by the height of the building, number of fixtures and the specific requirements of the local code. The connections are individually vented to indicate the maximum in code requirements. The individual vent lines may be either 1½-inches or 2-inches as called for in the local code. The particular venting arrangement is for example, and other arrangements can be used with equal success.

For the pedestal and women's urinals, 4-inch closet bends are used. These connect with a 4-inch horizontal waste line which in turn connects with the 4-inch soil stack. The individual vents are connected to the closet bend as shown. Again, the individual vents can be 1½ or 2-inches as required by the local code. Other arrangements of roughing favored by the engineer may be used with equal success.

Throughout this chapter an exposed or concealed flush valve of more or less standard nature is illustrated. Foot operated push button or automatic flush valves may also be used. There is a broad selection from which to choose.

CHAPTER 53

Wall Mounted Urinal Installations

With the emphasis on a high degree of sanitary cleanliness in the rest room, there is a current trend to install fixtures mounted on the wall above the floor. Wall mounted urinals can be placed in this category. When fixtures are set above the floor, cleaning the floor is made easier.

Types of Urinals

Although there are a number of types and styles of urinals from which to select, the basic types may be classified as illustrated in Fig. 1: integral flush urinal, flushing rim urinal, and women's urinal.

The name "integral flush urinal" has been devised by this writer to define the unit shown in Fig. 1. The name has been

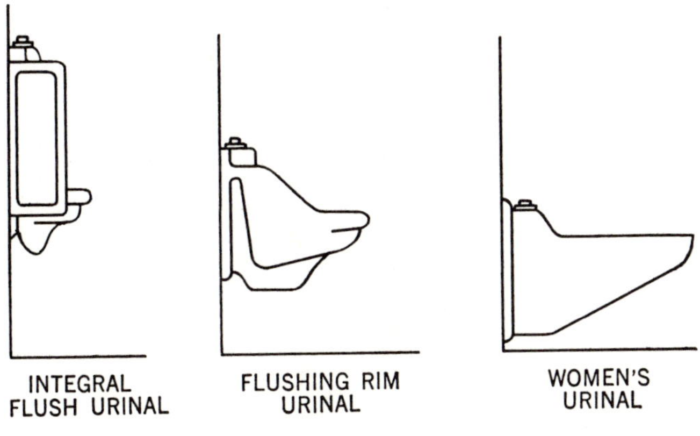

Fig. 1. Three basic types of wall mounted urinals.

borrowed from the integral flush spreader device that is incorporated in the urinal. Water may be supplied by an elevated flush tank, an exposed flush valve or a concealed flush valve and this applies to all three types mentioned. The urinal can be supported on one of several different devices that are available for the purpose.

The flushing rim urinal obtains its name from the manner in which the flushing action takes place in the fixture.

The women's urinal, as its name implies, is intended primarily

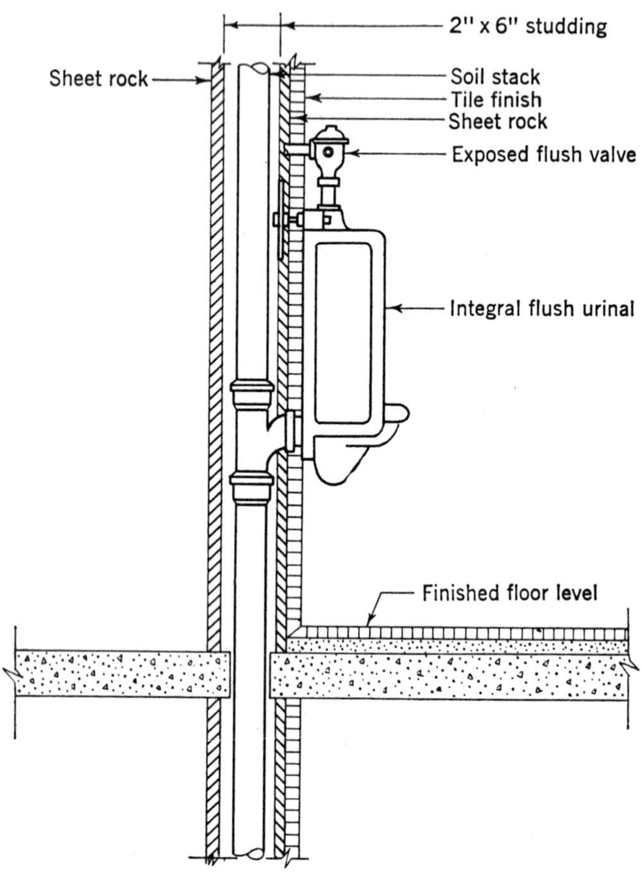

Fig. 2. Typical installation of an integral flush urinal with the roughing contained in a space of minimum dimensions.

for use by women. This type has become increasingly popular in recent years.

Although elevated flush tanks are mentioned, only flush valve installations are discussed in this chapter.

Integral Flush Urinals

Figure 2 shows a typical installation of this urinal using minimum dimensions. The construction consists of 2 x 6-inch studding faced by sheetrock and tile on one wall and sheetrock on the

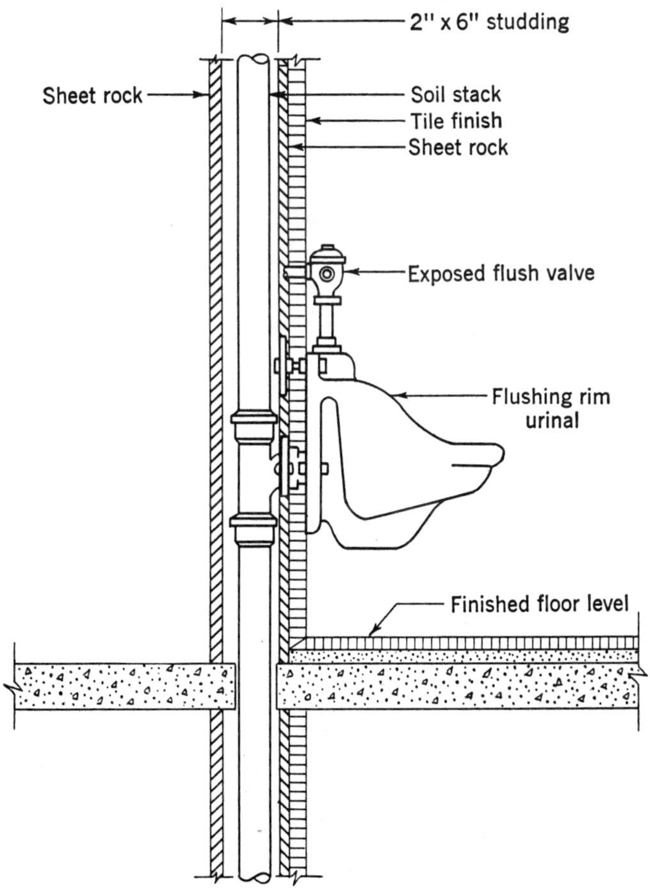

Fig. 3. Typical installation of a flushing rim urinal. Roughing is contained in a minimum pipe space.

other wall. The waste outlet connection from the urinal is generally 2 inches. The soil stack to which the waste is connected may be 2, 3 or 4 inches as required by the local code. Except in Fig. 8, cast iron hub and spigot pipe is used in all the illustrations to indicate maximum physical dimensions for the piping. Galvanized steel, galvanized wrought iron or copper piping may also be employed.

The urinal is shown as being supported on a steel plate which is, in turn, secured to the studding. Other types of devices are also available for the support of the urinal.

Figure 2 shows the urinal installed as a single fixture in the room. If other fixtures, as water closets and lavatories, are installed on the same wall, the urinal roughing would then be but one component of the roughing for all the fixtures.

Flushing Rim Urinal

Figure 3 shows the flushing rim urinal mounted on the same type of stud construction described in Fig. 2. The urinal is supported on two steel carrier plates which are secured to the stud construction. Threaded rods, attached to the steel plates, extend out and engage the urinal for support. Other types of supporting devices are available.

The urinal waste outlet is 2 inches, although a 3-inch outlet may be specified. The soil stack may be 2, 3 or 4 inches as required by the urinal outlet connection or the local code. Here again, the urinal has been shown as a single installation. When other fixtures are installed on the same wall, the urinal roughing will be an installed component with the rcughing of the other fixtures.

Women's Urinal

Figure 4 shows how a women's urinal may be installed using minimum proportions. The construction is similar to that of Fig. 2. The urinal waste outlet and stack are 4 inches. The urinal is supported on a cast iron or steel bracket which in turn is attached to a foot that is rigidly set in the floor construction. Threaded rods extend out from the plate to engage and anchor the urinal. If other fixtures are installed on the same wall, the urinal roughing will become a part of the roughing for the other fixtures.

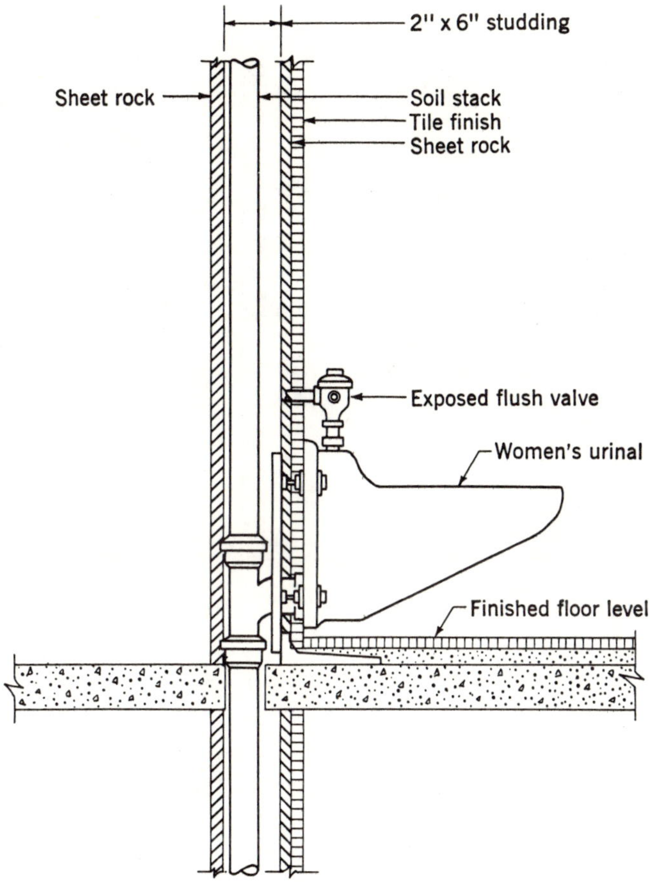

Fig. 4. Installation and roughing for a women's urinal.

Battery of Integral Flush Urinals

When the integral flush type urinal is contained in a battery installation, the roughing would be similar to that shown in Detail *A* of Fig. 5.

The waste from the urinal connects with a horizontal waste line which connects with the vertical soil stack. It is the arrangement of the vertical soil stack, vent stack and cold water riser that generally prescribes the amount of pipe space that must be allotted. In this case, the pipe space is 12 inches, which may be further reduced to 10 inches, with a special arrangement of the

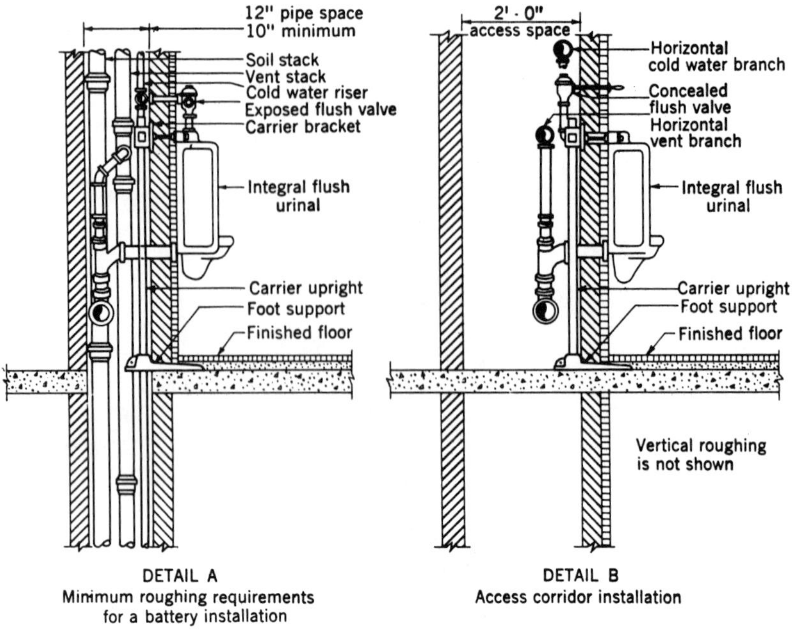

Fig. 5. Two installations for integral flush type urinals.

vertical lines. The floor in the pipe space has been deleted from the drawing to clarify the illustration of the vertical lines.

The urinal is supported on a carrier assembly consisting of feet, upright piping and a plate attached to the uprights, as shown. Threaded rods extend from the plate to engage and secure the urinal in place.

Detail *B* shows how the roughing would be arranged when an access corridor is contemplated. The 24-inch pipe space offers sufficient room for the maintenance man to perform his duties.

Battery of Flushing Rim Urinals

Roughing for the flushing rim urinal is shown in Fig. 6. The 2 or 3-inch urinal connection shown in Detail *A*, ties in with the horizontal waste line which connects with the vertical soil stack. Here again, the pipe space is 12 inches, with a possible reduction to 10 inches only when a special arrangement of vertical piping is utilized. The floor slab in the pipe space has been deleted from the drawing to clarify the arrangement of the

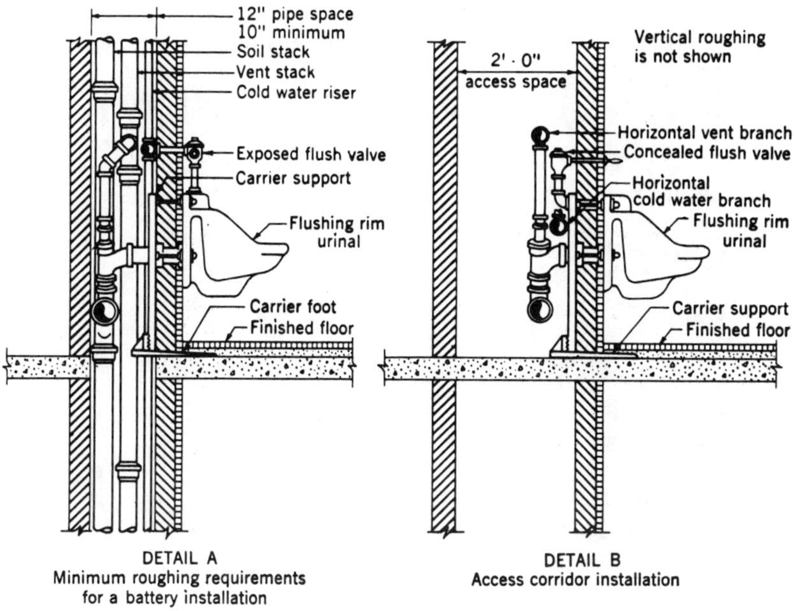

Fig. 6. Two installations for flushing rim type urinals.

necessary vertical piping.

The urinal is supported on a heavy cast iron carrier assembly which is held upright by a foot embedded in the floor construction. Threaded rods extend out from the assembly to engage and secure the urinal.

Detail *B* shows the typical arrangement that may be employed when an access corridor is used. The pipe space is 24 inches wide to offer freedom of movement for the maintenance man.

Battery of Women's Urinals

Figure 7 suggests the type arrangement that might be followed in the roughing for a women's urinal.

Detail *A* of this figure shows the carrier fitting combination that is used to support the women's urinal. This combination is comprised of an adjustable fitting, an upright support plate which is rigidly attached to a foot that is embedded in the concrete slab. Threaded rods extend out from the carrier face plate to engage the urinal and secure it in place.

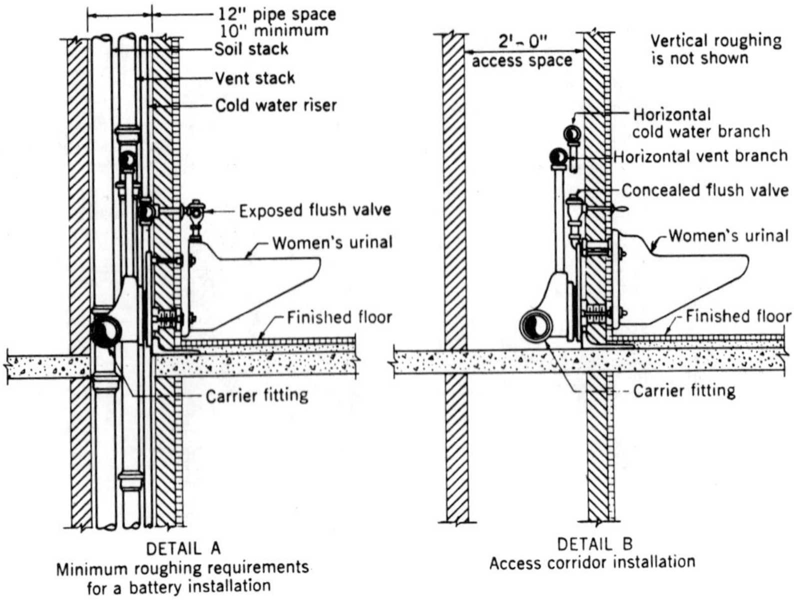

Fig. 7. Two installations for women's type urinals.

The standard pipe space is 12 inches. Cast iron hub and spigot piping has been used to indicate the maximum in physical piping dimensions.

Detail *B* shows that particular arrangement of roughing that may be followed when an access corridor of 24 inches is considered. Here a concealed flush valve is used. Access corridors should be well lighted and ventilated to serve a useful function. An access corridor permits adjustments to be made to the flush valves without disturbing the occupants of the rest room.

Back-to-Back Urinal Installations

Back to back urinal installations were not considered in this article since it is a rare occasion that urinals are generally placed in such a manner. It might only occur in a building strictly devoted to occupancy by men. Should this condition prevail, then the minimum pipe space for the roughing would range from 15 to 18 inches dependent upon the urinal used and the particular arrangement of the roughing.

Installation of Roughing

The specific methods of roughing that may be used with the different urinals covered in this article are shown in Fig. 8.

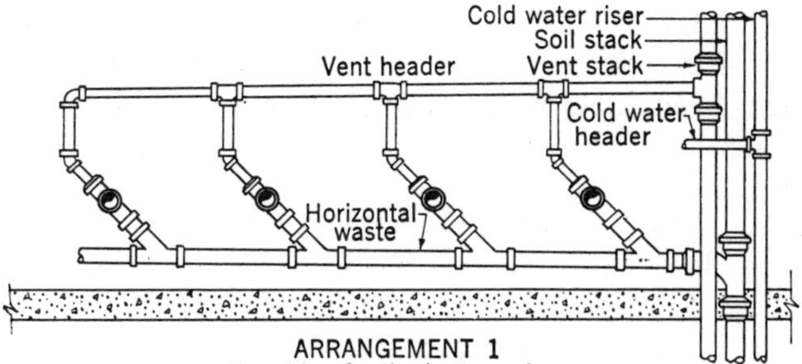

ARRANGEMENT 1
Roughing for the battery of urinals shown in Fig. 5 and 6

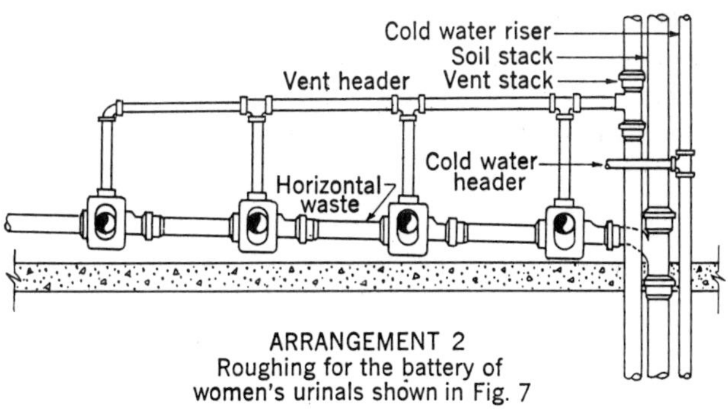

ARRANGEMENT 2
Roughing for the battery of women's urinals shown in Fig. 7

Fig. 8. Two arrangements of roughing for wall mounted urinals.

Arrangement 1 shows the typical installation of the roughing for the integral flush type urinal and the flushing rim urinal. Threaded pipe and fittings have been indicated in this case to show a variation over the hub and spigot cast iron piping used for example throughout this article. The waste piping has been placed at 45°. Each waste connection is individually vented. Cast iron hub and spigot piping has been illustrated for the vertical piping. The engineer can use any particular arrangement of roughing which he may favor, for there certainly are several different ways in which this part of the installation can be accomplished.

Arrangement 2 shows the typical roughing arrangement that might be employed for the women's urinal. Here the closet carrier combination has been used for the support of the urinals as well as to receive the discharge from the urinals. The combination fittings are adjustable in that they compensate for the pitch in the horizontal waste piping. The slot shown in the face of the fitting offers the range of adjustment needed for the pitch in the line.

The combination fittings shown are of the caulked type although threaded type fittings are also available for the function. The foot and face plate have been deleted from the detail for clarity in showing the fitting itself. The face plate will contain the threaded rods which extend out to engage the urinal.

The vent header is of threaded piping and the vertical vents are caulked into the combination fitting. The vertical piping is of the cast iron hub and spigot type. As mentioned earlier in this article, piping materials of galvanized steel, galvanized wrought iron or copper may be used with equal success as desired by the engineer. Another arrangement of roughing other than that shown in Arrangement 2 may be used, as favored by the engineer.

CHAPTER 54

Lavatory Installations

There are lavatories of every size and description to satisfy particular needs and space conditions. These units range from the small compact corner lavatory to special lavatories that are built into a room space. Various types of trim are available and the lavatories can be equipped with soap dispensers or other devices. However, this discussion is confined to the common lavatory types installed in both large and small buildings.

Types of Lavatories

The three basic types of lavatories, Fig. 1, are: Enameled lavatory; vitreous china high back lavatory; vitreous china deck type lavatory. These will now be described.

Enameled lavatory. This type is constructed of cast iron or steel and is furnished in several shapes and many colors. It is supported on a steel or cast iron bracket that is placed on the finished wall and is secured to the construction or other assemblies placed behind the construction. Lugs cast at the back of the lavatory engage slots in the bracket for support.

Vitreous china high back lavatory. This, too, comes in several shapes and sizes and many colors. It is supported on an exposed

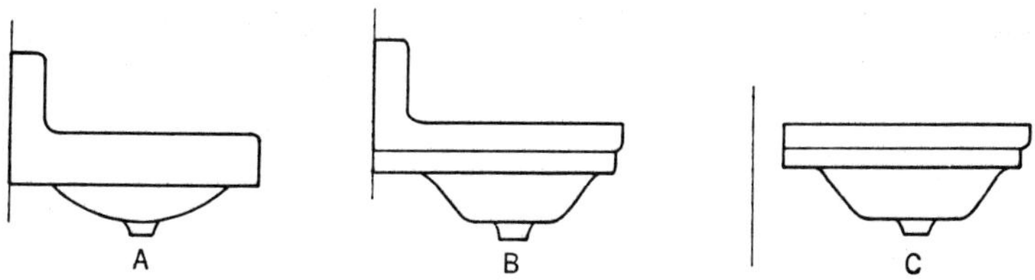

Fig. 1. Three basic types of lavatories: A—enameled lavatory; B—vitreous china high back; C—vitreous china deck type.

bracket which is secured to the construction, or to devices set behind the construction. This lavatory can also be supported on concealed arms secured to the construction or to devices set behind the construction.

Vitreous china deck type lavatory. Constructed of vitreous china, it can be furnished in several sizes and shapes and comes in many colors. This lavatory can be supported on concealed arms that are secured to the construction or to devices that are set behind the construction. The lavatory can also be supported on exposed type brackets although the concealed arms are more commonly used.

Enameled Lavatory

Roughing for the enameled lavatory easily fits minimum pipe spaces. Figure 2 shows two typical roughing installations. Arrangement 1 shows the lavatory installed as a single unit. The construction is 2-inch x 6-inch studding, faced on one side with a sheet rock finish and on the other side with a sheet rock and tiled finish. The 2 x 6-inch studding is shown because this is

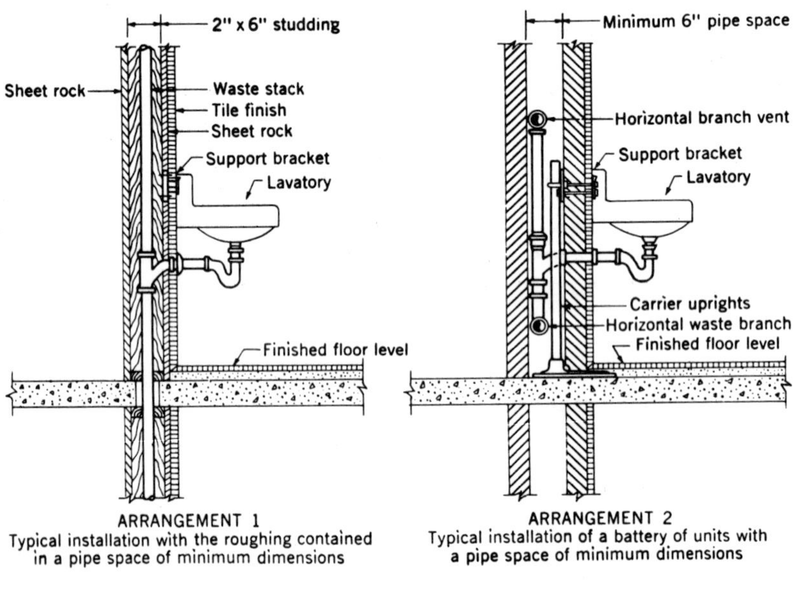

Fig. 2. Two installations of enameled lavatories.

the construction that would be required for a water closet that generally accompanies the lavatory.

The lavatory in this case is supported on a steel or cast iron bracket which in turn is secured by screws to the wood plank that is fastened to the stud construction. A steel plate is attached to the wood construction wih threaded rods that extend out to engage and secure the bracket.

Arrangement 2 illustrates the type construction that is encountered in larger buildings where batteries of lavatories are involved. It consists of glazed tile block, cinder or concrete construction. The wall surface in the rest room would be finished with tile. Any one of several other types of modern construction may be used with equal success. The pipe space is a minimum of 6 inches, since the roughing for a battery of lavatories is generally 2-inch horizontal piping with the individual connections 1½ or 2 inches.

The factors that create a need for the increase in size of the pipe space is the size of the vertical waste stack, vent stack and cold water risers. As these lines increase in size, the pipe space must be of wider proportions. If the lavatories are installed on the same wall with water closets or urinals, then the lavatory roughing will connect with the roughing for the other fixtures; the required pipe space in this event would be the minimum 12-inch standard.

The device used to support the lavatory consists of an assembly of two feet, two uprights and a steel plate rigidly fastened to the uprights. Threaded rods extending from the steel plate engage and secure the steel or cast iron bracket that is placed on the finished wall to hold the lavatory.

Supporting Devices for Enameled Lavatories

Several types of supporting devices that may be used with enameled lavatories are illustrated in Fig. 3.

Detail *A* represents the steel or cast iron bracket set on the finished wall and secured to the construction by threaded rods that extend through the construction and engage the steel backup plates set behind the construction.

Detail *B* shows the steel or cast iron bracket set on the finished wall with the threaded rods that extend through the construction to engage a steel plate placed behind the wall construction. This arrangement is suitable for wall construction of a flimsy nature

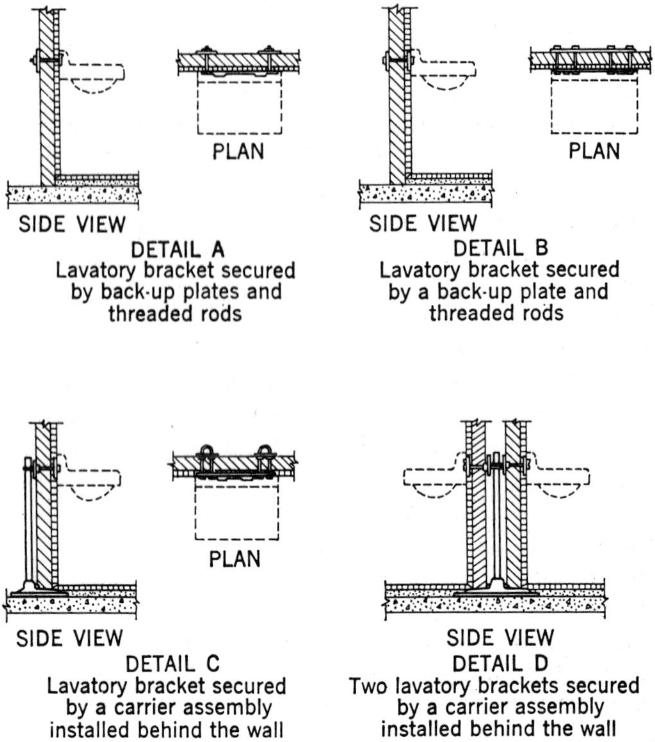

Fig. 3. Various supporting devices that may be used with enameled lavatories.

since the steel plate distributes the bearing load over a greater wall surface. Sometimes a continuous steel plate is used when a battery of lavatories is mounted on wall surfaces of lightweight construction.

Detail *C* illustrates the method of support described in Arrangement 2, Fig. 2.

A back-to-back installation of lavatories is shown in Detail *D*. For support, there is an assembly consisting of feet, uprights and steel plates.

Vitreous China High Back Lavatories

The roughing for the vitreous china high back lavatories, Fig. 4, easily fits a minimum pipe space.

Arrangement 1, Fig. 4, shows how the roughing is placed in

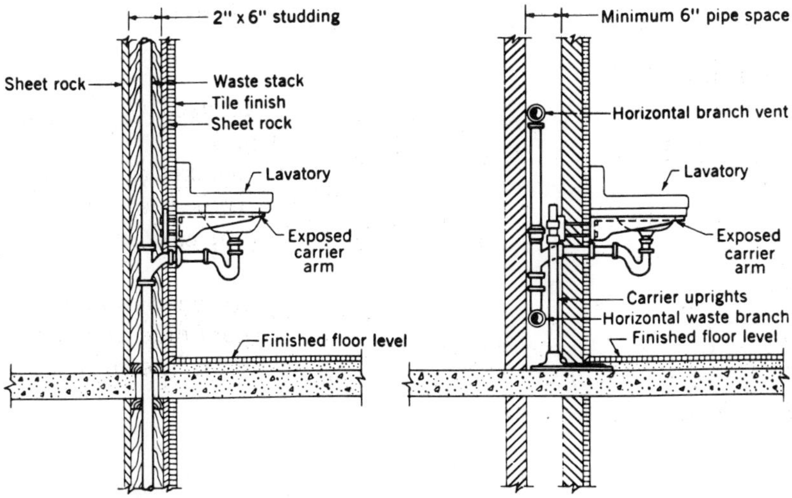

ARRANGEMENT 1
Typical installation with the roughing contained in a pipe space of minimum dimensions

ARRANGEMENT 2
Typical installation of a battery of units with a pipe space of minimum dimensions

Fig. 4. Installations of vitreous china high back lavatories with the roughing contained in a pipe space of minimum dimensions.

the pipe space. The construction consists of 2 x 6-inch studding as would be required for water closets that are normally contained on the same wall.

The lavatory, in this instance, is supported on exposed bracket arms that are in turn attached by means of threaded rods to a steel plate rigidly secured to the studding. The lavatory is anchored to the exposed bracket arms by toggle bolts or other devices.

Arrangement 2 indicates how the roughing for a battery of lavatories can be installed in a pipe space of minimum proportions. Here again, the pipe space is 6 inches but is subject to an increase, dependent upon the size of the waste stack, vent stack and cold water risers. If the lavatory roughing is a component part of the roughing for water closets, then the pipe space would be 12-inch standard. The exposed bracket arms in this case are secured to an assembly of two feet, two uprights and adapter couplings that are secured to the uprights. Threaded rods extend out from the adapter plates to engage and anchor the bracket arms in place. Once again, toggle bolts or other devices are utilized to secure the lavatory to the exposed bracket arms.

Supporting Devices

Several types of supporting devices may be used with vitreous china high back lavatories, as detailed in Fig. 5.

Detail *A* of the figure shows the exposed bracket arms secured to the construction by means of threaded rods that fasten the exposed bracket arms to an adjustable plate of cast iron or steel placed behind the wall.

Detail *B* shows the exposed bracket arms attached to an adjustable plate placed in the construction at the room side and secured by means of threaded rods to steel plates set at the back of the wall construction.

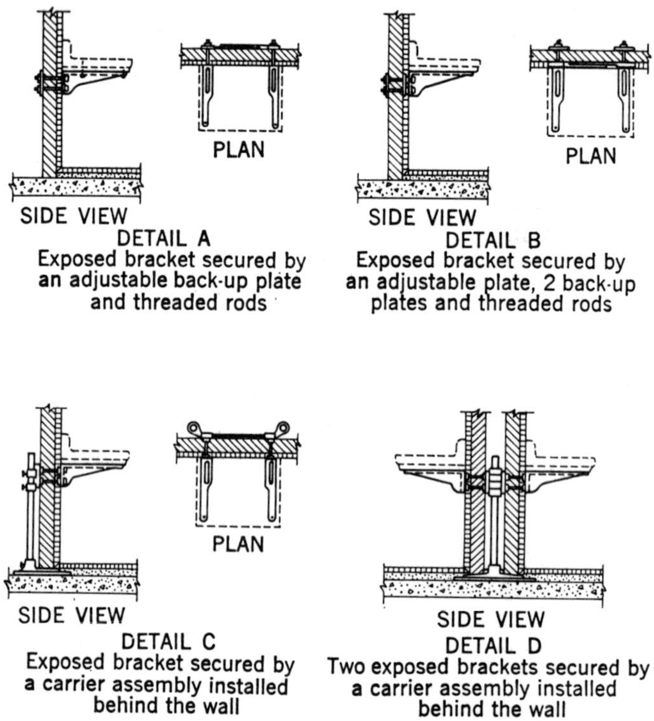

Fig. 5. Various supporting devices for vitreous china high back lavatories.

Detail *C* illustrates the method of support described in Arrangement 2 of Fig. 4.

Detail *D* shows a back-to-back installation of lavatories and for support, an assembly consisting of feet, uprights, adapter couplings and exposed bracket arms.

Vitreous China Deck Type Lavatories

Roughing for the vitreous china deck type lavatories easily fits minimum pipe space dimensions as detailed in Fig. 6.

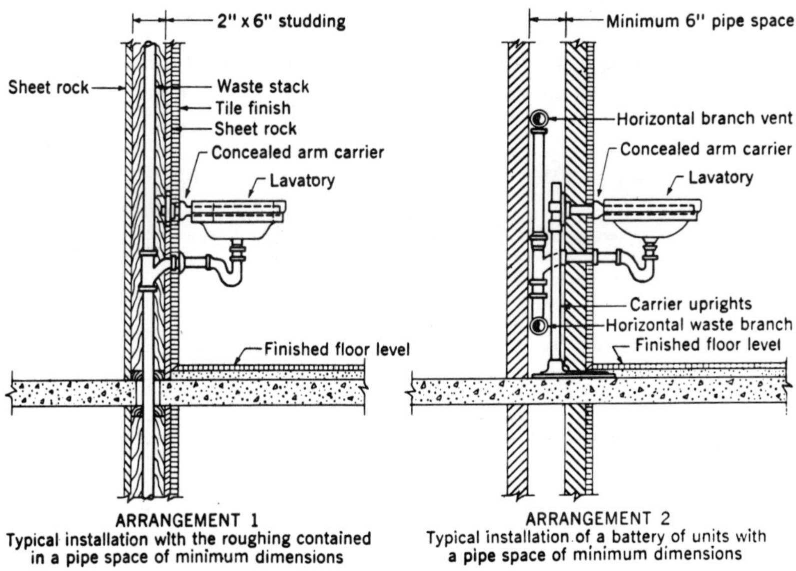

Fig. 6. Installations of vitreous china deck type lavatories with the roughing contained in a pipe space of minimum dimensions.

Arrangement 1 of Fig. 6 shows the roughing installed in the minimum space offered by 2 x 6-inch stud construction. Here the lavatory is shown as a single unit. The 2 x 6-inch construction has been shown since this is the space required for water closet roughing usually placed on the same wall as the lavatory.

In this case the lavatory is supported on concealed arms which are in turn secured to an adapter plate that is rigidly affixed to the stud construction. The concealed arms slip into holes that have been provided at the back of the lavatory. Leveling screws are incorporated on the concealed arms to align the lavatory. Also a locking device is located on the concealed arms to anchor the lavatory in place. Access to the leveling screws and anchoring device is provided by holes in the underside of the lavatory. The deck type lavatory is set approximately 2 inches away from the wall. Thus, the portion of the concealed arms which would normally be exposed is hidden by a chrome plated escutcheon that fits between the wall and the back of the lavatory.

Arrangement 2 shows the vitreous china deck type lavatories installed in a battery arrangement. The pipe space is 6 inches wide as it would be used for the battery installations. This pipe space is subject to an increase dependent upon the size of the vertical waste stack, vent stack and cold water risers that are used when water closets or urinals are installed on the same

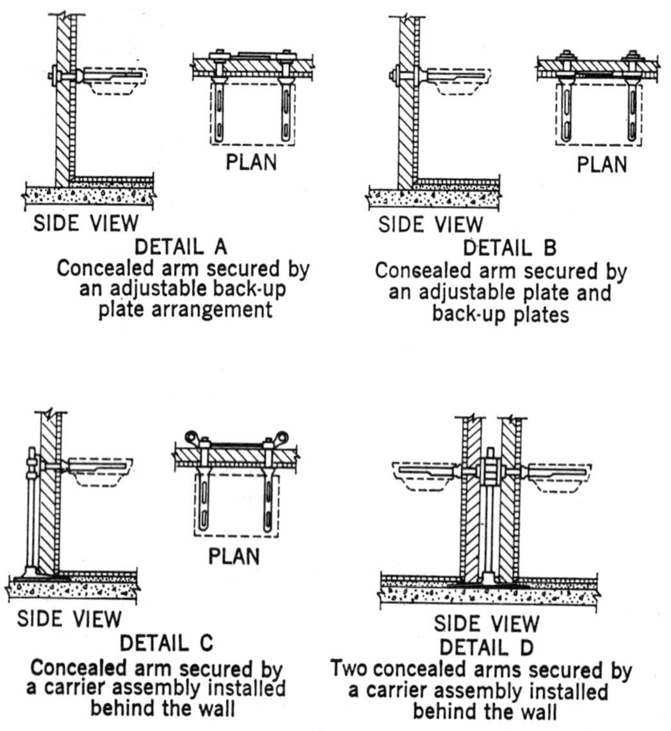

Fig. 7. Various supporting devices for vitreous china deck type lavatories.

wall. The required pipe space dimension would then be the standard 12 inches.

Here, each of the lavatories in the battery arrangement are supported on concealed arms which are in turn secured to an assembly of two feet, two uprights and two adapter couplings.

Supporting Devices for Deck Type Lavatories

Several devices that may be utilized to support vitreous china deck type lavatories are shown in Fig. 7.

Detail *A* of this figure shows how concealed arms can be secured to the construction by an adjustable adapter plate set behind the wall construction. A lock nut at the finished wall anchors the concealed arm.

Detail *B* shows the concealed arms secured to the construction by means of an adjustable adapter plate set in the finished wall with steel back-up plates at the rear of the construction.

Detail *C* illustrates in detail the method of support described in Arrangement 2 of Fig. 6.

Detail *D* shows how back to back lavatories may be supported on concealed arms which are in turn, fastened to an assmbly of two feet, two uprights and adapter couplings.

The many manufactured devices that are available make possible a very broad selection for the engineer from which to choose.

Installation of Lavatory Roughing

Two typical arrangements of lavatory roughing are illustrated in Fig. 8.

One shows a typical arrangement of roughing that might be employed for the upper floor levels in a building. Here the roughing is installed above the floor slab. The roughing with the exception of the vertical waste and vent stack, is of threaded pipe and fittings. The horizontal waste piping is run above the slab and connects with the vertical waste stack as shown. The individual vent lines are caulked into the fittings that receive the waste from the lavatories. This arrangement can be used with all the lavatories described.

The other is a typical installation of roughing at the ground or lowest floor level in the building. Here the horizontal waste is below the floor slab and is of cast iron hub and spigot pipe and fittings. The piping up to and including the lavatory waste

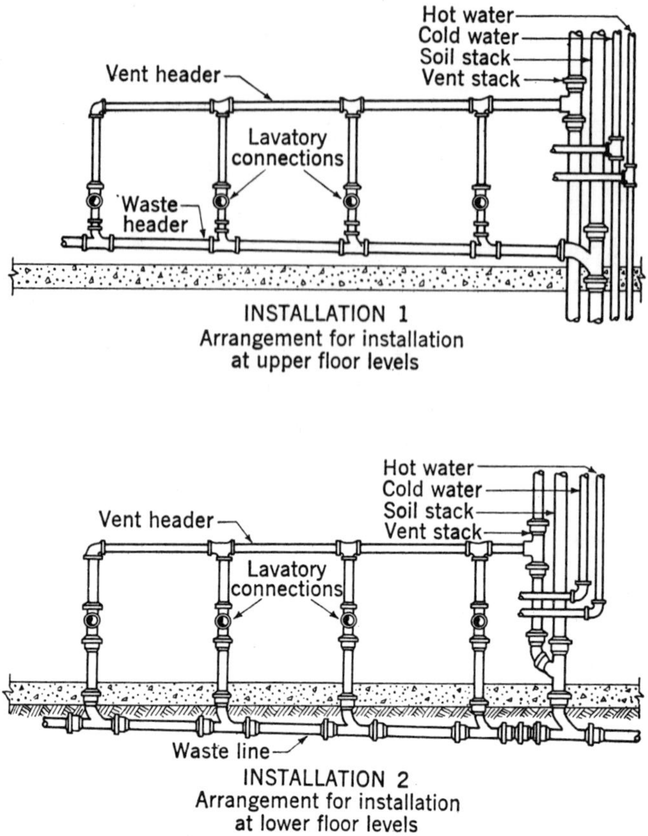

Fig. 8. Two arrangements of roughing for a battery of lavatories.

connection, is also of the hub and spigot pipe and fittings. The vent header and individual vents are of threaded pipe.

Galvanized steel pipe, galvanized wrought iron pipe, copper pipe or cast iron hub and spigot piping all may be used for the installation. The engineer will select the materials as well as the particular method of roughing arrangement he favors.

Part 6
Laboratory Services and Facilities

CHAPTER 55

Water Service for Research Laboratories

Research laboratories are a special type of building and are designed primarily to house the various kinds of equipment and apparatus used by scientists, engineers and chemists in their work. Research laboratories may differ in size and shape to suit the needs of the particular experiments to be performed; the specific type of occupancy equipment used, will determine the design of the building itself. In general, a research laboratory is that form of structure which is built around the exacting demands of the men and their equipment.

Regardless of the type, shape or function of the research laboratory, an abundant supply of cold and hot water is always required. It is important to consider a number of factors in the design of an adequate water supply system for the structure.

Distribution of Water Piping

Water for the laboratory is generally supplied from either an existing street service or from an elevated or underground storage tank, erected on the site. After the water enters the building it is furnished to the various fixtures and equipment by means of a distribution system. A typical arrangement of the piping for a research laboratory is shown in Fig. 1.

For the purpose of example, Fig. 1 shows a rectangular shaped building, devoted to chemistry research. Within the building, the water supply is divided into two separate systems. One system is known as the potable cold water, while the other is called process cold water.

The potable cold water serves the drinking fountains, kitchen, plumbing fixtures, emergency showers, eye wash fountains, air conditioning and heating equipment as well as the other domestic needs of the building. The process cold water is supplied to all laboratories, including outlets and equipment related to experimental purposes.

Since the scope of this chapter is confined only to the process cold and hot water system, the potable water system will not be emphasized even though it is mentioned. The process cold and hot water system of piping might be arranged in the corridor of the basement, Fig. 1, and branches taken from the cold and hot water mains shown are run up through the service shaft where take-offs are made for the laboratories and equipment.

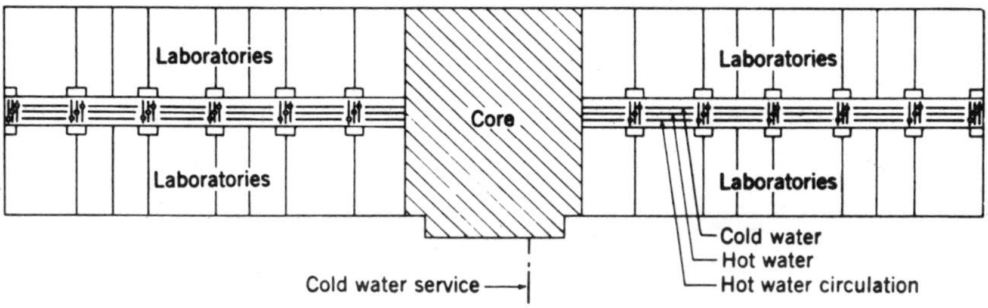

Fig. 1. Laboratory building showing typical distribution of process, cold and hot water services.

Separation of Water Supply

It is very important that a separate system be furnished for the process cold and hot water for this prevents the danger of backflow into the potable water system in the event of a pressure drop in the outside water supply system. In some instances, backflow can suck liquids used in experiments, into the process water piping and thus into the potable water system unless adequate protection is afforded. Where potable water is concerned, every precaution must be taken to avoid its contamination.

In order to provide a potable and process water system, it becomes necessary to separate the water supply at a convenient point in the building. There are several different methods that can be employed to separate the water supply and the method shown in Fig. 2 illustrates one means.

As indicated in Fig. 2, separation in the building is accomplished at a point near the water supply entry. Assuming that water metering is required, the water supply passes through a control valve, strainer, valve, water meter, test tee and another valve. The water meter is so arranged that it can be easily removed and another meter quickly installed in its place. The valved bypass permits uninterrupted water service while the replacement is being made. If pressure reducing valves are required to handle an excessive water pressure, they can be placed either near the water meter or on both the process and potable water supply branches. In any event, the water meter and control must be installed to conform with the requirements of the local water department or plumbing authorities, whichever holds jurisdiction in this matter.

Next, the supply is separated as shown. At this point, both the potable and process water services are each equipped with: A valve,

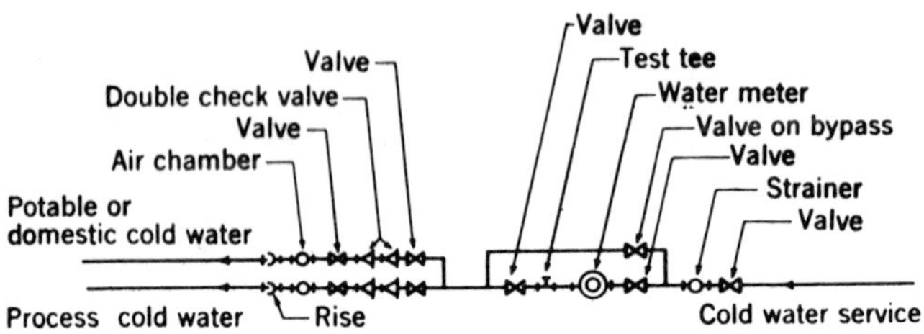

Fig. 2. Arrangement of water supply lines to provide separation of services.

double check valves and another valve. The purpose of the double check valves is to ensure against any possibility of backflow in the event of a pressure drop in the water service or main.

Adequate protection is afforded in the laboratory in the form of vacuum breakers to prevent back siphonage of water through laboratory faucets or outlets. However, the physical separation including double check valves is an added safety factor. When complete and positive separation is desired because of the dangerous nature of the experiments being performed, then it becomes necessary to isolate the process cold water by means of a water tank located on the roof of the building.

Hot Water Systems

When both potable and process cold water systems are provided in a research laboratory building, then the potable and process hot water systems should also be separate. Thus an individual means of generating hot water should be included for both systems. A hot water circulating line in each case will assure hot water at the turn of the tap. The type of hot water generating equipment selected should be of the size and capacity best suited for the demand.

Support of Distribution Piping

The process cold and hot water distribution piping may be located in the basement of the building, in the top floor or attic space of the building or in other areas as selected by the engineer and the architect. In any event, it is imperative that the piping be adequately supported to avoid strain to the piping and to prevent its vibration. Figure 3 represents a suggested method of support.

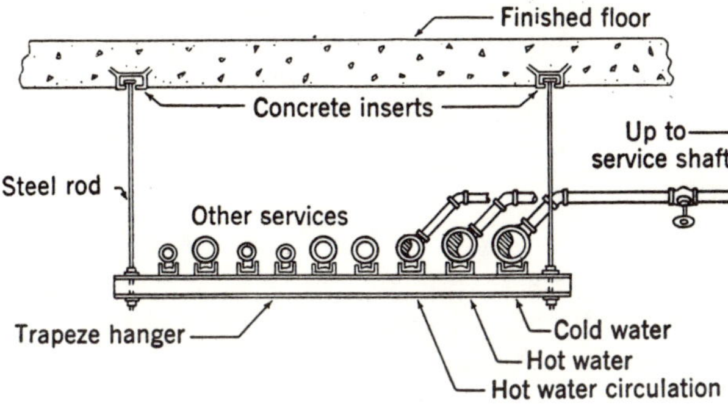

Fig. 3. Arrangement of process, cold and hot water circulation on distribution rack in corridor.

Referring to Fig. 3, it will be noted that a trapeze hanger is employed which consists of two channels fastened together and secured in place with steel rods, hung from concrete inserts. The piping in turn rests on roller supports that are bolted to the trapeze. The rollers permit some movement of the piping, especially in the case of the hot water lines.

Since expansion is a factor that must be considered with hot water lines, proper expansion devices must be utilized. The expansion devices are usually loop offsets provided in the piping or are patented items made by several manufacturers. When these devices are used, the trapeze hanger that follows the expansion device is generally anchored. This is accomplished by angular bracing of the trapeze hanger at either side, to concrete inserts set in the underside of the concrete slab. The hot water lines are then anchored or clamped to this particular trapeze to limit the extent of expansion or to confine the expansion between anchored trapezes. Continuous strip concrete inserts are used in many cases to offer a greater flexibility in locating the steel rods at the most convenient spot for the piping.

Trapeze hangers are placed at prescribed distances apart, dependent upon the piping material used. Since copper tubing is used rather extensively, the trapeze hangers are set rather close together. When carefully installed and the piping properly arranged, the trapeze hanger will give many years of trouble-free service.

To prevent sweating or condensation of the cold water lines, the piping is covered with suitable anti-sweat covering. To retain the hot water temperature suitable covering or insulation is applied to the piping. There are several devices manufactured, to set into the underside of the insulation at the point where it rests on the roller supports and thus prevent damage or flattening of the insulation at this point.

Service Shafts

The service shaft is a cut in the concrete that allows for the passage of the piping risers. Figure 4 suggests a typical arrangement of a laboratory room and its component service shaft.

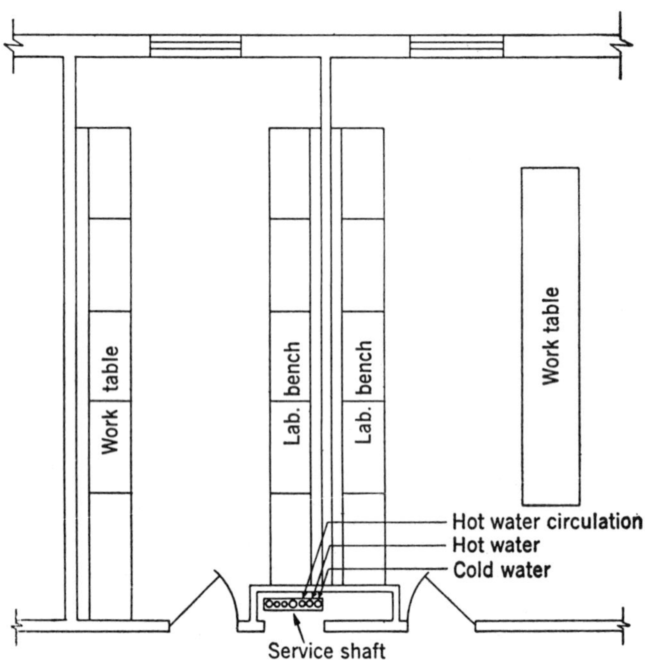

Fig. 4. Typical arrangement of piping in service shaft.

Branches are taken from the hot, cold and circulation mains in the corridor and are installed as vertical risers to supply one, two or more floors of laboratories. Control valves are placed at each floor to shut off the branches to the laboratory equipment or outlets contained in the laboratory bench. Often, valves are placed at the base of the risers in the corridor to control the riser. The slot in the service shaft should be large enough to permit each service contained to be adequately supported.

The service shaft is a rather large space which often contains in addition to the laboratory piping, exhaust ducts, electric controls, cylinder gas bottles and other equipment needed for the laboratory benches.

Service shafts are designed in different sizes and shapes to suit the particular design needs of the building. Moreover, the service shaft may be located as shown in Fig. 4, or can be placed in other positions to suit special needs.

Laboratory Benches

Laboratory benches usually contain a sink and a number of table top outlets for the different services. Figure 5 is a typical example of how the various services are installed under the benches.

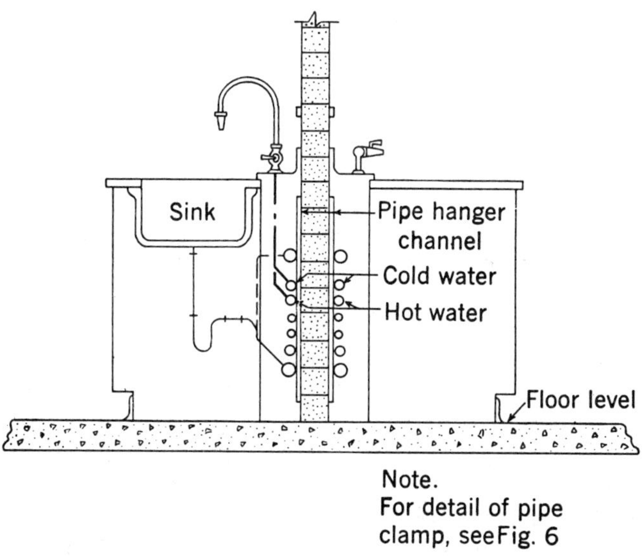

Fig. 5. Arrangement of process, cold and hot water for laboratory bench.

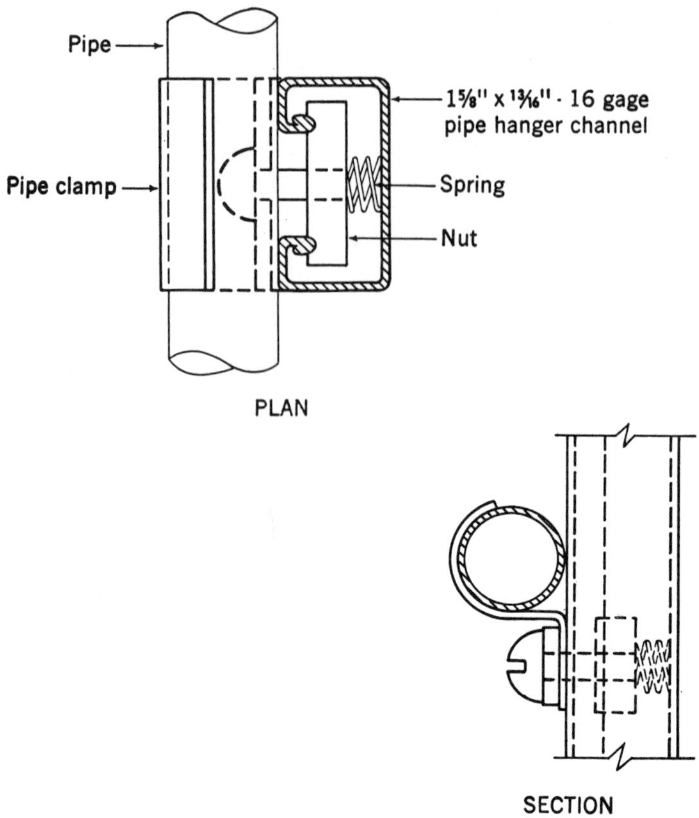

Fig. 6. Pipe clamp arrangement for support of process, cold and hot water services under laboratory bench.

In this instance, a bench is included on either side of the wall. Although, there are a number of ways in which the piping can be supported, Fig. 5 shows a method that is frequently used. Since only the process cold and hot water is featured in this discussion, the accent is placed on the support of these lines and it will be noted that the water lines are supported in a certain prescribed way as detailed. Figure 6 shows how the piping is supported.

Pipe hanger channel strips are placed vertically every few feet and pipe clamps are attached. These clamps assure an adequate support for the piping and yet offer a very flexible arrangement for a group of services.

When a process cold and hot water system is carefully designed and installed, it will give many years of trouble-free performance and if the piping is properly supported, undue strain on the piping itself caused by pipe movement will be avoided. Expansion devices compensate for the expansion of the hot water piping. Should dissimilar metals be joined together in the piping system, electrolysis can be prevented by means of patented isolator fittings or couplings. Properly sized and placed water hammer eliminators will prevent the clatter and noise normally associated with water hammer. Equally important, vacuum breakers installed in suitable locations are assurance against back siphonage.

CHAPTER 56

Emergency Devices for Protection of Laboratory Personnel

Since World War II, there has been a significant increase in the construction of research laboratories. This has brought with it, special engineering problems in meeting the services required. In research laboratories, considerable flexibility is offered in the design of water, waste and gas facilities.

Research is a dangerous type of effort because it involves experimentation with untried materials, compounds and procedures. Therefore, certain safety factors must be incorporated into the design of the building. Thus, prescribed safety factors include structural provisions for buildings in which blasts or explosions can occur; specialized fire protection for buildings where spontaneous combustion or flash fires is a potential hazard; where highly corrosive liquids are handled, attention must be given to the proper design and specific piping materials and equipment that can safely handle and dilute such liquids.

Laboratory personnel, too, must be protected against the hazards connected with their experiments. Accidental splashing or dousing by a highly concentrated form of acid or caustic solution can cause severe and sometimes fatal injury to the technician involved. There are a number of emergency devices designed as emergency safeguards for laboratory personnel should such an accident occur.

Emergency Measures

Many different types of acids, chemicals, alkalis, radiological, bacteriological and biological materials, as well as other contaminants, are regularly used in laboratory work. Although, extreme care is generally exercised in the handling of this material, accidental spilling or splashing can still occur. There is always the possibility that a mixture of ingredients will erupt and splash the technician.

When this happens, it is necessary to act quickly in removing such substances from one's clothes and body before it can react on the skin. As the first step of treatment must be taken immediately, the devices for emergency protection must be close at hand. Water in generous quantities is regarded as the most favorable means for diluting or washing away the dangerous chemicals and solutions. The device furnishing this water is called an emergency shower. It should be located as close as possible to the laboratory for convenience and to speed emergency treatment.

Emergency Shower

An emergency shower is a large single shower or a combination of small shower heads, so arranged, that when it is activated, it will quickly douse a person with water. The shower is usually located in a position where a person that has been splashed, can find it with his eyes closed.

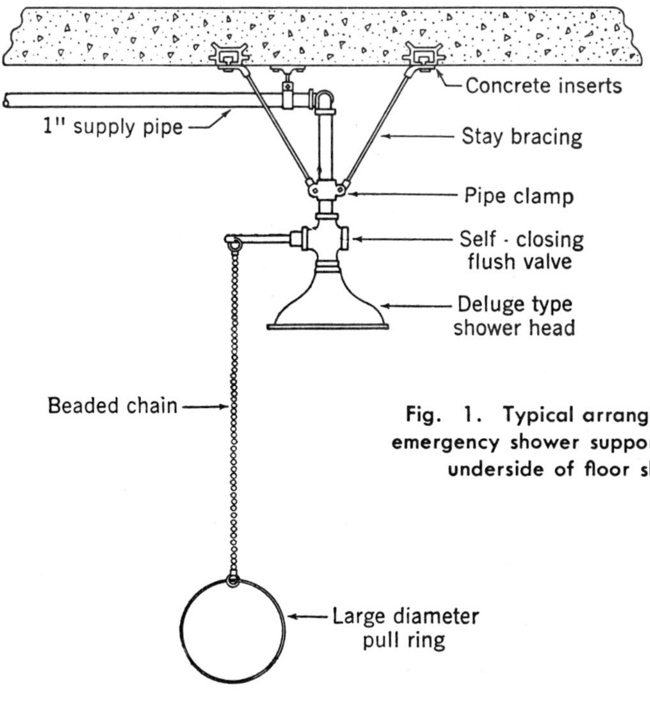

Fig. 1. Typical arrangement of emergency shower supported from underside of floor slab.

Figure 1 illustrates a basic type of emergency shower. It shows a single large diameter shower head known as a deluge type head for it permits a large volume of water to be discharged.

The piping is rigidly fastened to the underside of the slab or ceiling construction. A large diameter ring, which a person can hook with his arm is used to activate the shower. Many different arrangements with regard to installation are possible with this unit. Likewise, several different methods for activating the shower may also be utilized.

Location of Emergency Showers

The emergency shower should be placed as close as possible to the point of contact or contamination by the dangerous materials. Prescribed locations are either in the laboratory itself or directly outside the laboratory, near the door.

Figure 2, suggests locations where an emergency shower might be placed. One recommended position is in the laboratory as shown

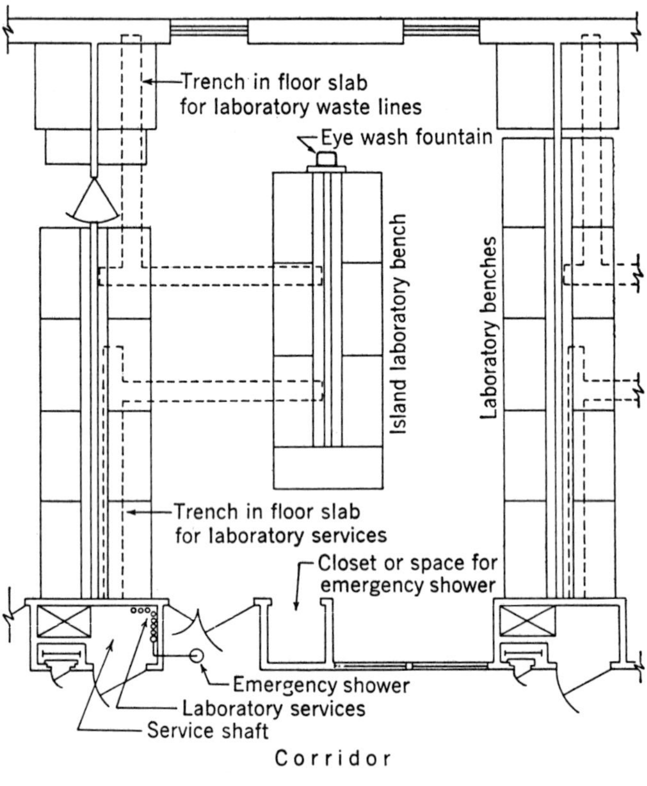

Fig. 2. Two-module laboratory room showing location of the emergency shower and eye wash fountain.

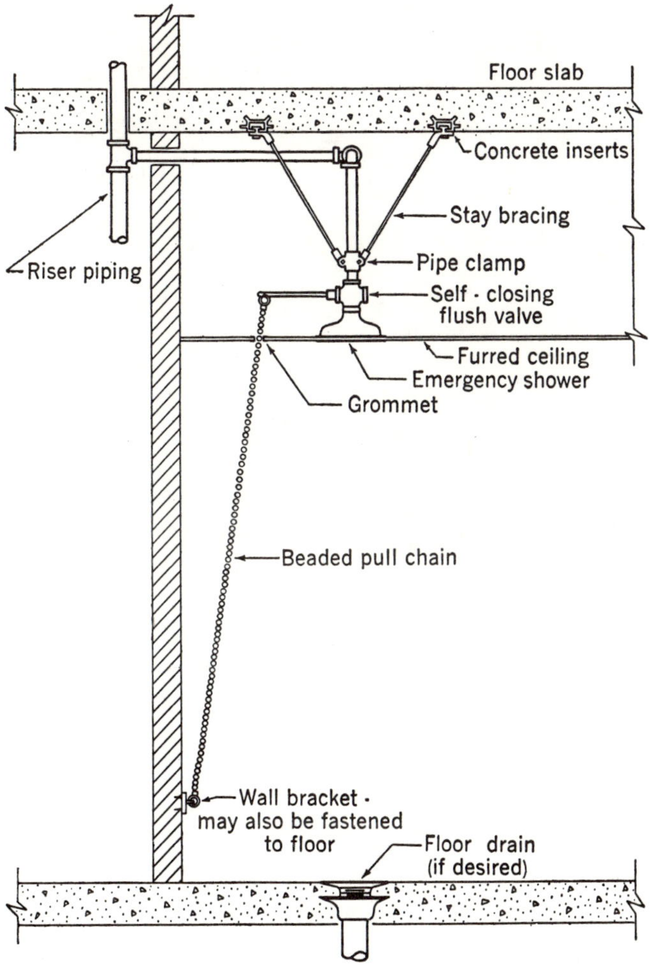

Fig. 3. Arrangement of emergency shower for condition shown in Fig. 2.

while another position is directly outside the laboratory. The laboratory plan illustrated is a rather typical layout of a 2-module laboratory and is presented only for an example. The emergency shower is placed outside the laboratory when, because of the hazardous nature of the work being conducted, splashing would intensify the danger to personnel. In either of the locations, the showers are easily reached by a person, in need of its benefit. The eye wash fountain has also been included in Fig. 2, and will be discussed later.

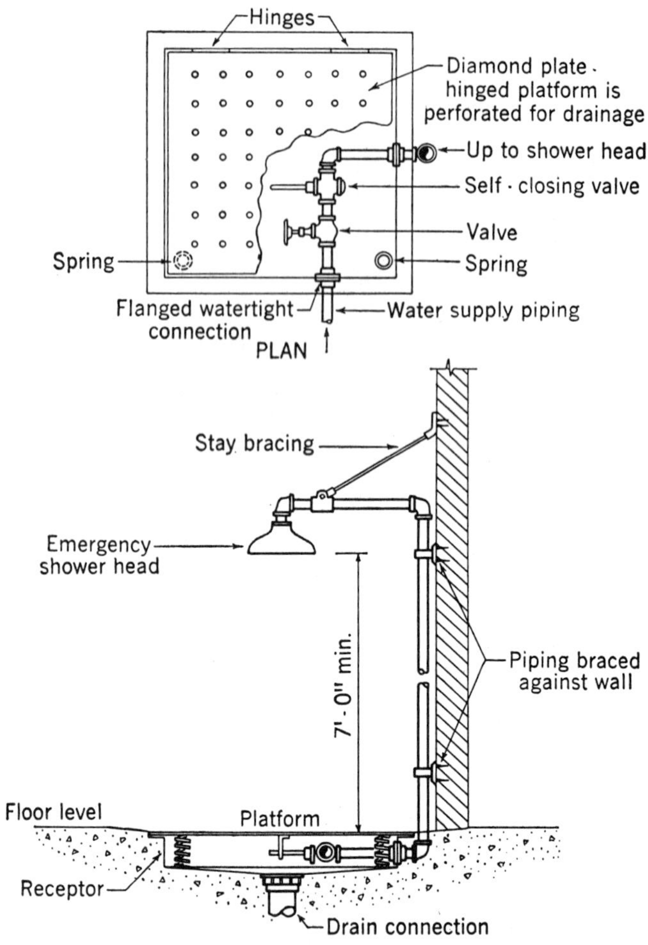

Fig. 4. Typical arrangement of a platform type emergency shower.

Types of Emergency Showers

There are several different types of emergency showers currently being used with a great deal of success in research laboratories. The following are basic and are shown to point out (1) the application of this emergency protection and (2) to suggest the range of specific needs to which the showers may be adapted. The basic types are deluge, platform, swing gate and swing gate with partition.

Deluge Type. This type of protection, Fig. 3, can be employed inside or outside of the laboratory, as desired. Here, it is assumed

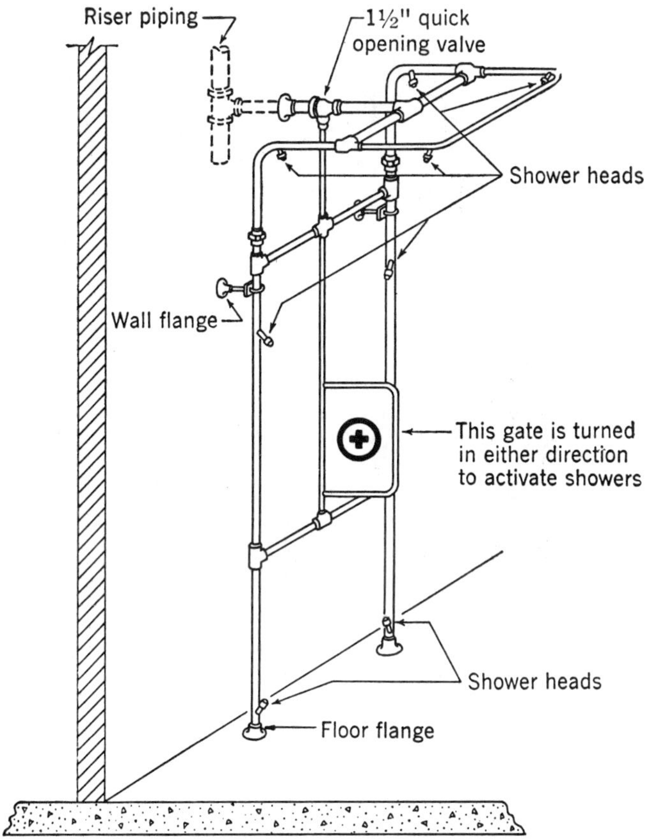

Fig. 5. Arrangement of an emergency shower unit that utilizes individual shower heads.

the position is outside the laboratory, in a similar manner to that shown in Fig. 2. The shower head is a deluge type that can be mounted flush with a furred ceiling. The valve and piping in this case are all concealed. As a person leaves the room, he has only to pull on the chain and the necessary water protection is available. A floor drain may be added or omitted at the option of the owner or engineer. Generally, a floor drain is not included since the shower is only for emergency purposes and therefore when it is used, the water dispensed on the floor is easily mopped up. Moreover, odors from a floor drain can be a hazard in itself unless the trap water seal is frequently replenished.

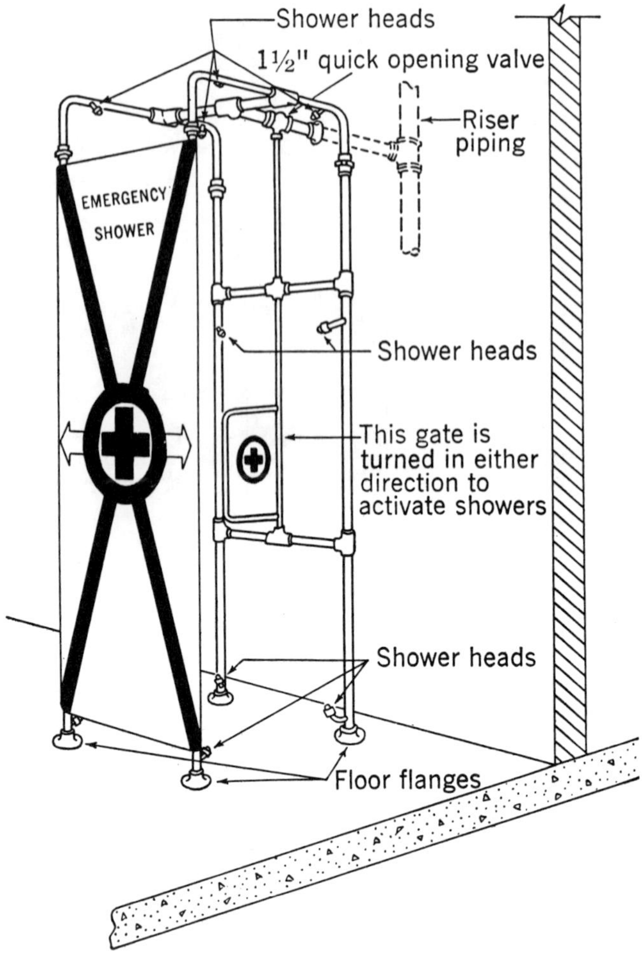

Fig. 6. Arrangement of a shielded emergency shower unit that utilizes individual shower heads.

Platform Type. Here is a shower, Fig. 4, that might be included in the laboratory. It is only necessary to step on the platform to activate this shower. This emergency shower consists of a drain connected receptor with a perforated platform that is hinged on one side while the other side rests on springs. The activating valve and the piping and control valve are installed in the receptor space for ease of inspection and maintenance. If required, a shield or partition may be placed in front of the unit to prevent or limit the area that will be splashed with water. This type of an emergency shower operates in a very favorable manner and offers the complete protection sought.

Swing Gate Type. This is an emergency shower that can also be installed in or outside of the laboratory. Figure 5 represents a detail of its method of operation. The unit illustrated consists of 8 shower heads, so arranged to completely envelop a person's body with water. It is necessary only to swing the gate in either direction to activate the showers.

Swing Gate Type with Partition. This type unit, Fig. 6, operates in a manner equal to that shown in Fig. 5, with the exception that a partition or shield has been added in front of the shower unit to prevent the water from splashing beyond a confined area. Entrance to the shower is gained from both sides and the shower is activated by swinging the gate in either direction. A floor drain of course, can be added or omitted as desired.

Water Supply

Emergency showers and their location have been discussed and a few types illustrated. Now it is important that some attention be given to the piping system supplying water to the showers.

It is always preferable to utilize a potable water supply for the showers. Very often, in research laboratories, the water supply is separated into two systems, a potable water system and a process water system. The potable supply is used for all the domestic needs of the building while the process water is used for all the experimental work in the laboratories.

Let us assume that our laboratory building has both systems and that the water for the emergency showers is supplied from the potable system. A typical piping system would then be similar to that shown in Fig. 7.

The system shown supplies both the emergency showers and the eye wash fountains. The sizing of the pipe is based on the particular type of actuating valve used and the volume of water needed by the shower heads. Sizing tables included in the manufacturers' catalogs offer recommended piping sizes. The size of the general piping need not be excessive since it is rare that more than one emergency shower will operate at one time.

A water hammer arrestor is included at the top of each riser line to absorb the hydrostatic shocks that water hammer in the line can generate. Self closing and quick opening valves contribute to water hammer and, therefore, the arrestors will not only absorb the shock

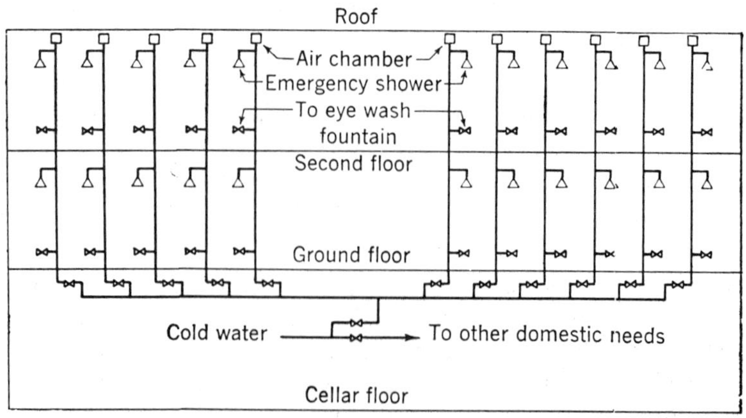

Fig. 7. Arrangement of water risers supplying emergency showers and eye wash fountains.

but will also prevent the piping clatter associated with the shock. With careful design and installation, this system can be counted upon to function in a most favorable manner when the need arises.

Temperature of Water

When the installed piping system is similar to that shown in Fig. 7, the water contained in the lines will warm up to room temperature. Even though a limited number of plumbing fixtures are supplied by the line, the temperature of the water will still remain rather warm for use.

Sometimes, however, for reasons of economy or otherwise, the emergency showers are connected to the process cold water line serving the laboratories. In this event the water is continually running through the lines and the water temperature is cold. When one is splashed with a dangerous liquid to some extent there is personal shock. A dousing by the emergency shower at this time, although it does remove the caustic material, can also intensify this shock. Cold water can be especially startling to a person utilizing an eye wash fountain.

Frequently, the laboratory staff requests that water of a controlled temperature be maintained in the piping system for the emergency showers. If the installed system is similar to that illustrated in Fig. 7, there is no particular problem. If the water is supplied from lines that are continually running, special attention is necessary to temper the water. There are a number of ways that tempered water can be furnished in a practical and economical way. One way is to provide a separate riser that supplies only the emergency showers and eye wash fountains. This method offers a riser through which water will not circulate, thus allowing the water to reach room temperature. When the emergency shower is used, the first good dousing will be with warm water before the cold water can reach the point of discharge. The engineer designing the system can devise a method that best suits his needs.

Eye Wash Fountains

Whenever emergency showers are needed, eye wash fountains are also generally required. Eye wash fountains are intended primarily for protection of the eyes of the laboratory workers.

Should an acid or contaminant accidently splash into a person's eyes, every second counts in the speed at which the matter can be washed from the eyes. There are different styles of eye wash fountains. Basically, they all function in the same manner. The units are equipped with two jets or sprays that create a gentle flow of water. The person has only to lower his head into the fixture and open his eyes, the gentle flow of the water will thoroughly wash away the dangerous material. These units are often equipped to offer both hand and foot activation. This provision is important because it is often necessary to assist a person to the unit, and to hold his face in position. In this instance, the foot pedal is employed by the person doing the assisting.

Although this chapter was confined to research laboratories, the units illustrated have a much wider application inside and outside the building in many industries.

Research work with radioactive materials has opened an ever greater need for this type of protection. In this field, water is used for decontamination purposes, and its immediate application is often as important as for the other hazardous materials previously mentioned.

CHAPTER 57

Distilled Water Systems

While water as supplied for human consumption is pure and palatable, a water of still higher quality is required for laboratory tests and process work. In such cases, distilled water may be necessary.

Distilled water is used in many fields, a few of which follow: Research laboratories, food processing, electronics, automotive industry, bottling plants, pharmaceutical industry, chemical processes, textiles, paper and pulp industry, hospitals and universities.

Fundamentally, to produce distilled water, tap water is gently boiled in a flask. The vapors thus generated, rise and enter a condenser where they are cooled by a flow of condensing water. The vapor returns to a liquid state and is collected in a container for future use. This water is free from minerals, dissolved gases and other impurities and basically it can be produced by the apparatus shown in Fig. 1, which is generally employed for classroom demonstrations only. We shall attempt to cover the manufactured apparatus available for distilled water production, and the system of piping that will convey distilled water to its point of application or consumption in the system.

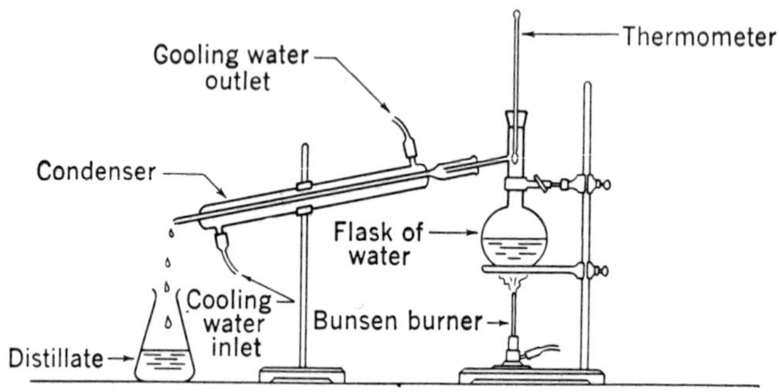

Fig. 1. Laboratory apparatus used in the distillation of water.

Research Laboratories

A research laboratory is used, for example, since varying amounts of distilled water are used — from a few gallons a day for some laboratories to many more gallons for others.

Distilled water for laboratory research work must meet certain exacting purity requirements depending on the nature of the experiment being performed. For a higher rate of purity, double or triple distilled water apparatus will ensure the quality of water needed.

Distilled water plays an important part in many experiments because it is a neutral liquid and does not impart impurities to the experiment as would tap water. Since this water is measured in ohms of purity, the laboratory personnel must inform the design engineer of the degree of purity needed. Once the desired degree of purity is known, the engineer is able to select and specify the type of equipment to maintain the standards of operation that were selected.

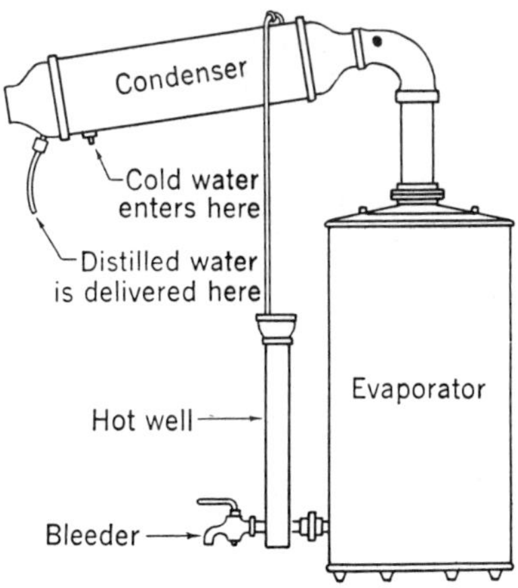

Fig. 2. A typical water distilling unit.

Individual Distilling Units

When but a small quantity of distilled water is needed in a laboratory, this amount can be furnished by a small water distilling unit installed in the laboratory. Such a typical unit, Fig. 2, is capable of generating 1 to 10 gallons per hour. The still can be heated by gas, electricity, steam or oil. Several different types of units are offered by the manufacturers, and the one shown has been used for example only. Because such units are compact and durable, they will produce distilled water economically for many years.

Larger Capacity Units

One may desire to have a still of greater capacity and centrally located so that the laboratory personnel can draw distilled water in the quantities needed for their experiments. A typical unit of this type is depicted in Fig. 3. This unit is manually operated and uses a storage tank in conjunction with the still. The storage tank holds a reserve of distilled water for ready use. The unit shown is a steam heated still and tank combination but can also be obtained with automatic controls.

The still and storage tank are sized in accordance with the demand requirements of the laboratory. For example, the still may be capable

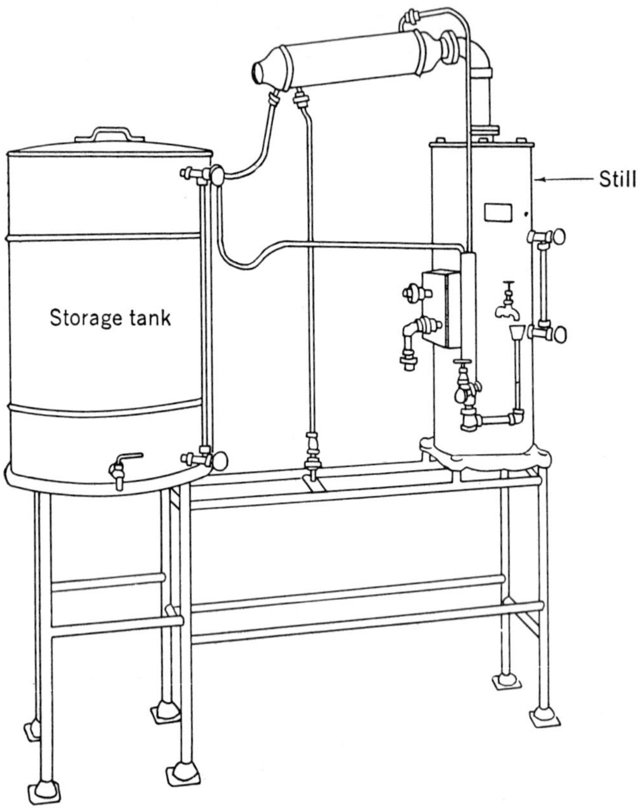

Fig. 3. Detail of a manually operated steam heated still and tank combination.

of producing 30 gal. of distilled water each hour while the tank will store say 50 gal. After the contemplated demand of the laboratory is known, it is always advisable to consult with the equipment manufacturer before selecting the type and capacity of apparatus best suited for the service.

Storage Tanks

For laboratories that require either a large amount or a flexible demand of distilled water, a storage tank is provided. The interior of the tank is lined with the specific material that will best maintain the degree of purity desired. Where small capacity storage tanks are needed, wooden boxes or small metal tanks are available. For larger capacities, tanks can be made in any shape or form to meet the precise standards that might be needed. Several of the tanks more commonly used are shown in Fig. 4.

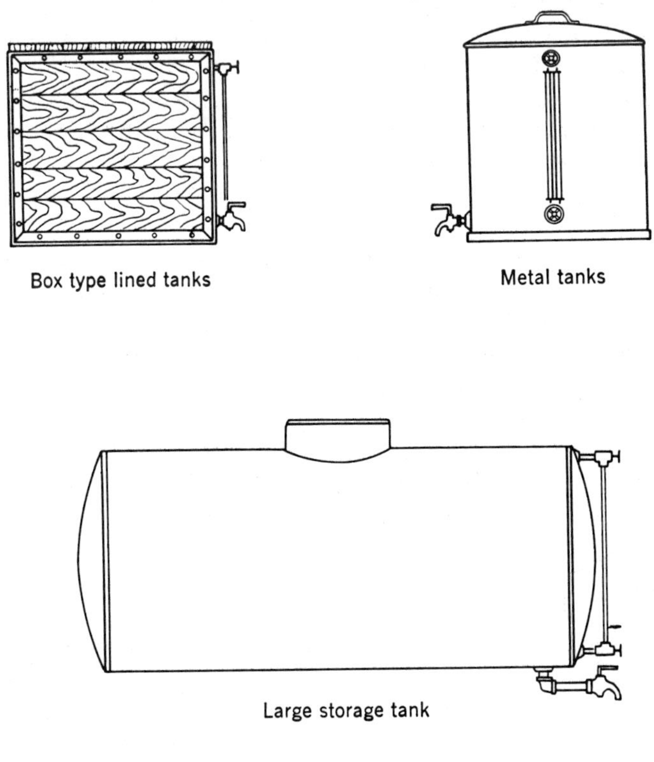

Fig. 4. Typical storage tanks.

Box type, lined tanks, constructed of 2 in. selected woods are securely stay-bolted, lined with copper, and heavily coated with pure tin. This tank is equipped with a snug fitting cover that is also lined with copper and heavily coated with tin. Although the box type tanks are more generally used for small capacity needs, it can be furnished in capacities ranging from 50 to 1000 gal.

Cylindrical or metal tanks of the laboratory or hospital type are constructed of heavy copper with an extra thick coat of pure tin on all interior surfaces. The hospital type tank contains a few additional features that further preserve the purity of the water stored.

Special tanks of either the vertical or horizontal type can be built to meet specific job requirements, pertaining to shape or capacity. Such tanks can be furnished in aluminum, stainless steel or other special materials.

A unique device termed a vent guard, can be furnished for the storage tanks to further safeguard the purity of the distilled water stored. This device is equipped with a replaceable filter element and contains a special combination of chemicals which will not permit impurities to enter the tank. The vent guard not only filters the air, it also prevents traces of organic vapors, alkali, gases, acid gases and other elements from entering the tank where they would be readily absorbed by the distilled water.

Double and Triple Distilled Water Outfits

Where the experimental work is of an exacting nature and water of high distillate purity is needed, double or triple distilled water is furnished. Water generated by such means represents the purity required for intravenous solutions. To produce water of such purity, either two or three stills are used, with the water from the last still running into the storage tank.

In multiple distillation, distilled water from the first still is automatically redistilled by the second still and when desired, by the third still. Water of such purity should satisfy the most exacting requirements.

Central Piping System

When numerous outlets for distilled water are contemplated in a laboratory building, a central system is used for the distribution of distilled water. A typical example of this type of system is detailed in Fig. 5, which shows in diagram a two-story laboratory building including a basement. Since distilled water is required in each of the laboratory rooms, a gravity system is employed. In this instance,

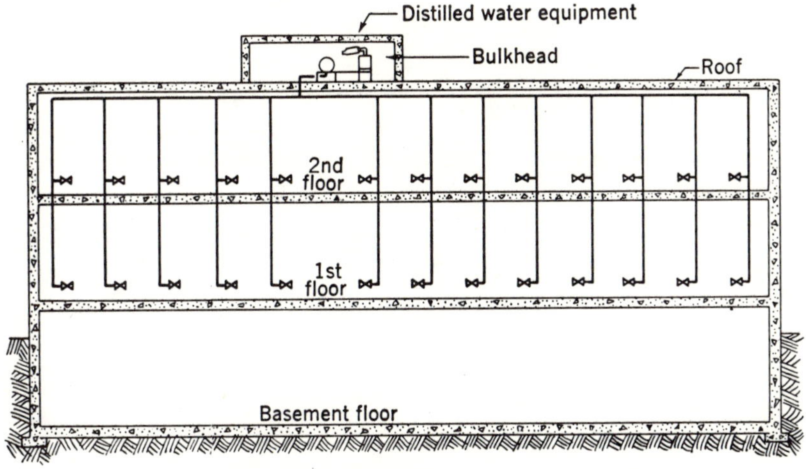

Fig. 5. Arrangement of distilled water distribution piping in a research laboratory.

the distilling apparatus is placed at a high point in the building, specifically the bulkhead over the laboratory core as shown. Distribution mains run at or in the second floor ceiling. The vertical drops of piping are placed in the service shaft enclosures, adjoining each laboratory room. Branch lines for the laboratory benches are extended from the vertical lines. The arrangement of branch piping to the laboratory outlet would be similar to that shown in Fig. 6, and these branches are supported under the bench in the manner shown. Supplies are taken from the branch piping to feed the distilled water outlet. Since there are a number of different types of outlets available for the purpose, the unit shown is for the purpose of example only. Although not shown, a cup drain is installed in the laboratory bench and directly under the water outlet to catch any spillage.

Figure 7 illustrates the type of equipment that might be furnished in the space above the second floor to produce the quantity of distilled water needed. In this case, steam is used as the heating medium and the controls are fully automatic. A horizontal storage tank has been used to hold the constant reserve of distillate needed.

Maintaining Distillate Purity

To preserve the purity of the distilled water through the distribution system, special piping materials must be used such as: Piping lined with block tin or glass, aluminum pipe, plastic pipe, stainless steel or other piping materials adapted for this purpose. However,

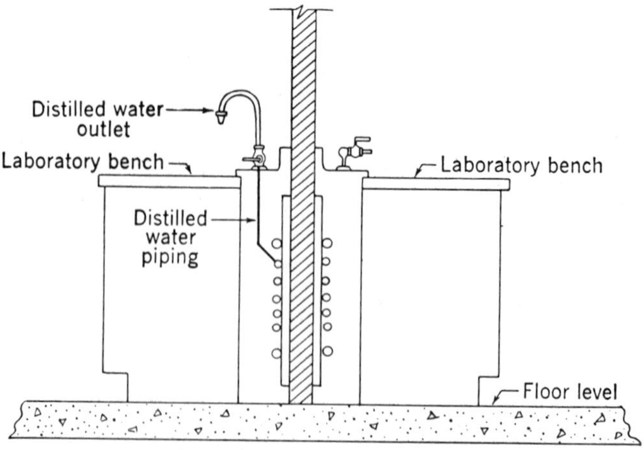

Fig. 6. Arrangement of distilled water piping at laboratory bench.

regardless of the material or lining used, all materials will contribute a certain amount of contamination to the distilled water. Therefore, if a prescribed rate of water purity is to be maintained, it becomes necessary to distill the water to a higher degree of purity than is

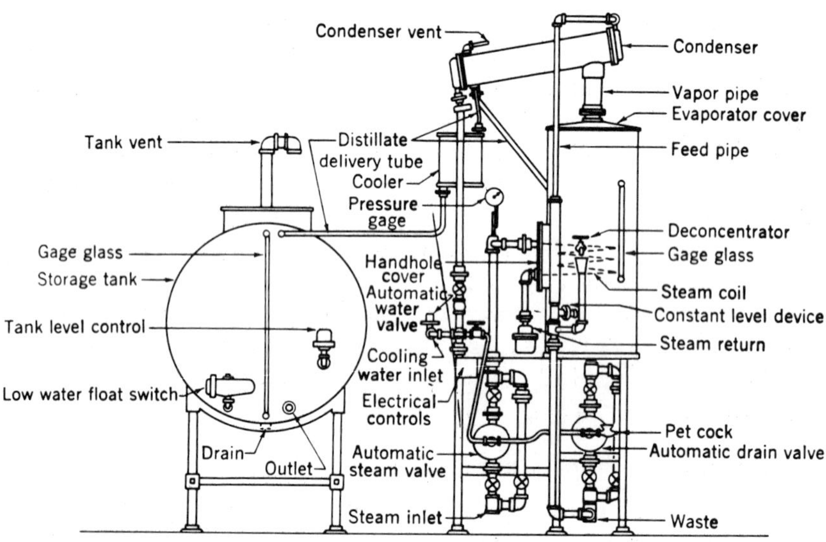

Fig. 7. Typical arrangement of still, storage tank and piping.

actually needed in the laboratory rooms. Thus, by the time the water has traveled the distance from the storage tank to the laboratory outlet, it will have lost some of its purity but yet will be within the prescribed limits of purity needed in the laboratory.

During the average work day, the water in the storage tank and in the piping will be used up but will be continually replaced by the still. There are times, nevertheless, when the water may not be used for several days and will remain dormant in the storage tank and piping. A typical example of this is a long holiday weekend when the water may not be used for two or more days. In this span of time, the water will lose its purity to an extent that it will not meet the laboratory requirements. In this event, it will be necessary to draw-off the entire supply of distilled water in the storage tank and piping and generate a fresh supply.

Demineralized Water

There are times when it is desired to remove certain substances from the water before it is distilled and consideration is therefore given to demineralizing equipment. The demineralizing process removes only the ionizable impurities contained in the water. The difference between distilled water and demineralized water follows. Distillation removes all types of impurities, ionizable and non-ionizable including minerals, organics, gases and bacteria. Water purified by chemical ion-exchange is called demineralized or de-ionized water. The organic, bacterial and non-ionizable content of the water remains essentially unchanged.

Since the generation of demineralized water is a low cost process, its benefit can be realized in countless applications where sterility and freedom from organic impurities are of little importance.

When faced with the problem of selecting distilled water equipment of adequate capacity and performance, the manufacturer of the equipment favored should be consulted.

CHAPTER 58

Water and Drainage Facilities for Animal Quarters

As medical research expands, so does the need for adequate quarters to house the animals used for experimental study.

Animals play a large part in this increasing quest for scientific knowledge. Experiments are performed with the animals quartered under sanitary conditions to prevent the spread of disease from one test animal to others. Weeks of study would thus be nullified.

Since it is necessary to house the experimental animals in the confines of research laboratories, adequate quarters must be provided that are convenient for treating, observing and caring for the animals. In this chapter, data are presented for the proper water and drainage facilities required in the animal cages. For example, only the proper housing for monkeys and animals of comparable size will be considered. Only the larger size cages are considered because they offer greater freedom of motion for the animals and permit closer study of the physical conditions of the animals.

Adequate Quarters

Since the monkeys must be properly housed, careful thought should be given to both the cages in which they are to be kept, and to the necessary facilities for keeping the cages clean and sanitary. A typical arrangement of cages would be similar to that shown in Fig. 1.

An arrangement of six cages has been used for this example, with a floor drain in each cage. Serving the group of six cages is a tempering valve that supplies water of the correct temperature for washing this group of cages.

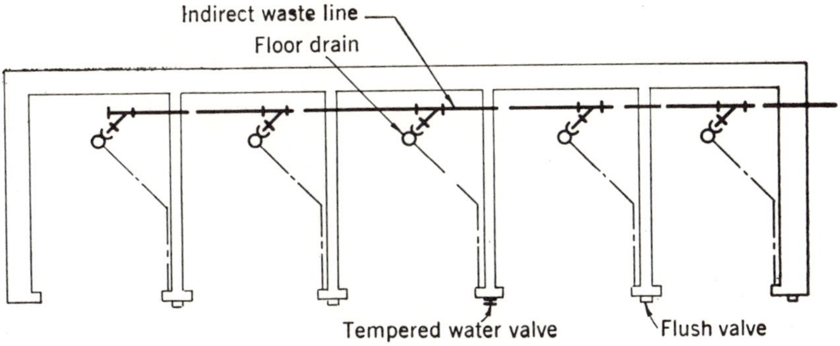

Fig. 1. Plan of a typical arrangement of water and waste facilities in cages.

Cleaning the Cages

Since cleanliness is an important factor and must be strictly maintained, it is necessary to provide the means for facilitating good sanitation.

A tempering valve regulates the temperature of the water to a degree suitable for cleaning the cages. Droppings from the monkeys as well as bits of food are the debris that must be removed. Figure 2 shows an arrangement that will permit proper cleaning.

As shown in Fig. 2, the tempered water is furnished by mixing cold and hot water to the desired temperature. Warm water is more effective in cleaning the droppings that stick to the floor of the cage. A hose is used to clean the cage floor, and water is applied to the floor in a gentle manner. By means of a squeegee or a scraper, all the matter is swept into the drain. This cleaning can usually be accomplished from outside the cage and this is quite an advantage.

Frequent and proper cleaning of the cages is an important factor and is quite necessary to insure true results from the experiments.

This manner of cage cleaning is recommended over the method of just flushing the cage with a hose. In addition, this method avoids wetting the monkeys or splashing the sides of the cage with water. The result is that the cage is quickly cleaned in an efficient manner and the floor will dry rather quickly because radiant heating coils are generally used to maintain the correct temperatures in the cages. Vacuum breakers must be installed on the water supply line to avoid possible contamination of the water supply through backflow. This reversal of flow results from a drop in water pressure in the

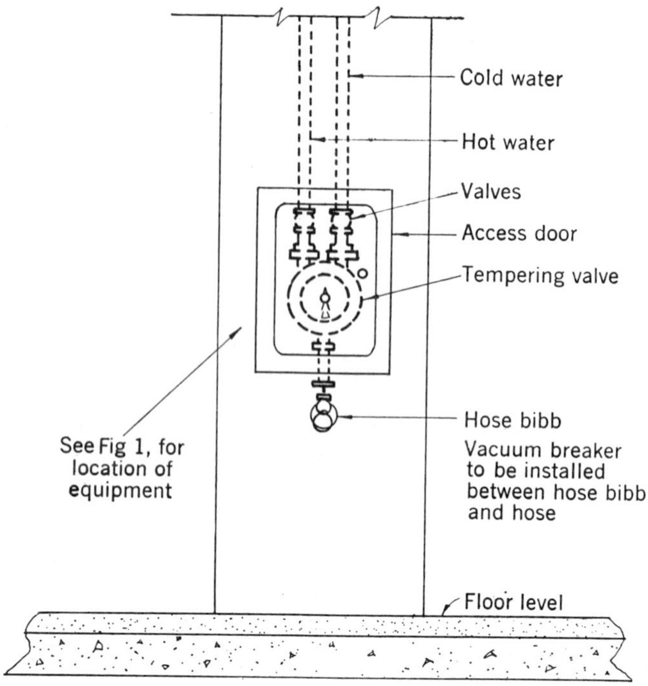

Fig. 2. Arrangement of water supply for washing cages.

main that feeds the supply line for the cages. The drop of pressure can be caused by any number of reasons.

Sterilization of the Cages

Sterilization of the cages is periodically required to assure clinically clean conditions for the varied experiments. Either live steam or strong disinfectants are used for this purpose. The entire cage can be thoroughly cleaned by these processes.

Since strong possible corrosive, solutions of disinfectants may be used, consideration should be given to both the floor drain and the waste piping to insure that they are of materials that will withstand attack by the solution.

Proper Drainage

Proper cleaning of the cage requires an approved drainage system. Water used in washing can be intercepted and drained through the floor drain in each cage. The drainage system, however, must be carefully designed to serve the needs of the cages. The floor drain is

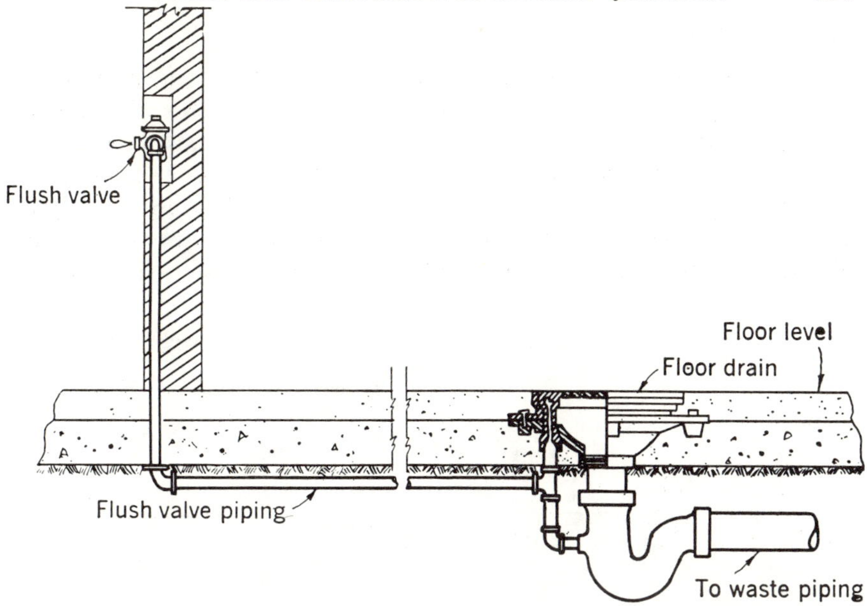

Fig. 3. Section through a cage to show arrangement of flush valve and floor drain.

a special type and this plus the drainage system are detailed in Fig. 3.

The special floor drain includes a secured grate to prevent its removal by the monkeys. Furthermore, the interior of the drain is enameled to provide the highest sanitary conditions. The enameled interior of the drain offers a surface that is smooth and, therefore, foreign material is prevented from clinging to it.

Droppings from sick or infected monkeys or other animals are contaminated and even though the cage has been cleaned by washing this matter into the drain, the drain itself must be so designed to be sure that the waste matter is discharged in a positive manner into the drainage system for the cage.

This is accomplished by a flush valve in a suitable location outside the cage. After the cage is cleaned, the valve is flushed to thoroughly wash out the drain with fresh water. This flushes out all matter remaining in the drain. The flush valve should be of the type that includes a vacuum breaker to avoid the contamination that backflow could create in the water supply line. The method described offers a positive way of keeping the drain clean.

It will be noted that flashing has been installed with the drain so that there is a watertight installation in the floor slab.

Waste System

Although every consideration must be given to the cages to maintain the highest standards of sanitation, sufficient thought should also be given to the waste line to which the drains are connected. This system must be so installed that a stoppage in another part of the sanitary system will not back up and overflow in the cages. The surest way to accomplish this is to provide an indirect system of waste piping. A suggested piping arrangement is shown in Fig. 4.

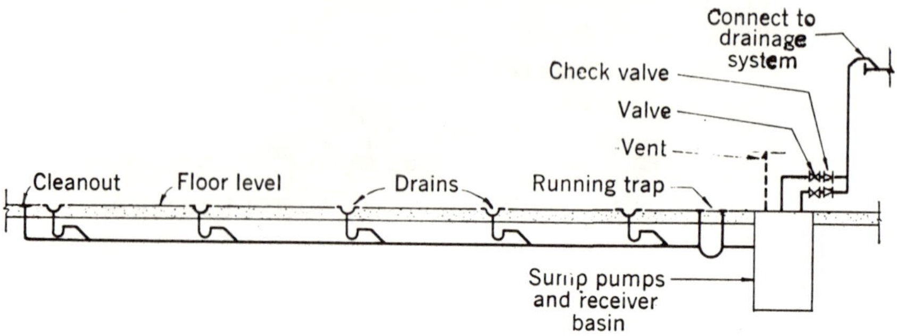

Fig. 4. Diagram to show how waste facilities and fittings are placed.

This piping receives the discharge from each of the drains and in turn discharges into a sump receiver. The purpose of a sump receiver tank is to allow for the introduction of disinfectants to the waste to be certain that the waste matter is properly treated before it enters the sanitary system. This chemical treatment of the waste is often required before this matter enters the sanitary lines in the building, and it will prevent the possible spread of this contamination to other portions of the sanitary system.

Not only must consideration be given to the cages, but in an equal manner, utmost care must also be given to the design of the waste system and the manner with which it connects with the sanitary system.

Other Applications

Although the information presented covers the water and waste facilities required for animal cages in a laboratory, the same consideration may also be extended to similar cages in a zoo.

There is an entirely new concept of thinking in the planning of zoos, including quarters that represent more closely the natural habitat of the animals. Glass is now used for certain cages where bars were formerly used. Moats or other specific barriers are now employed to restrict the activities of certain larger animals. Such innovations offer zoo patrons an unobstructed view of the wildlife in its natural habitat. In addition, these considerations insure the health of the animals because the animals enjoy sufficient exercise and do not experience the feeling of confinement. To satisfy this new concept in zoos, a more elaborate water supply system and drainage facilities will be required.

Water supply, for example, will be required for drinking, spray ponds, pools, watering as well as for washing the animal quarters. Therefore, it must be abundant in supply and so installed that it will not be subject to freezing. Control valves must be accessible at all times.

The drainage lines must be adequately sized for the loads imposed. Drains and catch basins have to be carefully selected for the function intended. Leaves, dirt and other foreign material poses problems that only a carefully designed drainage system can handle.

Winter quarters for the animals also require careful consideration with regard to the design of the water and waste facilities.

The food preparation rooms, kitchens and baking areas, as well as the food storage lockers, also need a more comprehensive system of facilities. Areas used for raising rodents or other special food needed in the diets of some animals and reptiles also demand certain prescribed design requirements. This is a subject by itself.